ACS SYMPOSIUM SERIES **646**

Environmental Immunochemical Methods

Perspectives and Applications

Jeanette M. Van Emon, EDITOR
U.S. Environmental Protection Agency

Clare L. Gerlach, EDITOR
Lockheed Martin Envrionmental Systems

Jeffre C. Johnson, EDITOR
U.S. Environmental Protection Agency

Developed from a symposium sponsored
by the U.S. Environmental Protection Agency
at the National Immunochemistry Summit IV,
Las Vegas, Nevada, August 2–3, 1995

American Chemical Society, Washington, DC

Library of Congress Cataloging-in-Publication Data

Environmental immunochemical methods: perspectives and applications /
Jeanette M. Van Emon, Clare L. Gerlach, Jeffre C. Johnson, editors.

p.   cm.—(ACS symposium series, ISSN 0097–6156; 646)

"Developed from a symposium sponsored by the U.S. Environmental
Protection Agency at the National Immunochemistry Summit IV, Las
Vegas, Nevada, August 2–3, 1995."

Includes bibliographical references and indexes.

ISBN 0–8412–3454–X

1. Immunoassay—Congresses. 2. Pollutants—Analysis—Congresses.
3. Environmental monitoring—Technique—Congresses.

I. Van Emon, Jeanette M., 1956–   . II. Gerlach, Clare L., 1948–
III. Johnson, Jeffre C., 1957–   . IV. United States. Environmental
Protection Agency. V. National Immunology Summit (4th: 1995: Las
Vegas, Nev.) VI. Series.

QP519.9I42E56   1996
628.5—dc20
                                                    96–28179
                                                        CIP

This book is printed on acid-free, recycled paper.

Advisory Board

ACS Symposium Series

Foreword

THE ACS SYMPOSIUM SERIES was first published in 1974 to
provide a mechanism for publishing symposia quickly in book
form. The purpose of this series is to publish comprehensive
books developed from symposia, which are usually "snapshots
in time" of the current research being done on a topic, plus
some review material on the topic. For this reason, it is necessary that the papers be published as quickly as possible.

Before a symposium-based book is put under contract, the
proposed table of contents is reviewed for appropriateness to
the topic and for comprehensiveness of the collection. Some
papers are excluded at this point, and others are added to
round out the scope of the volume. In addition, a draft of each
paper is peer-reviewed prior to final acceptance or rejection.
This anonymous review process is supervised by the organiz-
er(s) of the symposium, who become the editor(s) of the book.
The authors then revise their papers according to the recommendations of both the reviewers and the editors, prepare
camera-ready copy, and submit the final papers to the editors,
who check that all necessary revisions have been made.

As a rule, only original research papers and original review papers are included in the volumes. Verbatim reproduc-
tions of previously published papers are not accepted.

ACS BOOKS DEPARTMENT

Contents

Preface... ix

OVERVIEW

1. **Environmental Immunochemistry: Responding to a Spectrum
   of Analytical Needs**.. 2
   Jeanette M. Van Emon and Clare L. Gerlach

   RESEARCH AND TECHNOLOGY DEVELOPMENT

2. **Assay of Heavy Metals Using Antibodies to Metal–Chelate
   Complexes**.. 10
   Diane A. Blake, Gary N. Dawson, Pampa Chakrabarti,
   and Frank M. Hatcher

3. **Enzyme-Linked Immunosorbent Assay for the Detection
   of Mercury in Environmental Matrices**... 23
   Craig Schweitzer, Larry Carlson, Bart Holmquist,
   Mal Riddell, and Dwayne Wylie

4. **A New Approach to Electrochemical Immunoassays
   Using Conducting Electroactive Polymers**................................... 37
   Omowunmi A. Sadik

5. **Environmental Immunosensing at the Naval Research
   Laboratory**... 46
   Lisa C. Shriver-Lake, Anne W. Kusterbeck,
   and Frances S. Ligler

6. **Enzyme Immunoassay Analysis Coupled with Supercritical
   Fluid Extraction of Soil Herbicides**.. 56
   G. Kim Stearman, Martha J. M. Wells, Scott M. Adkisson,
   and Tadd E. Ridgill

7. **Development and Application of an Enzyme-Linked
   Immunosorbent Assay Method for the Determination of Multiple
   Sulfonylurea Herbicides on the Same Microwell Plate**................... 65
   Johanne Strahan

8. Immunoaffinity Extraction with On-Line Liquid
   Chromatography—Mass Spectrometry 74
   Jeanette M. Van Emon and Viorica Lopez-Avila

9. Sensitive Analyte Detection and Quantitation Using
   the Threshold Immunoassay System.. 89
   Kilian Dill

10. Optical Sensing Technology for Environmental
    Immunoassays.. 103
    Stephen L. Coulter and Stanley M. Klainer

APPLICATIONS AND EVALUATIONS

11. Recent Developments in Immunoassays and Related Methods
    for the Detection of Xenobiotics 110
    Ingrid Wengatz, Adam S. Harris, S. Douglass Gilman,
    Monika Wortberg, Horacio Kido, Ferenc Szurdoki,
    Marvin H. Goodrow, Lynn L. Jaeger, Donald W. Stoutamire,
    James R. Sanborn, Shirley J. Gee, and Bruce D. Hammock

12. A Status Report on Electroanalytical Techniques
    for Immunological Detection 127
    Omowunmi A. Sadik and Jeanette M. Van Emon

13. A First Application of Enzyme-Linked Immunosorbent Assay
    for Screening Cyclodiene Insecticides in Ground Water.................. 148
    Tonya R. Dombrowski, E. M. Thurman, and
    Greg B. Mohrman

14. Maumee Area of Concern Sediment Screening Survey,
    Toledo, Ohio.. 155
    Thomas J. Balduf, Jeff Wander, Philip A. Williams,
    Brent Kuenzli, and Patrick J. Heider

15. Validation of an Immunoassay for Screening Chlorpyrifos-
    methyl Residues on Grain.. 161
    Brian A. Skoczenski, Titan S. Fan, Jonathan J. Matt,
    J. Terry Pitts, and J. Larry Zettler

16. An Evaluation of a Microtiter-Plate Enzyme-Linked
    Immunosorbent Assay Method for the Analysis of Triazine
    and Chloroacetanilide Herbicides in Storm Runoff Samples........... 170
    Michael L. Pomes, E. M. Thurman, and Donald A. Goolsby

17. Evaluation of an Automated Immunoassay System
for Quantitative Analysis of Atrazine and Alachlor
in Water Samples 183
    Barbara Staller Young, Andrew Parsons, Christine Vampola,
    and Hong Wang

18. An Evaluation of a Pentachlorophenol Immunoassay Soil
Test Kit 191
    Alan Humphrey

COMMUNICATION AND DATA INTERPRETATION

19. An Immunochemistry Forum: A Proposal for an
Immunochemistry Web Site 216
    Donna W. Sutton and Jeanette M. Van Emon

20. Screening Tests in a Changing Environment 227
    Richard L. Ellis

21. Considerations in Immunoassay Calibration 240
    Thomas L. Fare, Robert G. Sandberg, and David P. Herzog

22. Quality Assurance Indicators for Immunoassay Test Kits 254
    William A. Coakley, Christine M. Andreas,
    and Susan M. Jacobowitz

23. Maximizing Information from Field Immunoassay
Evaluation Studies 265
    Robert W. Gerlach and Jeanette M. Van Emon

HUMAN EXPOSURE ASSESSMENT

24. Biomonitoring for Occupational Exposures Using
Immunoassays 286
    Raymond E. Biagini, R. DeLon Hull, Cynthia A. Striley,
    Barbara A. MacKenzie, Shirley R. Robertson,
    Wendy Wippel, and J. Patrick Mastin

25. Application of an Immunomagnetic Assay System
for Detection of Virulent Bacteria in Biological Samples 297
    Hao Yu and Peter J. Stopa

26. Application of Enzyme-Linked Immunosorbent Assay
    for Measurement of Polychlorinated Biphenyls
    from Hydrophobic Solutions: Extracts of Fish and Dialysates
    of Semipermeable Membrane Devices 307
        James L. Zajicek, Donald E. Tillitt, James N. Huckins,
        Jimmie D. Petty, Michael E. Potts, and David A. Nardone

27. Immunochemical Methods for Fumonisins in Corn 326
        Mary W. Trucksess

Author Index 333

Affiliation Index 334

Subject Index 334

Preface

THE ENVIRONMENTAL ANALYTICAL COMMUNITY has responded with enthusiasm to the need for reliable, inexpensive, field-transportable methods for the characterization of hazardous waste sites by adapting antibody-based technologies. Immunochemical methodology, with a strong record in clinical applications, was a natural for the analytical challenges faced by environmental scientists. There is a need for economical methods that can detect trace levels of hazardous compounds in complex media such as soil, sludge, oil, and food products. Immunochemical methods are capable of achieving very low detection levels, are adaptable to a variety of matrices, and require little in the way of heavy or expensive instrumentation. In addition to these features, cost savings can be achieved for studies requiring repetitive analysis or high sample loads.

The 1980s saw a boom in immunochemical research and development for environmental applications. Researchers at university laboratories worked with scientists at federal and state regulatory agencies to develop better methods for immunochemical analysis. Entrepreneurs based new businesses on the desirability of immunoassay test kits for site monitoring and human exposure assessment. Major chemical manufacturers, recognizing the cost savings in immunochemical analyses, addressed issues such as the need to quantify cross-reactivity and minimize matrix effects in order to lower the cost of registration and re-registration of pesticide products. In the 1990s, the U.S. Environmental Protection Agency officially began to accept certain immunochemical procedures for the analysis of several compounds. Back in the laboratories, new and exciting refinements were being discovered. Research and regulatory interest escalated. The race was on.

In 1992, the U.S. Environmental Protection Agency laboratory in Las Vegas (at that time the Environmental Monitoring Systems Laboratory, now the National Exposure Research Laboratory, Characterization Research Division), seeing the need for intra- and interagency communication, sponsored the first Immunochemistry Summit Meeting. Since then, the summit has been held annually, growing to an attendance of about 150 in 1995 at Immunochemistry Summit IV. Summit IV featured speakers from university research centers, government laboratories, major chemical companies, and commercial producers of immunochemical reagents, equipment, and immunoassay test kits. Representatives from local, state, and federal government laboratories participate in the summits and provide regulatory insight as well as case studies that illustrate the success or failure of immunochemical methods for a variety of applications.

This volume is intended to inform novice readers about the strength and versatility of immunochemical methods for a wide spectrum of analytical environmental applications and to provide experienced researchers with a selection of readings in various areas of applied environmental immunochemistry. Authors from regulating agencies present the latest uses, both accepted and innovative, of immunochemical and related methods as well as quality assurance guidelines. Authors from universities present updates on breakthrough research. Authors from manufacturing companies provide insight into recently developed products and procedures.

The U.S. Environmental Protection Agency is proud to have initiated this recognized forum for technical communication in the area of immunochemistry. The expansion of the work begun at the summit meetings will continue on several fronts: leveraged research by participants, joint projects, and alignment behind common goals. Future summit meetings may be held jointly with professional societies and technical organizations, like the American Chemical Society, with an eye toward recruiting more practitioners of traditional analytical chemistry into the exciting field of immunological methods. An electronic immunochemistry forum within the Environmental Protection Agency's National Exposure Research Laboratory home page will be available to technical personnel with Internet access. Digital communication will bring the networking of the summit meetings to the next level of technical interaction. We hope that a combined force of publications, conferences, focused work groups, and electronic forums will result in an expanded awareness of the power of immunochemical technologies and a continued cooperation between the developers and users of innovative scientific methods.

JEANETTE M. VAN EMON
JEFFRE C. JOHNSON
Characterization Research Division
National Exposure Research Laboratory
U.S. Environmental Protection Agency
P.O. Box 93478
Las Vegas, NV 89193–3478

CLARE L. GERLACH
Lockheed Martin Environmental Systems
980 Kelly Johnson Drive
Las Vegas, NV 89119

May 3, 1996

OVERVIEW

Chapter 1

Environmental Immunochemistry: Responding to a Spectrum of Analytical Needs

Jeanette M. Van Emon[1] and Clare L. Gerlach[2]

[1]Characterization Research Division, National Exposure Research Laboratory, U.S. Environmental Protection Agency, P.O. Box 93478, Las Vegas, NV 89193–3478
[2]Lockheed Martin Environmental Systems, 980 Kelly Johnson Drive, Las Vegas, NV 89119

The field of environmental immunochemistry brings together several specialties, including analytical chemistry, biochemistry, molecular biology, and environmental engineering. This multidisciplinary nature is both a benefit and a confusion to practitioners, rewarding a mastery of several scientific skills with the excitement of innovative technology. Environmental immunochemistry can be as simple as a disposable immunoassay test kit or as complex as an integrated system that employs immunochemical techniques as a component of a multistep process. The growing regulatory and user acceptance of immunochemical methods for dozens of regulated compounds ensures the continued growth in this technology – at the lab bench, at the hazardous waste site, and beyond. Applications are widespread, including determination of agricultural runoff, assessment of human and ecological exposure, quantification of food and pharmaceutical purity, and groundwater monitoring.

Environmental immunochemistry has grown dramatically in the academic, commercial, and regulatory areas over the past fifteen years. Before the 1980s, immunochemical methods were widely used in clinical applications (1) and their success in these critical studies led environmental scientists to consider immunoassay use for screening of hazardous compounds in various media. Body fluids are complex media, but new challenges were presented by soils, sludges, food, and agricultural products. Methods were developed, test kits were designed and manufactured, and many comparison studies were initiated to evaluate the performance of these environmental analytical newcomers.

It is appropriate at this juncture in the development and use of environmental immunoassays, to review the success of the methods, assess the status of regulatory acceptance, and welcome the next tier of practitioners to the arena of immunochemical and related technologies. This volume presents in-depth research reports, environmental applications studies, data interpretation subtleties, and commercial success stories. It is hoped that the scope of this monograph will make it interesting to the experts, give new applications ideas to researchers, and provide a strong technical basis for novice users. The multidisciplinary character of immunochemical technology is one of the strengths of the Immunochemistry Summit Meetings, upon which this volume is based. The popularity of the Summit Meetings is due to the recognized value of inter-agency and intercorporation exchange of ideas.

All immunoassays are based on the interaction between an antibody and a target analyte. Antibodies are produced in response to an immunogen by a complex mechanism (*2*). Recently, as analytical chemists have become more interested in the technology, better quantitative methods have been developed, and the user community has benefitted (*3*). Commercial manufacturers of immunoassay test kits have contributed to the availability of more and better analytical tools. University researchers continue to push the technical envelope, extending immunochemical capabilities well beyond their status in 1990. The combined enthusiasm of these groups is apparent in this volume and is palpable at the Summit Meetings.

Research in the area of human exposure monitoring is described in this edition. R. E. Biagini and fellow researchers at the National Institute for Occupational Safety and Health (NIOSH) present work that demonstrates the efficiency of immunoassay test kits for measuring alachlor metabolites in urine. Absorption, partitioning, and excretion of toxic compounds is reliant upon several factors and this multivariate character presents a challenge in data interpretation as well as in analytical procedures. Examples of the research at NIOSH are presented, including the use of circulating antibodies and antibody techniques to monitor exposure in the urine of exposed workers. Research at NIOSH demonstrated that circulating antibodies to morphine can be present in the absence of urinary analytes. In another NIOSH study, an immunoassay test kit, originally developed for alachlor analyses in groundwater, was found to be 4-5 times more sensitive in detecting the primary human metabolite of alachlor, alachlor mercapturate, than in detecting the parent molecule. For some compounds, immunoassay techniques are orders of magnitude more sensitive than traditional gas chromatographic/mass spectrometric techniques. Benefits such as increased sample throughput, reduced cost, simpler sample preparation and no derivitization steps make this type of analysis very attractive.

V. Lopez-Avila and J. M. Van Emon discuss their work in the coupling of immunoaffinity chromatography (IAC) extraction with on-line liquid chromatography and mass spectrometry (MS). The IAC technique is based on the ability of antibodies to separate a target analyte from the complex matrices that often challenge environmental analytical chemists. On-line analysis is done by high-performance liquid chromatography (HPLC) and electrospray MS. The system is particularly useful for the analysis of compounds that are water-soluble, nonvolatile, thermally labile, or highly polar. Sample throughput of the MS is increased by integrating IAC with the analytical instrumentation, providing automated, streamlined sample preparation.

I. Wengatz and co-workers at the University of California-Davis are developing immunoassays to assess human exposure to xenobiotics, such as pesticides (4). Metabolites of certain xenobiotics may serve as biomarkers of exposure in toxicity studies. The UC-Davis group is developing immunoassays for trace levels of triazine herbicides and their metabolites, and is using cross-reactivity information to enable antibody use for screening classes of analytes. Human exposure research is increasingly important as regulatory agencies move from prescriptive methods based on absolute contaminant concentration to risk-driven guidelines based on the bioavailability of the contaminant and its threat to human and ecological health.

Other immunochemical-based technologies, such as biosensors, are being developed at the Naval Research Laboratory. L. Shriver-Lake and fellow researchers describe a continuous flow immunosensor that can be used to measure small molecules in discrete samples or in monitoring process streams. A fiber optic biosensor, based on a competitive immunoassay being performed on the fiber core of a long optical fiber, is also being studied. Response is measured by the change in the fluorescent signal. Electrochemical immunoassays are based on modifications of enzyme immunoassays with the enzyme activites being determined potentiometrically or amperometrically. O. Sadik provdes a status report on electrochemical immunosensors based on conducting electroactive polymers. Immunosensors provide the analytical advantages of conventional immunoassay methods, as well as the option of obtaining real-time monitoring measurements. An electrochemical immunosensor is also described for the analysis of polychlorinated biphenyls.

The promise of these methods is in the eventual development of a sensor that can be used remotely, gathering information without an operator and sparing personnel the possible exposure associated with some environmental work.

S. Coulter and associates discuss a solid state system that combines the advantages of optical sensing and competitive immunoassays. This sensing package comprises a light source which provides the output through the waveguide, the sensing chemistry, and the appropriate detector. This chip-based sensor is easily manufactured and has a sensing arm and a reference arm. By combining fiber optic chemical sensor technology with immunoassay, these systems enlarge the panorama of analytical tools available to environmental scientists. Research that reduces the number of steps in an analytical procedure will be appreciated as the environmental analysis emphasis moves from the laboratory to the hazardous waste site.

Immunoassay test kits and other immunochemical procedures are now used almost routinely to monitor the purity of food and drugs, and drinking water. Immunoassay test kits are increasingly used at hazardous waste sites regulated under Comprehensive Environmental Response, Compensation and Liability Act (CERCLA) and Resource Conservation and Recovery Act (RCRA). An important step was taken in 1993 when the EPA's Office of Solid Waste and Emergency Response (OSWER) accepted nine immunoassay methods for its compendium, SW-846 (5). This regulatory acceptance means that these methods can be used for certain RCRA applications. The EPA's Office of Water and Office of Drinking Water are utilizing field methods such as immunoassays to determine the safety of the nation's water supplies for drinking, agriculture, and recreational use (6). The U.S. Geological

Survey uses immunoassays to analyze water samples (7). The U.S. Food and Drug Administration (FDA) uses immunochemical methods to determine the purity of processed foods and manufactured drugs. The U.S. Department of Agriculture uses immunoassays to measure the levels of pesticides in crops and their byproducts and in meat and poultry inspection.

M. Trucksess and associates describe work done at the FDA to determine the levels of fumonisins in corn. Fumonisins are mycotoxins that have demonstrated toxicity in horses and swine, and have been implicated in certain cancers in humans. Immunochemical methods are now preferred for mycotoxin monitoring in foods because results are obtained much faster than with traditional methods, such as thin-layer, liquid, and gas chromatography. Enzyme-linked immunosorbent assay (ELISA) methods are commercially available and the FDA study compares and evaluates these technologies.

The need for inexpensive, easy-to-use methods has been well documented (8,9). With this low-cost, however, users may sometimes relinquish maximum analytical performance, such as extremely low detection limits, very high precision, and even analyte-specific identification. But not necessarily. Many quantitative immunochemical methods are now available that achieve extremely low detection levels and rival their traditional laboratory counterparts. Often there is a need for both screening methods (e.g., in characterizing hazardous waste sites) and higher cost analytical procedures.

Special quality assurance (QA) considerations are needed in immunochemical methods. W. A. Coakley of the U.S. EPA's Environmental Response Team and a technical support team from Roy F. Weston, Inc., describe a QA system that focuses on generic and core indicators of confidence. Generic indicators assess the reliability of the total sampling and analysis scheme. Core indicators are specific to the mode of analysis, in this case, immunoassays. Several features are key to the interpretation of immunoassay results: temperature, analyte specificity, non-analyte interference, moisture content, and dilution factors. Understanding the entire process and the potential effects of these and other factors is essential to the quality of information obtained with these innovative methods.

The first role of immunoassay test kits was in screening applications where they provided a welcome addition to the field-portable instruments commonly used in hazardous waste site characterization. Their results are comparable to those from gas chromatography/mass spectrometry (GC/MS). Though immunoassays are frequently more sensitive than GC/MS, high immunoassay results are still routinely confirmed by laboratory procedures. In this aspect, technical strength has outpaced regulatory and user acceptance. Early uses at hazardous waste sites were conducted by EPA's Superfund Innovative Technology Evaluation (SITE) program (10). In 1988, the first SITE demonstration of a measurement technology evaluated immunoassays for pentachlorophenol (11). Subsequent SITE demonstrations evaluated immunoassay test kits for benzene/toluene/xylene (12) and polychlorinated biphenyls (13).

The use of immunoassays to obtain quantitative results has escalated in the past several years. With this increased reliance upon sensitivity and specificity come several challenges in the area of data interpretation. T. L. Fare and fellow researchers at Ohmicron Corporation detail the importance of correct calibration techniques. Three basic approaches are discussed: empirical, semi-empirical, and equations derived from the Law of Mass Action. Types of error are described and recommendations are made regarding the processing of immunoassay data. R. W. Gerlach and J. M. Van Emon discuss a variety of data analysis and interpretation issues identified as a result of their work in field evaluations of environmental immunoassays. Analysis of multiple estimates for parameters such as false positive and false negative rates suggest that interval estimates are often better than point estimates. Response factors which control false negative and false positive rates are identified. The effect of explicit and implicit experimental design factors on data interpretation and their impact on the use of advanced non-linear calibration analysis are also reviewed. These papers are critical in understanding the strengths and weaknesses of various statistical procedures for interpreting data from an immunochemical study in the laboratory or in the field.

The success of early environmental immunoassay studies led to increased research, publication, and commercial development. More immunoassays were developed for a wider number of compounds of environmental concern. This research effort resulted in test kits that were capable of achieving more reliable results and lower detection limits, with less cross-reactivity. Immunoassays are available for individual compounds and for groups of related compounds. For example, one can use immunoassay test kits to monitor a specific triazine, such as atrazine, or a group of closely related triazines. In some cases, if the ratio of cross-reactivity for specific compounds is known, monitoring for the group and multiplying by the correct factor can be a time-saving and inexpensive method for characterizing a hazardous waste site.

Novel innovations are also expanding the range of commercially available detection systems. K. Dill describes the Threshold Immunoassay System, a commercial sandwich immunoassay detection system, that is based on a silicon chip with eight identically etched sites. The system reduces the distortions due to solid-phase/liquid-phase interactions by using solution-phase binding and is capable of detecting a wide range of molecular weights, from pesticides to DNA. The normal sandwich immunoassay format was modified to indirectly detect the herbicide atrazine. This paper presents an excellent example of industrial response to technical market requirements. The market drives the research into faster, less expensive, and versatile analytical methods.

The primary use of environmental immunoassays is in field-screening procedures because of the relative low cost and the ease-of-use in hazardous waste site environments, but research based on the high sensitivity of immunochemical methods has elevated the technology to a strong competitor in the quantitative laboratory as well. This next step to acceptance as a quantitative analytical procedure is critical. Pragmatic acceptance dictates regulatory acceptance. Regulatory acceptance stimulates commercial interest. Commercial interest results in more candidate methods. These new methods may then gain pragmatic acceptance. Thus, the circle of research and development is perpetuated, resulting in better procedures for chemists, better data for end users, and a better environment for everyone.

By linking immunochemical procedures to other analytical and sample preparation steps, analysts are exploring new avenues for technical advancement of immunologic analytical procedures. In this volume, work is presented that describes supercritical fluid extraction with ELISA, electroimmunochemical processes, and the use of metal chelates in certain environmental applications. There are opportunities for research in teaming GC/MS methods with immunoassays. Capillary electrophoresis with laser-induced fluorescence offers another linking option for immunochemical methods. The results of the research so far indicates considerable promise in these hyphenated techniques and research is ongoing at university, private, and government laboratories.

The future of environmental immunochemical technologies is very promising. Ongoing research and continued regulatory interest set the stage for an expanding technological base – in field and laboratory applications, in human exposure monitoring, and in food and agricultural uses. The editors wish to thank the hundreds of participants in the Summit Meetings whose interest and input have made the meetings successful and made this volume possible. The editors gratefully acknowledge the contributions of all the authors and the reviewers for their valuable comments in preparing this book.

Acknowledgments

The editors are indebted to Allan W. Reed for his tireless efforts in format, layout, and production matters. Al Reed is a Senior Environmental Employement Program enrollee, assisting the U.S. EPA under a cooperative agreement. The U.S. Environmental Protection Agency (EPA), through its Office of Research and Development (ORD), funded and collaborated in the research described here. It has been subjected to the Agency's peer review system and has been approved as an EPA publication. Mention of trade names or commercial products does not constitute endorsement or recommendation for use.

Literature Cited

1. Berson, S.A.; Yalow, R.S.; Bauman, A.; Rothschild, M.A.; Newerly, K. *J. Clin. Invest.* **1956**, 35, 170-190.
2. Van Emon, J.M.; Seiber, J.N.; Hammock, B.D. In *Advanced Analytical Techniques*; Sherma, J., Ed.; Academic Press: San Diego, CA, 1989, Vol. XVII.
3. Hammock, B.D.; Gee, S.J.; Cheung, P.Y.K.; Miyiamoto, T.; Goodrow, M.H.; Van Emon, J.M.; Seiber, J.N. In *Pesticide Science and Biotechnology*; Greenhalgh, R. And Roberts, T.R., Eds.; Blackwell: Oxford, 1986; 309.
4. Hammock, B.D.; Mumma, R.O. In *Pesticide Analytical Methodology*; Zweig, G., Ed.; ACS Symposium Series No. 136, American Chemical Society, Washington, DC, 1980; 321-352.

5.   Office of Solid Waste and Emergency Response. *Test Methods for Evaluating Solid Waste. Physical/Chemical Methods. Vols. I and II. Third edition*; EPA/SW-846;   U.S. Environmental Protection Agency: Washington, DC, 1995.

6.   Environmental Monitoring and Management Council. *Environmental Monitoring Methods Index, Version 1.0, Users Manual*; EPA/821-B-92-001; U.S. Environmental Protection Agency: Washington, DC, 1992.

7.   Goolsby, D.A.; Battaglin, W.A.; Thurman, E.M. *Occurrence and Transport of Agricultural Chemicals in Mississippi River Basin in July-August, 1993*; Circular 1120-C; U.S. Geological Survey: Washington, DC, 1993.

8.   Van Emon, J.M.; Lopez-Avila, V. *Anal. Chem.* **1992,** 64(2), 79A-87A.

9.   Van Emon, J.M.; Gerlach, C.L. *Environ. Sci. Technol.* **1995**, 29(7), 312-317.

10.   *Superfund Innovative Technology Evaluation Program, Technology Profiles, Sixth Edition*; EPA/540/R-93/526; U.S. Environmental Protection Agency: Washington, DC, 1993.

11.   Van Emon, J.M.; Gerlach, R.W. *Bull. Environ. Contam. Toxicol.*, **1992**, *48*, 635-642.

12.   White, R.J.; Gerlach, R.W.; Van Emon, J.M. *An Immunoassay for Detecting Gasoline Components*; EPA/600/X-92/116; U.S. Environmental Protection Agency: Washington, DC, 1992.

13.   Johnson, J.C.; Van Emon, J.M. *Development and Evaluation of a Quantitative Enzyme-Linked Immunosorbent Assay (ELISA) for Polychlorinated Biphenyls*; EPA/600/R-94/112; U.S. Environmental Protection Agency: Washington, DC, 1994.

RESEARCH AND TECHNOLOGY DEVELOPMENT

Chapter 2

Assay of Heavy Metals Using Antibodies to Metal–Chelate Complexes

Diane A. Blake[1], Gary N. Dawson[1], Pampa Chakrabarti[2], and Frank M. Hatcher[2]

[1]Department of Ophthalmology, Tulane University School of Medicine, 1430 Tulane Avenue, New Orleans, LA 70112
[2]Department of Microbiology, Meharry Medical College, 1005 D. B. Todd Boulevard, Nashville, TN 37208

The features of a model immunoassay designed to measure heavy metals are reviewed. The assay used a monoclonal antibody specific for indium-EDTA chelates in an antigen inhibition format. This report demonstrates how the sensitivity of the immunoassay can be modified by changing the nature of the soluble inhibiting antigen and suggests new assay formats which may increase assay sensitivity without reducing specificity. Those assay formats most likely to succeed as an on-site test for heavy metals are emphasized. A monoclonal antibody specific for complexes of cadmium and ethylenediaminetetraacetic acid (EDTA) isolated in the authors' laboratories was tested for its ability to detect cadmium-EDTA complexes in the presence of excess EDTA. Assay performance was insensitive to EDTA at concentrations as high as 230 mM. This antibody was subsequently used in an antigen inhibition immunoassay to distinguish between soil samples which were heavily and minimally contaminated with cadmium. These preliminary studies will be expanded to develop an on-site test for cadmium contamination.

The contamination of the environment with heavy metals poses a continuing and increasing threat to plant and animal life. At least twenty metals are known to be toxic and fully half of these, including arsenic, cadmium, chromium, copper, lead, nickel, silver, selenium, and zinc are released into the environment in sufficient quantities to pose a risk to human health (*1*). Upon entering the body, metals accumulate and impair several physiological functions (for a review, see *2*). Cadmium is toxic to a wide range of tissues including lung, liver, kidney, and testis (*3,4*). Even low levels of lead in blood cause developmental abnormalities in human fetuses and children (*5*). Although zinc is an essential trace element, overexposure can lead to impaired immune function (*6*). Chromium has been shown to cause

0097–6156/96/0646–0010$15.00/0

chromosomal damage and disrupt redox reactions within the cell; the hexavalent form is many times more toxic than the cationic form (7).

Existing technologies to measure heavy metals require complex instrumentation (atomic absorption or inductively coupled plasma emission spectroscopy) in a centralized facility. These instruments measure the total amount of a specific metal in an environmental sample, but provide no information about metal oxidation state or speciation. Immunoassays offer an alternate approach for metal ion detection and they have significant advantages over more traditional metal ion detection methods. Immunoassays are quick, inexpensive, simple to perform, and reasonably portable; they can be both highly sensitive and selective. A limited number of immunoassays are now available that measure environmental contaminants, including industrial pollutants (8) pesticides (9) and herbicides (10). Although most commercial immunoassays are directed toward halogenated or aromatic compounds, this method is theoretically applicable to any pollutant for which a suitable antibody can be generated. Our laboratories have recently demonstrated the feasibility of immunoassays for specific metal ions. The features of a prototype immunoassay have been published in *Analytical Biochemistry* (11). This assay used an antigen inhibition format and measured the soluble metal indium at concentrations from 0.005 ppb to 320 ppm. In this report we describe how the sensitivity of the metal ion immunoassay could be modulated by changing the structure of the inhibiting antigen. We also introduce an immunoassay that we are developing to detect cadmium in contaminated soil samples

Material and Methods

Ultrapure bovine serum albumin (BSA) was purchased from Boehringer-Mannheim Biochemicals (Indianapolis, IN). Atomic absorption grade standard cadmium, indium, copper, magnesium, and manganese were obtained from Perkin Elmer Corp. (Norwalk, CT). These standard metals were handled and disposed of according to the directions in the Materials Safety Data Sheets provided by Perkin-Elmer. Tissue culture medium, glutamine, antibiotics, goat anti-mouse IgG coupled to horseradish peroxidase and ethylenediaminetetraacetic acid (EDTA) were purchased from Sigma Chemical Co. (St. Louis, MO). Fetal bovine serum and Fetalclone supplement were products of Hyclone Laboratories (Logan, UT). 3,3',5,5'-Tetramethylbenzidine peroxidase substrate (TMB Microwell Substrate) was from Kirkegaard-Perry Laboratories (Gaithersburg, MD). All other reagents were the purest commercially available. ELISA plates were a product of Costar, Inc (Cambridge, MA). All water was purified by filtration through a Barnstead Nanopure II water purification system. Metal-free disposable pipette tips were a product of Oxford Labware, Inc. (St. Louis, MO). All glassware was mixed acid washed (12) and liberally rinsed with purified water and all plasticware was soaked overnight in 3 M hydrochloric acid and liberally rinsed with purified water before use. Monoclonal antibody directed towards indium-EDTA chelates (CHA255), p-isothiocyanate-benzyl-EDTA and p-nitrobenzyl-EDTA (L-isomers) were generous gifts of Hybritech, Incorporated (San Diego, CA). Indium-EDTA-BSA was available from a previous study (11). Isolation of hybridomas which produce monoclonal antibodies specific for cadmium-

EDTA chelates and the preparation of cadmium-EDTA-BSA will be described elsewhere (Blake et al., submitted). The monoclonal antibody directed towards indium-EDTA chelates (CHA255) was purified as described previously (*11*). A monoclonal antibody directed towards cadmium EDTA-chelates (2A81G5) was purified from ascites fluid by chromatography on a Protein G Superose column.

Indirect ELISA. The optimum metal-EDTA-BSA concentration for coating microwell plates and the best working dilution for the purified monoclonal antibodies was determined by indirect ELISA. Antigen (indium-EDTA-BSA or cadmium-EDTA-BSA, 1-100 μg/ml in PBS) was absorbed to ELISA plates overnight at 4° C. After this and each subsequent step the plates were washed 3 times with 0.05% Tween 20 in 137 mM NaCl, 3 mM KCl, 10 mM sodium phosphate, pH 7.4 (PBS). The wells were blocked with 100 μl of 3% BSA in PBS for 1 h at 37° C, then the purified antibody was serially diluted into the wells. After 1 h at 37° C, 50 μl of goat antimouse-IgG-HRP conjugate (1/4000 dilution in PBS +3% BSA) was added and the plates were incubated for an additional 1 h at 37° C. For color development, 50 μl TMB microwell substrate was added to each well. The reaction was stopped with 1 N HCl and the plates were read in the dual wavelength mode (450-650 nm) using a Vmax Kinetic Microplate Reader from Molecular Devices.

Antigen inhibition ELISA. Soluble metals were diluted into an excess of EDTA (5, 138, or 230 mM), or into 5 mM (p-nitrobenzyl)-EDTA and preincubated with diluted monoclonal antibody (1/4000 in PBS + 3% BSA) for 1 h at room temperature, then added to a microwell plate coated with metal-EDTA-BSA conjugate (usually 1-10 μg/ml in PBS). After 1 h at 37° C, the plates were washed, enzyme-labeled second antibody was added, and the plates were incubated an additional 1 h at 37° C. Wash steps, color development, and data collection were performed as described for indirect ELISA. EDTA and (p-nitrobenzyl)-EDTA were omitted from the preincubation buffer when metal-EDTA-BSA conjugates were used as the soluble inhibiting antigen.

Assay of soil samples for cadmium. Standard reference soil (SRM 2711, Montana soil) containing 41.7 ppm cadmium was prepared by the National Institute of Standards and Technology (Gaithersburg, MD) and a minimally contaminated control soil sample (50455) containing 3.70 ppm cadmium as assessed by inductively coupled plasma emission spectroscopy (ICP) was generously provided by Dr. R. Smith of Analytical Services, Inc. (Norcross, GA). The dried soil samples were suspended at 10 mg/ml in 230 mM EDTA and incubated for 1 h at 60° C. After centrifugation, the supernatants were serially diluted from 1:2 to 1:64 into 230 mM EDTA and incubated for 1 h at 25° C with a 1:4000 dilution of a purified monoclonal antibody directed against cadmium-EDTA chelates. The samples were subsequently placed into 96-well plates that had been previously coated with 1 μg/ml of a cadmium-EDTA-BSA conjugate. Antigen inhibition ELISA was carried out as described above. The optical densities obtained from the samples were compared to a standard curve prepared using atomic absorption grade cadmium diluted into 238 mM EDTA.

Results

In our initial efforts to develop an immunoassay, we immobilized metal-free EDTA-BSA conjugates onto microwell plates and monitored antibody binding after adding serially diluted soluble metal to the test plates. Although this assay format was incredibly sensitive, it was also very imprecise since trace metals in the laboratory atmosphere were sufficient to cause false positives in this assay format (data not shown). We subsequently chose an antigen inhibition assay format for immunoassay of soluble metal.

Antigen inhibition format for immunoassay of heavy metals. The antigen inhibition format was much more precise than the previous format. The sensitivity of this assay could be modulated by changing the nature of the inhibiting antigen, as shown in Figures 1, 2 and 3. When indium-EDTA-BSA was used as the soluble antigen, the assay reliably monitored metal concentrations from 0.001 to 2000 ppb (0.009-17,400 nM) (Figure 1). The limit of sensitivity, defined as two standard deviations above the lowest detectable concentration (*13*), was 5 ppt. When soluble indium was serially diluted into 5 mM (p-nitrobenzyl)-EDTA and then preincubated with the monoclonal antibody, the assay detected indium at concentrations from 0.11 to 120 ppm (0.96-1044 μM) (Figure 2). The antibody could also detect EDTA complexes of indium (shown in Figure 3). In the presence of 5 mM EDTA, the assay detected indium at concentrations from approximately 0.6 to 320 ppm (5.22-2784 μM). Free indium also inhibited color formation in the antigen inhibition format, but at concentrations 20-30 times greater than those required for the indium-EDTA complex (data not shown).

Specificity of the indium immunoassay for metal ions. Since trace metals in the environment often exist in combinations, we tested our immunoassay with two metals that might occur as contaminants in waste water samples containing indium. Manganese is a common divalent metal which binds strongly to EDTA (*14*) and copper is solubilized with indium during the recycling of computer chips (*15*). Soluble indium in the presence of 5 mM EDTA decreased absorbance in the antigen inhibition immunoassay in a dose-dependent fashion (see Figure 4, solid circles). Copper or manganese complexes with EDTA inhibited color formation very weakly (shown by the squares and triangles, respectively, in Figure 4). As shown in the figure, the concentration curve for antigen inhibition shifted by more than 100 fold when Cu-EDTA or Mn-EDTA were used as the inhibiting antigens. Indium was also assayed in the presence of a large excess of magnesium (500 ppm) during other studies with bacterial solubilization of indium (*11*) and this level of magnesium did not effect the immunoassay's ability to detect indium (data not shown).

The effect of EDTA on the cadmium immunoassay. Our laboratory has recently isolated several hybridomas which produce monoclonal antibodies specific for cadmium-EDTA chelates (Blake et al., submitted). The most promising clone was expanded and the IgG1 secreted by the hybridoma was prepared by purification on a Protein G affinity column. This purified antibody was subsequently used in an

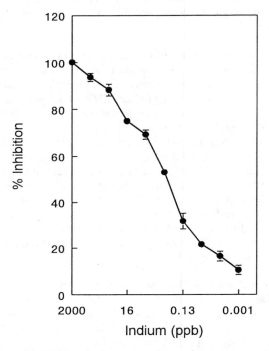

Figure 1. Effect of inhibiting antigen structure on immunoassay sensitivity. An antigen inhibition immunoassay was performed as described in **Methods**, using an indium-loaded EDTA-BSA conjugate as the inhibiting antigen. Each point represents a five-fold dilution of the antigen. The concentration of protein-bound, chelated indium in the assay was determined from the moles of EDTA conjugated to the carrier protein. Values are mean $\pm$SD of four to six determinations. Figure 1 is reprinted from (*11*) with permission from Academic Press, Inc.

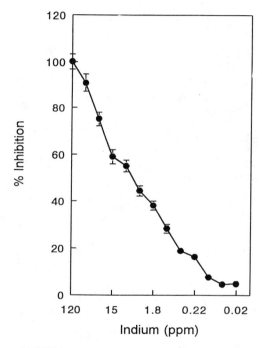

Figure 2. Effect of inhibiting antigen structure on immunoassay sensitivity. An antigen inhibition immunoassay was performed by diluting the indicated concentrations of indium into an excess (5 mM) of p-nitrobenzyl-EDTA . Each point represents a two-fold dilution of the antigen. Values are mean ±SD of four to six determinations.

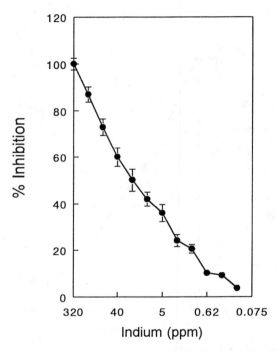

Figure 3. Effect of inhibiting antigen structure on immunoassay sensitivity. An antigen inhibition immunoassay was performed by diluting the indicated concentrations of indium into an excess (5 mM) of EDTA. Each point represents a two-fold dilution of the antigen. Values are mean ±SD of four to six determinations.

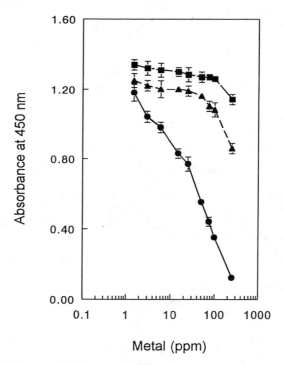

Figure 4. Metal ion specificity of indium immunoassay. Antigen inhibition immunoassays were performed using indium-EDTA (•), copper-EDTA (▲), or manganese-EDTA (■) as the inhibiting antigen. Each point represents the mean ±SD of four to six determinations.

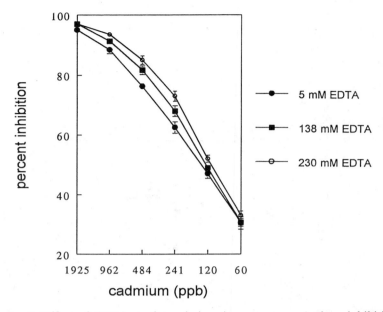

Figure 5. Effect of EDTA on the cadmium immunoassay. Antigen inhibition immunoassays were performed using cadmium diluted into 5 mM EDTA (•), 138 mM EDTA (■), or 230 mM EDTA (o) as the inhibiting antigen. Each point represents the mean ±SD of four to six determinations.

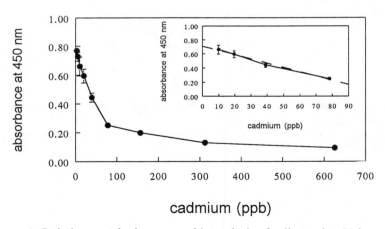

Figure 6. Cadmium standard curve used in analysis of soil samples. Values are mean ±SD of six determinations. *Inset*, linear portion of the curve used to quantify cadmium concentrations in soil extracts. The dotted line shows the best fit obtained by linear regression.

antigen inhibition format to measure samples of cadmium at different concentrations of EDTA. As shown in Figure 5, when atomic absorption grade cadmium was serially diluted into 5 mM EDTA, the assay measured cadmium at levels from 1925 to 60 ppb. Increasing the EDTA concentration to 138 and 230 mM had a minimal effect on assay performance at high cadmium concentrations and virtually no effect at the lower concentrations tested in these experiments. Since high concentrations of EDTA should facilitate extraction of cadmium and other metals from soil and sediment samples, 230 mM EDTA was used for all subsequent standard curves and in the assay of cadmium in a standard reference sample.

Assay of soil samples for cadmium content. The supernatants obtained from a EDTA extraction of a standard reference soil sample and a minimally contaminated control sample were assayed for cadmium content by indirect ELISA; the accompanying standard curve, prepared using atomic absorption grade cadmium, is shown in Figure 6. The linear portion of this curve (inset, 10-80 ppb cadmium) was used to analyze the cadmium content of the soil extracts. The cadmium content of both the standard reference material and the minimally contaminated control sample as assessed by ELISA were within two-fold of that reported using more conventional techniques of analysis. Standard reference material 2711, which was reported by NIST to have a cadmium content of 41.5, was determined to contain 71.5 ppm cadmium when assessed in the immunoassay. The minimally contaminated soil sample, which was determined by ICP to have a cadmium content of 3.7 ppm, gave a value of 8.0 ppm by immunoassay.

Discussion

The size and structure of a low molecular weight hapten can have a profound effect on antibody binding affinity. In any immunoassay, the ultimate limit of detection is directly related to this binding affinity. In an antigen inhibition immunoassay, the soluble antigen must compete with the antigen immobilized on the assay plate, which in the present assay consists of an indium-EDTA covalently bound to a carrier protein via a benzyl linker arm. Since the affinity of an antibody for its antigen is related to the number of molecular interactions that occur in the binding pocket, changing the structure of the soluble inhibiting antigen could make it a more effective inhibitor and thus increase the limit of sensitivity in the immunoassay. The 3-dimensional structure of the anti-indium antibody used in these studies has been determined by x-ray diffraction crystallography to 2.2-Å resolution (*16*). This study revealed that the antigen binding pocket accommodated not only the indium-chelate complex but also the benzyl linker arm and some of the peptide backbone of the carrier protein. We reasoned that including this benzyl linker arm as part of the soluble antigen would increase binding to the soluble antigen and make the compound a better inhibitor in the assay. Our results indicated that changing the soluble inhibiting antigen from indium-EDTA to indium-EDTA containing a covalently-linked benzyl group increased the sensitivity of the assay at least five-fold.

The use of the entire indium-EDTA-BSA molecule as the inhibiting antigen also significantly increased the sensitivity of the assay; however, there are technical drawbacks to the use of a metal-EDTA-protein conjugate as the inhibiting antigen. If an EDTA-protein conjugate were the only chelator present, then the concentration of metal binding sites in the assay would be limited by the extent of substitution and the solubility of the carrier protein. Other heavy metals, if present in excess, could conceivably fill all available sites on the limited quantity of carrier protein, thus interfering with the performance of the assay. In contrast, when indium-EDTA or indium-(p-nitrobenzyl)-EDTA are used as the soluble inhibiting antigen, the metal ion is diluted into a large excess of chelator and essentially all of the metal ions in the test solution exist as metal-chelate complexes, which are recognized most efficiently by the monoclonal antibody. The use of a low molecular weight chelator as the inhibiting antigen would be the preferred format for a field test, since the presence of excess chelator would insure that all the metal in the test solution would exist in a complex with chelator.

The crystallographic studies suggest another method to increase the sensitivity of our immunoassay. The original anti-indium antibody used in these studies was raised against a protein conjugate which contained a covalently-linked L-benzyl group (*17*). Love et al., (*16*) reported that the stereochemistry of this benzyl group is critical for the maximum number of interactions with the antibody. The L-benzyl-EDTA derivatives used in the crystallographic studies made a significant number of molecular contacts within the antigen binding pocket; however, if D-benzyl-EDTA-indium complexes were modeled into the antigen binding pocket in a manner that preserved all the interactions between the indium-EDTA complex and the protein, then the D-linked benzyl group extended out into the solvent. Thus the use of an immobilized metal-EDTA-protein conjugate containing a benzyl group linked in the D-configuration on immunoplates should increase the ability of soluble antigens to compete for the antibody binding site without significantly effecting the metal ion specificity of the assay. Studies are underway to determine whether this new approach will increase the sensitivity of our immunoassays.

The metal ion specificity exhibited by our immunoassay was qualitatively comparable to the metal binding properties of the antibody. The anti-indium antibody used in these studies (CHA255) was reported by Reardan et al. (*17*) to bind to indium-chelate complexes with an association constant of 4.9×10^9. The binding constants for manganese- and copper-chelates were reported to be approximately 1000 fold lower (2.8×10^6 and 1.7×10^6, respectively). In our studies, EDTA chelates of manganese and copper inhibited color formation in antigen inhibition immunoassay, but the concentration required for comparable inhibition was approximately 250-350 fold higher with each metal than it was for indium.

Although the indium assay has proven to be a valuable model for the development of immunoassays for heavy metals, environmental contamination with indium does not pose a health risk in the United States. New research has therefore been directed towards developing monoclonal antibodies to metals which have been designated as priority pollutants. A hybridoma which elaborates a monoclonal antibody directed towards cadmium-EDTA complexes has recently been developed by our laboratories (Blake et al., submitted). We have used this monoclonal antibody to begin the development of an assay for cadmium in soil and sediment samples. Existing protocols for the extraction of heavy metals from soil involve digestion with strong acid (*18*). Preliminary experiments demonstrated that high concentrations of EDTA (>100 mM) were as effective as strong acid digestion in extracting heavy metals from soil samples (Blake and Dawson, unpublished data). We therefore tested our anti-cadmium antibody to determine if these high EDTA concentrations had a deleterious on the enzyme immunoassay. The data in Figure 5 demonstrate that the anti-cadmium antibody is relatively unaffected by EDTA concentrations as high as 230 mM. Treatment with 230 mM EDTA was therefore used to extract cadmium from soil samples for immunoassay analysis. Two soil samples were tested in these preliminary analyses. The cadmium immunoassay was able to correctly differentiate between a heavily contaminated and a minimally contaminated soil sample, although the actual values for cadmium obtained in the immunoassay were approximately twice as high as those determined by other methods of analysis. This ability to differentiate between lightly and heavily contaminated soil samples will form the basis of a field assay for cadmium. Further experiments are in progress to optimize the field assay and to understand more about the positive interferences that we presently observe in the cadmium immunoassay.

Acknowledgments

This work was supported by a grant to D.A. Blake from the Environmental Protection Agency (R 824029) and by a grant from the Department of Energy (DE-FG01-93EW53023) to the Tulane-Xavier Center for Bioenvironmental Research. Grants from the National Science Foundation (RII-8804780 and RII-8714805) supported the Hybridoma Research Facility at Meharry Medical College.

Literature Cited

1. Ireland, M.P. In *Biological Monitoring of Heavy Metals*, Dillon, H.K.; Ho, M.H., Eds. John Wiley and Sons: New York, 1991; pp. 263-276.
2. Clarkson, T.W. In *Handbook on the Toxicology of Metals*, Friberg, L.; Nordberg, G.F.; Vouk, V.B., Eds. Elsevier: Amsterdam, 1986, Vol. 1; pp. 128-148.
3. Friberg, L.; Elinder, C.-G.; Kjellstrom, T.; Nordberg, G.F. *Cadmium and Health: A Toxicological and Epidemiological Appraisal;* CRC Press: Boca Raton, FL, 1986.
4. Friberg, L.; Kjellstrom, T.; Nordberg, G.F. In *Handbook of the Toxicology of Metals,* Friberg, L.; Nordberg, G.F.; Vouk, V.B., Eds. Elsevier: Amsterdam, 1986, Vol. 2; pp. 130-143.
5. Davis, J.M.; Svendsgaard, D.J. *Nature* **1987**, *329*, 297-300.
6. Schlesinger, L.; Arevalo, M.; Arredondo, S.; Lonnerdal, B.; Stekel, A. *Acta Pediatrics* **1993**, *82*, 734-738.
7. Baruthio, F. *Biological Trace Element Research* **1992**, *32*, 145-153.
8. Fleeker, J.R. *J. Assoc. Off. Anal. Chem.* **1987**, *70*, 874-878.
9. Van Emon, J.M.; Seiber, J.K.; Hammock, B.D. In *Analytical Methods for Pesticides and Plant Growth Regulators*, Academic Press: New York, 1989; Vol. XVII. pp. 217-263.
10. Thurman, E.M.; Meyer, M.; Pomes, M.; Perry, C.A.; Schwab, A.P. *Anal. Chem.* **1990**, *62*, 2043-2048.

11. Chakrabarti, P.; Hatcher, F.M.; Blake, II, R.C.; Ladd, P.A.; Blake, D.A. *Anal. Biochem.* **1994**, *217*, 70-75.
12. Thiers, R.C. *Meth. Biochem. Anal.* **1957**, *5*, 273-335.
13. Maggio, E.T. *Enzyme Immunoassay;* CRC Press: Boca Ratan, FL, 1980.
14. Sillen, L.G.; Martell, A.E. *Critical Stability Constants of Metal Ion Complexes;* Chemical Society: London, 1964.
15. Blake, II, R.C.; Bowers-Irons, G. In *Biocorrosion and Biofouling: Metal/Microbe Interactions*, Videla, H.A.; Lewandowski, Z.; Lutey, R.W., Eds. Buckman Laboratories International: Memphis, TN, 1993; pp. 162-170.
16. Love, R.A.; Villafranca, J.E.; Aust, R.M.; Nakamura, K.K.; Jue, R.A.; Major, J.G.; Radhakrishnan, R.; Butler, W.F. *Biochemistry* **1993**, *32*, 10950-10959.
17. Reardan, D.T.; Meares, C.F.; Goodwin, D.A.; McTigue, M.; David, G.S.; Stone, M.R.; Leung, J.P.; Bartholomew, R.M.; and Frincke, J.M. *Nature* **1985** *316*, 265-268.
18. Stoeppler, M. (1988) In *Biological Monitoring of Toxic Metals*, Dillon, H.K.; Ho, M.H., Eds. John Wiley and Sons: New York, 1991; pp. 481-497.

Chapter 3

Enzyme-Linked Immunosorbent Assay for the Detection of Mercury in Environmental Matrices

Craig Schweitzer, Larry Carlson, Bart Holmquist, Mal Riddell, and Dwayne Wylie

BioNebraska Inc., 3820 Northwest 46th Street, Lincoln, NE 68524

An immunoassay for the detection of mercury in environmental samples is described. The assay allows for real-time, user-friendly analyses at a fraction of the cost of traditional methods. The assay is available in two formats, a microplate format for large volume, quantitative laboratory analysis of samples and a tube format for rapid semi-quantitative analysis in the field.

Heavy metal contamination in the environment is recognized as a serious danger to humans and wildlife. Accordingly, the use of toxic heavy metals has become more strictly regulated; but careless practices in the past have led to massive deposits of these toxins in the environment. Mercury, one of the most toxic heavy metals, bioaccumulates up the food chain causing severe behavioral, reproductive and developmental problems (*1*). Environmental mercury largely derives from natural degassing of the earth's crust (*2*) and the burning of wastes and fossil fuels with additional lesser contributions from mining, smelting, chloralkali industries, electrical equipment, paint industries, military applications, agriculture, and medicine (*3*).

Analytical tools that can measure environmental contaminants in the field are central to the ability to regulate, manage, and decontaminate sites. Conventional analytical methods, such as atomic absorption, are generally precise, but can be used only in a laboratory setting. Immunochemical techniques provide sensitive and specific methods capable of measuring analytes of interest in complex biological matrices. The medical laboratory community has long recognized these qualities in immunoassays, but only in the last few years have they been adapted for use in detecting environmental contaminants.

BioNebraska has developed the BiMelyze immunoassay for the detection of mercury in environmental samples (*4,5*). The assay exists in two formats: a plate format which is quantitative and best suited for laboratory analyses, and a tube format which can be used for semi-quantitative

measurements in the field.   The immunoassay can specifically and
quantitatively detect mercuric ions in several different environmental matrices.
Analyses of laboratory and field samples using either format gives results that
are in good agreement with those obtained by more conventional analytical
methods, such as cold-vapor atomic absorption (CVAA), neutron activation
analysis (NAA), and X-ray fluorescence (XRF). The assay is not affected by
other metals at concentrations likely to be encountered in environmental
samples.
        The environmental sample, typically 5 g of soil or sediment, is
extracted for ten minutes with a mixture of hydrochloric and nitric acids. The
sample is buffered, filtered by means of a dropper bottle with an enclosed
filter, and then analyzed by the immunoassay. After extraction, the actual
immunoassay takes less than twenty-five minutes employing supplied
solutions of monoclonal antibody specific for mercuric ions, a secondary
enzyme-conjugated antibody specific for the monoclonal, and a chromophoric
substrate which is oxidized by the enzyme.   Mercury concentrations are
determined relative to supplied reference standards by means of a battery-
operated, field-portable, differential photometer available from BioNebraska.
The dilutions and additions of reagents are facilitated by dropper bottles
provided in the kit. The assay allows rapid and easy field testing of multiple
samples, resulting in lower evaluation and clean up costs at remediation sites.
Samples requiring analysis by slower, more expensive methods are therefore
minimized.   Based on in-house and independent field results, the assay
appears to be well-suited for low-cost, real-time, user friendly field screening
of mercury in the environment.   Other substances present in environmental
matrices do not appear to interfere with the assay. Results correlate well with
traditional analytical methods. In addition, preliminary data suggests that the
immunoassay can be applied to the measurement of mercury in seafood and
animal tissues, so that potential problems resulting from biomagnification of
mercury can be identified before the contaminated food sources are
consumed.

Experimental

Monoclonal Antibodies.   The production and characterization of the
mercury-specific, mouse monoclonal antibodies used in these analyses were
described previously (4,5).

Extraction of Samples for Immunoassay. Mercury is extracted from the
environmental sample using a kit available from BioNebraska. The procedure
requires digesting a 5 g sample, representative of the area being tested, in a
solution of concentrated hydrochloric acid, concentrated nitric acid, and
water (2:1:1) for ten minutes with intermittent, gentle agitation. The acids for
the extraction (ACS reagent grade or better) are provided by the end user or,
alternatively, are available from an independent supplier. After extraction, the
sample is buffered, filtered, and diluted by means of filter-tipped dropper

bottles provided in the soil extraction kit, then analyzed by the immunoassay. Standards, included in the kit, are extracted at the same time as the unknown samples. Comparisons of the unknown samples to the standards allow for semi-quantitation of mercury in the field.

BiMelyze Mercury Assay. The assay has been developed in two formats: a quantitative 96-well, plate method for analyzing large numbers of samples in the laboratory, and a semi-quantitative, field-portable tube method. Both formats consist of sequential addition of four reagents, with a five-minute incubation period for each. The BiMelyze immunoassay is based on the initial binding of the mercuric ion from a sample to the thiol group of glutathione (GSH) that has been covalently linked to bovine serum albumin (BSA) and bound to a solid support (Figure 1). After a water rinse, the mercury specific antibody is added. This primary antibody binds to mercury which is bound to the BSA-GSH conjugate. The tubes/wells are washed with a detergent then rinsed with water to remove unbound antibody. The amount of bound mercury specific antibody is detected by binding horseradish peroxidase (HRP)-labeled rabbit secondary antibody to the mouse primary antibody. After washing as above, substrate, which is oxidized by the peroxidase-labeled antibodies, to produce a green chromogen, is added. The color intensity that develops is directly proportional to the amount of mercury in the initial sample. Color development is terminated by addition of stop solution, and the tubes are read within an hour. The absorbance of the color in the tubes/plate is measured at 405 nanometers. Microplate readers, present in most larger laboratories, are used to read the color development of the plate assay. A field-portable, battery-powered, differential photometer, available from BioNebraska, allows the absorbance in the tube assay to be quantified.

Metal Specificity. Standard assays were performed as above except that various concentrations of additional metal salts were added to the samples prior to assay, and their effect on color development in the assay was determined.

Results

The lower limit of quantitation of mercuric ions with the BiMelyze immunoassay in a buffered aqueous solution is 0.25 parts per billion (ppb) for both the tube and plate assays. Figure 2 shows a dose-response curve in which Mercury Standard Reference Material 3133 from the National Institutes of Standards and Technology was diluted to various concentrations in 0.1 M HEPES buffer, pH 7.0, and analyzed with the tube assay. The results demonstrate that the absorbance in the ELISA is linear and reproducible over the range from 0.25 to 25 ppb with all coefficients of variation below 8%. Similar results have been reported previously for the mercury-specific plate assay (5).

Principle of the Mercury ELISA

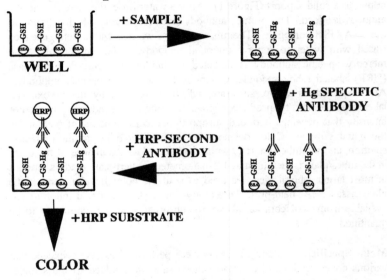

Figure 1. Basis of the BiMelyze Mercury Immunoassay ELISA. The schematic shows the major steps and the reagents used to perform the assay (see text).

Since the soil samples must be acid extracted and neutralized before assay, it was necessary to know the pH dependence of mercury binding to the ligand-coated tubes. This was tested with mercury concentrations of 0 and 1 ppb in buffers at pH 2, 4.75, 6, 7, 8, 8.75, 10, and 11.8 (Figure 3). The binding is essentially unaffected over the pH range 4.75 to 8.75. Consequently, the pH of the samples is adjusted to 7 for routine analysis.

Because many metals are ubiquitous in the environment, their effect on the reliability of the mercury-specific assay was characterized in detail using several approaches. First, a standard curve was constructed in which known concentrations of mercury were diluted in a multi-metal mixture and measured in the immunoassay. The composition of this mixture corresponded to that of the EPA Extract Metals Quality Control Sample formerly available from the Environmental Protection Agency. It contains 100 mg/L barium nitrate, 1 mg/L cadmium sulfate, 5 mg/L lead nitrate, 5 mg/L silver nitrate, and 5 mg/L chromium trioxide. Solutions of mercury at concentrations of 20, 2, 0.5 and 0.2 ppb were prepared in the metal mixture and used directly in the immunoassay without dilution. The results obtained with these samples were compared to a standard with mercury at the same concentrations in a buffer at pH 7.0 which did not contain the other metals. The standard curves obtained with mercury in these two diluents (Figure 4) are essentially identical, indicating that these concentrations of metals have no effect on the mercury immunoassay.

The potential interference by individual metals was examined over a wide concentration range, to a level higher than would normally be present in field samples (Figure 5). Standard curves were constructed in which mercury at 100, 50, 10, 5 and 0.5 ppb was diluted into solutions containing the indicated concentrations of these metals. According to the experimental design, for each metal there were five separate mercury-specific standard curves, each containing 1 mM, 10 mM, 100 nM, 10 nM and 1 nM of a potentially interfering metal. A control standard curve was also included in which the mercury was diluted to the same concentrations in an equal volume of metal-free buffer. The metal compounds examined were: arsenic trioxide, barium nitrate, cadmium chloride, chromium nitrate, cupric chloride, gold trichloride, iron sulfate, lead chloride, nickel chloride, silver nitrate, sodium bicarbonate, sodium chloride, strontium nitrate, thallium nitrate and zinc chloride.

Barium nitrate (Figure 5a), which gave results typical of most metals, shows no interference even at the highest concentrations employed. Only three of the metals tested affected mercury detection, but they did so only at high concentrations. Gold trichloride inhibited the response at the two highest concentrations (Figure 5b). The highest concentration of gold trichloride caused a purple precipitate when added to the tube in the first step of the assay (data not shown). Silver nitrate produced an increase in absorbance at the two highest concentrations (Figure 5c), which might be related to silver salt precipitation. An increase in signal was also seen with 1 mM chromium nitrate (data not shown).

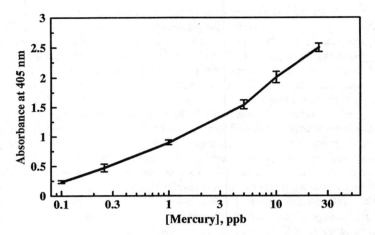

Figure 2.   Dose-response curve for mercury detection by the BiMelyze Mercury Tube Assay. Mercuric nitrate was diluted to final concentrations of 0, 0.25, 1, 5, 10, and 25 ppb in pH 7.0 buffer. The mercury solutions were then analyzed by ELISA as described in the Experimental section, with six replicates for each concentration.

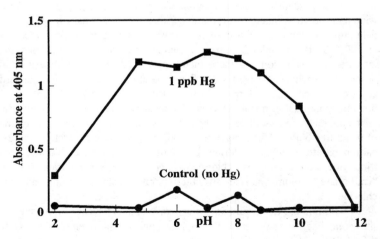

Figure 3. Effect of pH on the mercury binding to the ligand coated tubes as detected by ELISA. Solutions contained either 0 (●) or 1 (■) ppb mercuric nitrate in buffer adjusted to the indicated pH with either 1 N HCl or 1 N NaOH. The remaining steps in the assay are unchanged. Solutions were then used in the ELISA as described in the Experimental section.

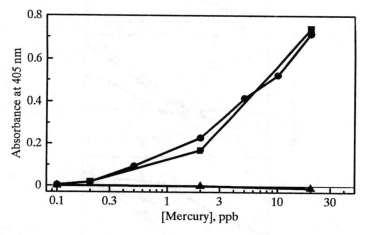

Figure 4. Detection of mercury in EPA Extract Metals Quality Control Sample. Mercuric nitrate was diluted to final concentrations of 0, 0.2, 2, or 20 ppb in a solution containing metals at the concentrations present in the EPA Extract Metals Quality Control Sample (■), as described in the Experimental section. A standard curve was then obtained by analysis of these solutions in the ELISA and compared to that obtained with mercury diluted to the same concentrations in 0.1 M HEPES buffer, pH 7.0 (●). A sample containing metals at the same concentrations as in the EPA quality control sample but without mercury was also included in the assay (▲).

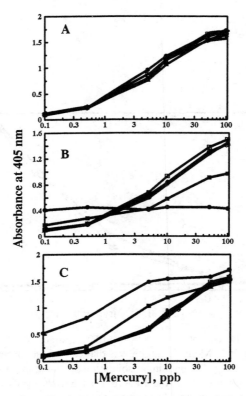

Figure 5. Effect of other metal salts on mercuric ion detection by ELISA. Standard curves of mercury concentrations ranging from 0.5 to 100 ppb were assayed in solutions containing potentially interfering metals at final concentrations of 1 nM (O), 10 nM (X), 100 nM (▲), 10 mM (■), and 1 mM (●). A control curve in which the mercury was diluted into pH 7.0 buffer was also included (□). These solutions were then used in the ELISA as described in the Experimental section. For each concentration of metal salt, a control point containing the same concentration of metal salt but with no added mercury was included and is represented by the bottom-left points. Figure 5a represents typical results obtained with the listed metals. Figure 5b shows the inhibition by gold trichloride, and Figure 5c shows the increase in signal with silver nitrate.

Finally, the mercury content of several soil samples containing certified amounts of various metals was measured. The descriptions of these samples and their metal compositions are shown in Table I. Triplicate samples were extracted according the BiMelyze protocol, and analyzed with the BiMelyze Mercury Tube Assay. By comparison with soil standards containing either 4 ppm or 15 ppm, the results were interpreted as less than 4 ppm, between 4 and 15 ppm, or greater than 15 ppm. As shown in Table II, the immunoassay correctly predicted the mercury concentrations of these samples in almost all cases. The only incorrect determination was in experiment #3 with soil sample #4.

The reproducibility of the soil assay was examined with 5 g aliquots of reference soils containing 0, 1.0, 2.0, 3.2, 4.0, 4.8, 6.0, and 8.0 ppm mercury, as determined by CVAA. Seven replicate analyses were done of each. The results (Table III) are presented as the differential absorbance of each mercury concentration relative to a 4 ppm standard. The data demonstrate the ability of the test to distinguish between small differences in mercury concentration in soils around a 4 ppm action level. They also show the reliability of the assay, since only one false-negative and one false-positive were obtained in these analyses.

To demonstrate the usefulness of the assay under field conditions, the BiMelyze Mercury Tube Assay was used to analyze ten environmental samples. Mercury concentrations were also measured by both NAA (n=10) and XRF (n=4). The results are shown in Table IV.

The general format of the study was to compare the unknown soil samples to the 5 and 15 ppm mercury-in-sediment standards. The data for the tube assay were interpreted as less than 5, 5-15, or greater than 15 ppm by comparison of the absorbance of each sample to that of the standards. The tube assay gave excellent agreement with the reference methods, differing from NAA in only two samples (#3 and #5). However, XRF analysis of sample #5 agreed with the immunoassay rather than with NAA. Sample #3 was not analyzed by XRF.

Another field study was conducted by an environmental testing company who collected samples and analyzed them with the BiMelyze Mercury Tube Assay and by CVAA. The description of the samples, along with the results of the analysis (Table V) indicate a good agreement by both methods, showing a disparity with only one sample. However, even with that sample, the difference was not large, since the CVAA value was 14 ppm and the BiMelyze results were greater than 15 ppm. This independent analysis tested matrices which have not been tested by BioNebraska (e.g., paint, cinderblock, and sludge), but suggests the versatility of the method. The matrices for which the BiMelyze assay is applicable appears to be limited only by the ability of the acid mixture to disassociate and oxidize mercury to the mercuric form.

Table I. Metal Compositions of Certified Reference Soils

Metal	Sample Metal Composition, ppm					
	1[a]	2[b]	3[c]	4[d]	5,6,7[e]	8[f]
Aluminum	4090	7.5%[g]	6.11%	6000	6.1%	4090
Antimony	<12	7.9	3.79	27.8	na[h]	<12
Arsenic	<2	17.7	23.4	67.7	7.3	<2
Barium	50.3	968	414	187	360	50.3
Beryllium	<1.0	na	na	57.5	na	<1.0
Cadmium	<1.0	0.38	3.45	110	na	<1.0
Calcium	1190	1.89%	2.6%	2040	na	1190
Chromium	6.63	130	135	189	88	6.63
Cobalt	<5.0	13.4	14.0	87.0	19	<5.0
Copper	<5.0	34.6	98.6	141	na	<5.0
Iron	8710	3.5%	4.11%	10800	2.6%	8700
Lead	8.01	18.9	161	100	80	8.01
Magnesium	1100	1.51%	1.2%	2050	na	1100
Manganese	167	538	555	294	1400	167
Mercury	<0.10	1.4	1.47	2.36	107	122
Molybdenum	na	2.0	na	124	na	na
Nickel	<8.0	88	44.1	79.6	na	<8.01
Potassium	1310	2.03%	2.00%	2130	1.5%	1310
Selenium	<1.0	1.57	1.12	99.1	na	<1.0
Silver	<2.0	0.41	na	124	5	<2.0
Sodium	<260	1.16%	0.55%	527	1200	<260
Thallium	<1.0	0.74	1.06	67.9	na	<1.0
Vanadium	15.4	112	95	84.8	60	15.4
Zinc	23.6	106	38	197	160	23.1

[a] Environmental Research Associates Inorganics Blank Soil.

[b] NIST Standard Reference Material (SRM) 2709 San Joaquin Soil.

[c] NIST SRM 2704 Buffalo River Sediment.

[d] Environmental Research Associates Priority Pollutant/CLP Lot # 216.

[e] NIST-SRM 8408 Mercury in Tennessee River Sediment.

[f] Environmental Research Associates Custom Mercury Standard.

[g] Weight percent basis.

[h] na=not analyzed.

Table II. Analysis of Certified Reference Soils Using BiMelyze Mercury Tube Assay

Soil Sample	[Mercury] ppm	A_{405}[a]			Interpretation
		Exp. 1	Exp. 2	Exp. 3	
1	<0.10	0.12	0.05	0.08	Neg[b]
2	1.40	1.01	0.64	0.47	<4
3	1.47	0.78	0.41	0.19	<4
4	2.36	1.54	0.84	0.93[c]	<4
5	4[d]	1.76	1.01	0.83	4[b]
6	15[d]	1.99	1.45	1.59	15[b]
7	50[d]	2.04	1.73	2.02	>15
8	122	2.55	2.55	2.55	>15

[a] Absorbance at 405 nanometers with the BiMelyze Differential Photometer
[b] Standard reference point, no interpretation
[c] Only value which gives incorrect conclusion
[d] NIST-SRM solid phase diluted from an initial concentration of 107 ppm.

Table III. Reproducibility of BiMelyze Mercury Assay Tube Kit with Extracted Soil Samples

Seven replicate 5 g soil samples were extracted with a 4 mL mixture of 2:1:1 hydrochloric, nitric acid and water. The samples were then analyzed by both the tube assay and CVAA. CVAA data represents an average of the seven analysis and the immunoassay data are presented as the difference relative to a 4 ppm standard.

[Hg] ppm	[AA] ppm	TUBE ASSAY EXPERIMENTS						
		1	2	3	4	5	6	7
0.0	0.0	-1.38	-1.32	-1.02	-0.57	-1.05	-0.98	-1.03
1.0	1.05±0.07	-0.63	-0.62	-0.57	-0.34	-0.69	-0.51	-0.41
2.0	2.08±0.12	-0.25	-0.38	-0.40	-0.16	-0.47	-0.37	-0.08
3.2	3.27±0.14	-0.06	-0.26	-0.17	+0.28	-0.22	-0.05	-0.04
4.0	4.16±0.24	-----[a]	-----	-----	-----	-----	-----	-----
4.8	4.93±0.31	+0.26	+0.05	+0.03	+0.46	+0.16	+0.15	+0.02
6.0	6.11±0.32	+0.23	-0.10	+0.07	+0.66	+0.21	+0.22	+0.21
8.0	7.97±0.36	+0.23	+0.15	+0.30	+0.92	+0.25	+0.37	+0.19

[a] 4.0 ppm used as standard

Table IV. Analysis of Mercury in Soils Using the BiMelyze Mercury Assay Tube Kit

Sample	Concentration by NAA ppm	XRF ppm	Absorbance at 410 nm	Interpretation
1	na[a]	na	0.077	Neg[b]
2	na	na	0.131	5[b]
3	na	na	0.216	15[b]
4	116	90-116	0.357	>15
5	<3.3	na	0.104	0-5
6	11	na	0.293	>15
7	<2.1	na	0.096	<-5
8	<5.3	22-52	0.292	>15
9	<1.5	na	0.127	0-5
10	<7.7	na	0.088	0-5
11	87	150-159	0.337	>15
12	19	na	0.259	>15
13	121	18-122	0.264	>15

[a] na=not analyzed
[b] Standard controls

Table V. Independent Analysis of Mercury in Samples Using the BiMelyze Mercury Assay Tube Kit at an Abandoned Battery Reclamation Site

SAMPLE DESCRIPTION	TEST KIT (ppm)	CVAA[a] (ppm)
Process Room	< 5	0.83
Dust from process room	< 5	> 4.5
Soil, alkaline	< 5	0.93
Sludge from tank	> 15	4,400
Sump sludge	5-15	14
Cinderblock	< 5	3
Cinderblock- duplicate	< 5	na[b]
Soil	5-15	14
Paint	> 15	34
Background cinderblock	< 5	1.4
Background paint	> 15	14
Debris from CO_2 blast	> 15	19
Groundwater	< 0.5 ppb	< 0.4 ppb

[a] Cold Vapor Atomic Absorption
[b] na=not analyzed

Discussion

The BiMelyze mercury tube immunoassay provides an accurate, reliable method for detecting mercury in a variety of matrices. The tube assay was designed as a field test for on-site evaluation of environmental samples. Under these conditions, the assay is semi-quantitative. The mercury concentration is determined by direct comparison to a standard with a known amount of mercury. The tube assay results agreed with those obtained by a reference method (either NAA or CVAA) in at least 20 of 22 various samples (Tables IV and V). The importance of determining the mercury concentration by comparison to a standard analyzed at the same time as the unknown samples must be emphasized. The mercury assay is an enzyme-linked immunosorbent assay (ELISA) whose interpretation depends on the absorbance obtained by enzymatic conversion of a colorless substrate to a colored product (Figure 1). The absorbance obtained with samples having the same mercury concentration can vary from day to day (Table II), since the enzyme activity of the assay is affected by ambient conditions. Analysis of a reference standard at the same time as the unknown samples controls for this variability.

With environmental samples, acid digestion is needed to extract total mercury from constituents of the matrix. This treatment oxidizes free mercury and mercurous forms to mercuric ions, for which the antibody is specific. Excellent correlation has been reported previously when the plate assay was used for analysis of environmental samples and compared to other analytical methods (6). The reliability of the results obtained in the previously reported study led to the inclusion of the BiMelyze Mercury Tube and Plate Assays into the Department of Energy Methods Compendium as Method MB 100. The method can accurately measure mercury in a variety of environmental samples. Independent organizations have tested cinder block and paint chips in addition to various soils, and it is likely to be applicable to other matrices which have not previously been assayed successfully (Table V). Methylmercury is not recognized by the mercury specific antibody. We are currently working on an efficient, rapid, and user-friendly extraction and oxidation protocol for application in biological tissues such as fish.

In the past, accurate testing of environmental samples for mercury has been limited by the availability of analytical methods, such as CVAA, that utilize expensive equipment requiring highly trained personnel for proper operation. Another disadvantage of these procedures is the lag time between sample collection and acquisition of the results, and problems arising from the instability of the sample, since, in most cases, the samples must be sent off to reference laboratories for analysis. In contrast, the BiMelyze mercury assay provides a convenient, cost-effective, rapid method for monitoring and surveying environmental sites for mercury that can be performed in the field by personnel with minimal training. Its use can thus reduce the number of samples that must be analyzed by more expensive, traditional methods. The rapid data acquisition reduces potential re-mobilization costs which can occur if initial remediation is insufficient. The method is ideal for measurement of mercury in remote areas where sample storage, inventory and transportation present logistical problems. The kit is stable for at least six months at 4°C and for shorter periods of time at elevated temperatures. The only instrumentation required is a field-portable spectrophotometer that is inexpensive (<$1,000) compared to the instrumentation needed for traditional analytical methods. The method has a high selectivity for mercury. Further, it is not affected by metals or other ions likely present in environmental samples, (Figure 5) as they are diluted more than 700 fold in the immunoassay.

With increased awareness on the part of both the general public and various governmental regulatory agencies concerning toxic chemicals in the environment, the demand for convenient, reliable methods for their detection will certainly increase. Although environmental immunoassay technology is relatively new, it provides an excellent way for regulatory agencies to implement effective monitoring programs on increasingly tight budget constraints. Immunoassays can be used in conjunction with traditional analytical methods to allow a larger number of samples to be analyzed at a lower total cost.

*Portions of this material were presented at the tenth annual Waste Testing & Quality Assurance Symposium, July 1994, sponsored by the American Chemical Society, Arlington, Virginia.

Literature Cited

1. Friberg, L.; Nordberg G.F.; & Voulk V.B. *Handbook on the Toxicology of Metals 2*, 1986; pp 389-396.
2. *World Health Organization Environmental Health Criteria 86, Mercury-Environmental Aspects*, WHO, Geneva, 1989.
3. Fitzgerald, W.F. *The World & I, "Mercury as a Global Pollutant"*; 1986; pp 192-199.
4. Wylie, D.E.; Lu, D.; Carlson, L.D.; Carlson, R.; Babacan, K.F.; Shuster, S.M.; & Wagner, F.W. *Proc. Natl. Acad. Sci. USA.* **1992**, *vol 89*, pp 4104-4108.
5. Wylie, D.E.; Lu, D.; Carlson, L.D.; Carlson, R.; Wagner, F.W.; & Shuster, S.M. *Anal. Biochem.* 1991, *vol 194*, pp 381-387.
6. *Department of Energy Methods for Evaluating Environmental and Waste Management Samples*; Goheen, S.C.; McCulloch, M.; Thomas, B.L.; Riley, R.G.; Sklarew, D.S.; Mong, G.M.; and Fadeff, S.K. 1993; Method MB100, "Immunoassay for Mercury in Soils", pp.1-9.

Chapter 4

A New Approach to Electrochemical Immunoassays Using Conducting Electroactive Polymers

Omowunmi A. Sadik[1]

Intelligent Polymer Research Laboratory, Department of Chemistry, University of Wollongong, New South Wales 2522, Australia

The problem of generating a rapid, sensitive and reversible electrochemical signal with antibody-antigen (Ab-Ag) interactions has previously been addressed. It was shown that the use of antibodies immobilized in conducting electroactive polypyrrole matrices, with pulsed amperometric detection, and flow injection analysis, provides a unique solution to this problem. A sub-ppm detection limit for the target protein thaumatin was obtained, and a high selectivity towards other non-target proteins was realized. These encouraging results have resulted in further scrutiny of the principle of the mechanism involved as reported in this paper. Evidence from electrochemical quartz crystal microbalance studies (EQCM) and cyclic voltammetry confirmed that a reversible mass increase was obtained in the presence of the antigen. The results showed that the application of alternating voltage waveforms induced changes in the conducting polymers such that a detectable interaction with a target analyte (antigen) was obtained in a reversible manner. Thus the detection method resulted in a reusable immunological sensor that responded within a time scale of minutes.

The use of antibodies in electrochemical sensing technologies promises a degree of selectivity previously unattainable (1,2). In practice, however, some difficulties arise which affect the generation of a sensitive analytical signal and the reversibility of the antibody-antigen (Ab-Ag) interaction. Several attempts at overcoming these problems include: the use of potential measurements, indirect amperometric immunoassay, as well as direct measurements of changes in capacitance at the sensor surface.

[1]Current address: National Exposure Research Laboratory, U.S. Environmental Protection Agency, P.O. Box 93478, Las Vegas, NV 89193–3478

In most of these cases, the procedures do not address the Ab-Ag reversibility issue.

Conducting electroactive polymers (CEPs) represent a new class of electrode materials which are polymeric and yet conductive. These polymers can be switched from a highly conductive state to a resistive state by controlling the electrode potential. This electrochemical conversion involves mass and charge transport in the polymer film. The immobilization of specific molecules capable of substrate recognition is carried out during polymerization (3). The resulting CEP-based biosensors have been proposed for direct and continuous detection of low concentrations of organic species in process streams, environmental samples, and biological fluids (3,4). The use of CEP-modified electrodes in the detection of simple inorganic ions, halogenated acetic acids, and other small organics has been demonstrated (5-7).

Recently, it has been shown that the use of pulsed amperometric detection (PAD) provides a sensitive and selective analytical signal for complex biological molecules such as proteins (8 -11). The fundamental idea is that a protein molecule can selectively bind to a specific biological molecule incorporated into a conducting polymer membrane assembly. The binding event leads to a change in the surface nature of the polymer matrix upon the application of pulsed potentials. Thus, the transduction can be quantitatively measured. One unique advantage of this approach is that it is generic enough to be applicable to the interactions of enzymes, antibodies, receptors, or cells. At the same time, it is sensitive enough to meet the analytical requirements for biosensors.

The immunological biosensor research described here utilizes a novel pulsed amperometric detection methodology for the generation of a useful analytical signal involving Ab-Ag interactions. The overall objective is the development of a simple antibody-based analytical tool which utilizes the intrinsic signal generation capability of the antibody to reversibly detect antigen (analyte) in real time without the use of enzyme or optical labels. In the course of this work, the incorporation of antibodies into conducting polymers was probed. The electrochemical control of the Ab-Ag interaction in effecting electrical signal generation was investigated. Furthermore, the issue of the effects of the applied pulsed potential producing changes in the structure of the polymer matrix, and the impact of the protein binding steps in giving rise to an analytical signal was also addressed.

Experimental

Polypyrrole (PP) electrodes were prepared by galvanostatic electropolymerization of pyrrole monomer from an aqueous solution containing antibody (Ab) as previously reported (8-10). The procedure can be represented by equation 1:

$$\tag{1}$$

The characterization of the PP/Ab-containing film was carried out by means of electrochemical methods such as cyclic voltammetry, and quartz crystal microbalance (EQCM). The PP/Ab films were used as detectors for the antigen by applying a unique pulsed waveform and measuring the resulting current signals.

Results & Discussion

The new sensing system requires that antibodies be incorporated into a conducting polymer matrix, and bioactivity be maintained. The polymer produced as such was conductive and electroactive as shown by the cyclic voltammograms recorded for the polypyrrole/anti-thaumatin (PP/ATHAU) electrode (Figure 1). The responses observed are due to oxidation/reduction of the polymer and there was no change in the voltammogram obtained when the polymer was exposed to the antigen.

This sensing element was then used in a flow injection analysis (FIA) system (Figure 2). The analysis involved the movement of analyte in a stream of eluent through the detection cell. The residence time in the detection cell was short (less than 1 min); hence the signal generation was fast. In order to achieve a rapid signal generation, a pulsed potential waveform was employed (Figure 3). E_1 and E_2 were chosen such that Ab-Ag interactions were encouraged at E_1 and then discouraged at E_2. The frequency (pulse width t_1 and t_2) was such that the Ab-Ag interaction did not reach a stage where it became irreversible. The electrical signal was obtained by repetitive sampling of the current at a specific current sampling time (at t_1). Pulsing to a more positive potential produced a small response which did not increase in magnitude, whereas the signal increased in magnitude as the potential was pulsed to more negative values. A well defined (non-tailing) and reproducible signal response was obtained during FIA using the pulsed potential routine as shown in Figure 4. The response obtained was very rapid (in minutes), sensitive (about 3-4 orders of magnitude) and reproducible current responses (± 3% over 10 injections) were observed.

The selectivity towards other test proteins was investigated by the injection of these proteins into the flow stream. As shown in Figure 5, responses to these other proteins were obtained at the PP/ATHAU electrode.

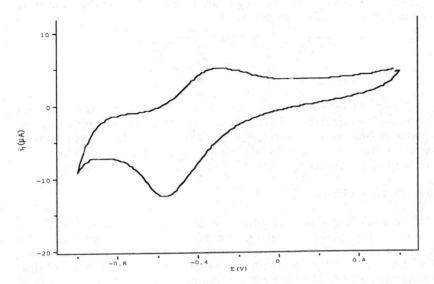

Figure 1. Cyclic voltammogram obtained using PP/ATHAU. Electrolyte was 0.05 M Phosphate buffer (pH 7.4). Scan Rate = 100 mV/sec. Polymerization solution contained 0.5 M Pyrrole and 100 mg/l Anti-thaumatin solutions made up in distilled water, current density = 0.5 mA/cm^2, growth time = 5 minutes.

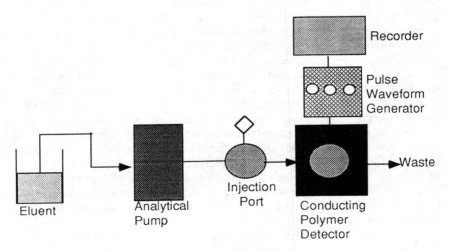

Figure 2. Flow Injection Analysis System

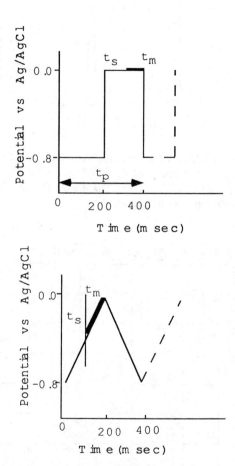

Figure 3. Typical potential waveforms employed: t_s = sampling time, t_m = sampling period, t_p = pulse width.

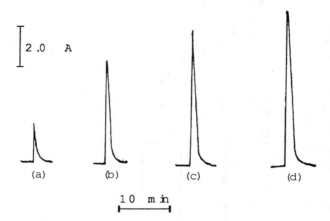

Figure 4. Typical FIA signals recorded for the injection of 50 mg/L thaumatin solutions using PP/ATHAU electrode at different pulse potentials. $E_1 = 0.00V$, E_2 = (a) 0.2 V, (b) 0.4 V, (c) 0.6 V, (d) 0.9 V .Other conditions as in Figure 1.

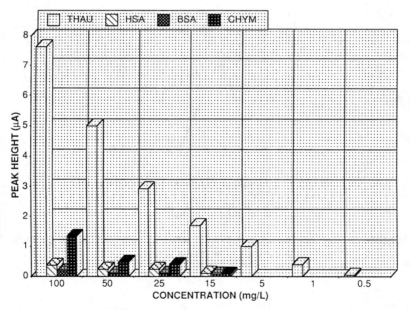

Figure 5. Response obtained by injecting various proteins. THAU = Thaumatin, BSA = Bovine serum albumin, HSA = Human serum albumin, CHYM = Chymotrypsin. Other conditions as in Figure 1.

However, the signal was much lower in magnitude than that recorded for the same concentration of thaumatin. It may be possible that these responses arose from non-specific interactions on the polymer backbone.

The reversibility of the Ab-Ag signal was also considered. This was confirmed with the use of EQCM. The PP/ATHAU electrode was coated on quartz crystal by galvanostatic electrodeposition. Changes in the mass of the quartz crystal following the oxidation/reduction of the attached electroactive polymer films can be given by the Sauerbrey equation (12) which relates changes in resonant frequency of the quartz crystal to mass changes in the film:

$$\Delta f = \frac{-2f_o^2}{\sqrt{\rho_\Omega \mu_\Omega}} \cdot \frac{\Delta m}{A} \tag{2}$$

where ρ_Ω is the density of quartz (2.68 g/cm^3), μ_Ω is the shear modulus of quartz (2.947×10^{11} dynes cm^{-2}, 1 dyne = 10^{-5}N), f_o is the resonance frequency of the unloaded quartz crystal (6 MHz), and A is the piezoelectrically active area of the quartz (cm^2).

The results indicated a mass increase on reduction, and a decrease on reoxidation (Figure 6). A slight increase in mass was recorded for the PP/ATHAU electrode coated on quartz crystal using only 0.05 M phosphate buffer (Figure 6a), but there was a notable increase in the mass recorded upon the addition of 100 ppm thaumatin (Figure 6 b). The observed mass increase, which was reversible, may be due to the interaction of thaumatin with the anti-thaumatin antibody as the polymer was reduced.

Conclusions

A rapid, sensitive and reproducible detection method for antigens based on the use of polypyrrole-antibody with pulsed amperometric measurements was developed for use with FIA systems. A sensitive analytical signal was obtained by using a unique electrical signal generation process available with conducting polymers and a pulsed potential waveform. The selectivity was enhanced by direct incorporation of antibody-based bio-recognition sites into the conducting polymer materials. The results obtained with EQCM experiments showed that the mass increase was reversible. Current studies are being focused on the determination of the mechanism of signal generation as well as the transfer of this biosensing technology onto microelectrodes. Immunosensor methods are therefore recommended for direct and continuous monitoring of environmental samples.

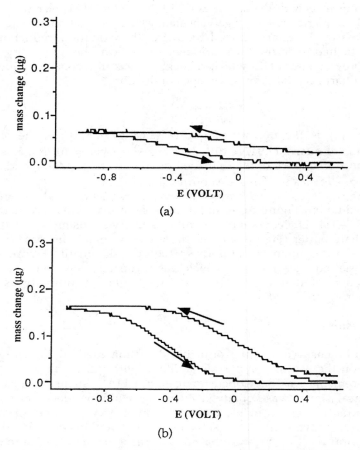

Figure 6. Mass *versus* potential profile using PP/ATHAU electrode in (a) 0.05 M phosphate buffer only, (b) 100 mg/L thaumatin in 0.05 M phosphate solution, scan rate = 20 mV/s. Other conditions as in Figure 1.

Acknowledgment

The author gratefully acknowledges the support of the Australian Government through the award EMSS fellowship. Invaluable discussions held with Gordon Wallace, Zhao Huigin, Lin Yuping, Chee on Too and Norm Barisci in the early stages of this work are also acknowledged.

Literature Cited

1. North, J.R., *Trend. Bio. Tech.*, 1985, 7, 180
2. Nagy, G., Pungor, E., *Bioelectrochem. Bioenerg.*, 1988, 20, 1
3. Bidan G., *Sensors & Actuators,* B6 (1992) 45.
4. Sadik O.A.,*Analytical Methods & Instrumentation* (In Press).
5. Omowunmi A. Sadik, Gordon G. Wallace, *Electroanalysis*, 5 (1993) 555.
6. Sadik O.A., Talaie A., Wallace G.G., *J.Intel. Mat. Syst. and Struc.*, 4 (1) (1993) 123.
7. Omowunmi A. Sadik, Gordon G. Wallace,*Electroanalysis*, 6 (1994) 860.
8. Sadik O. A., Wallace G. G., *Anal. Chim. Acta.*, 297 (1993) 209.
9. Sadik O. A., Wallace G. G., *Proc. 207th ACS National Meeting, March 13 - 18, 1994, San Diego, CA., USA.,* Vol.70 (1994) 178.
10. Barnett D., Laing D.G., Skopec S., Sadik O.A., Wallace G.G., *Anal. Lett.*, 27 (13), (1994) 2417.
11. Sadik O.A., John M.J., Wallace G.G., Barnett D., Clarke C., Laing D.G., *Analyst*, 119 (1994) 1997.
12. Sauerbrey G., *Z. Phys.*, 155 (1959) 206.

Chapter 5

Environmental Immunosensing at the Naval Research Laboratory

Lisa C. Shriver-Lake, Anne W. Kusterbeck, and Frances S. Ligler

Center for Bio/Molecular Science and Engineering, Naval Research Laboratory, Code 6910, 4555 Overlook Avenue Southwest, Washington, DC 20375–5348

In the Center for Bio/Molecular Science and Engineering at the Naval Research Laboratory (NRL), two different types of immunosensors are being developed for detection of environmental pollutants and monitoring of bacteria for bioremediation. Both biosensors are rapid, sensitive and easy to operate. The first sensor, the continuous flow immunosensor, is based on displacement of fluorescently-labeled antigen from antibodies immobilized on beads. Antigen is injected into a flow stream that passes over a 100 μL bed volume of antibody-coated beads saturated with fluorescently-labeled antigen. The displacement of the labeled antigen causes an increase in fluorescence, proportional to the antigen concentration, to be observed downstream. The other sensor, the fiber optic biosensor, utilizes long, partially clad optical fibers. Antibodies are immobilized onto the fiber core in the unclad region at the distal end of the fiber. Upon binding of antigen and a fluorescent molecule, a change in the fluorescence signal is observed. For small molecules, competitive immunoassays are performed in which a decrease in the fluorescence signal is observed which is proportional to the antigen concentration. For bacterial cells, sandwich or direct immunoassays are performed which generate an increase in the fluorescence signal proportional to the specific cell concentration.

Accurate monitoring of the environment for pollutants and other toxic chemicals has become increasingly important in recent years. Current technologies, such as gas chromatography/ mass spectrometry (GC/MS), are costly and time-consuming. A single sample may require up to a week to analyze at a cost of $1000-$2000. The development of antibodies specific for over 50 pollutants and organic compounds within the last few years now makes it possible to exploit immunoassay technology for the detection of environmental contaminants.(1)

To meet U.S. Navy and U.S. EPA guidelines for monitoring and controlling military/industrial pollution, NRL has developed two biosensors. The continuous flow immunosensor is a relatively simple device for repetitive analyses for small molecules. The fiber optic biosensor is technically more complex, but can be adapted to measure both small molecules and the bacteria capable of degrading them.

Continuous flow immunosensor

Continuous flow immunosensor detection of small molecules involves a sequence of events that takes place when the analyte of interest is injected into the system. The key elements of the sensor are: 1) an antibody specific for the analyte, 2) signal molecules similar to the analyte, but labeled so they are highly visible to a fluorescence detector, and 3) a fluorescence detector. A schematic of the immunosensor lab device is shown in Figure 1. To perform an analysis, the antibodies which specifically recognize the contaminants are immobilized on a solid support and the fluorescently-labeled signal molecules are bound to them, creating an antibody/signal molecule complex. The functionalized support is placed in a small column (typically 100 µL bed volume) and connected to a water stream. A sample is then introduced to the system. If the sample contains the target analyte, a proportional amount of the labeled signal molecule is displaced from the antibody and detected by a fluorometer downstream. Once the appropriate operating parameters have been determined, the displacement reaction is highly reproducible and predictable for a given antibody/analyte combination.

As seen in Table I, the system has been developed to detect a wide range of compounds, including drugs, explosives, and pesticides.(2,3) Using mathematical equations derived to describe the behavior of the sensor over time(4,5), dose response curves and detection limits can be determined for each assay. The detection limit for the small molecular weight analytes is typically in the low ng/mL (ppb) range. Figure 2 illustrates a typical dose response curve for the explosive trinitrotoluene (TNT) generated using a single antibody column.

A number of features make the continuous flow immunosensor well-suited to *in situ* monitoring. The instrument can be used for measuring either discrete samples containing small molecules in under five minutes per test or monitoring process streams at selected intervals. Because there are no incubation periods or reagent additions required, the analysis time is minimal, making repeat field measurements of a single sample possible. Alternatively, multiple samples can be examined rapidly.

Operational costs are also minimal. Samples collected in a water environment usually require no pretreatment or extractions. Unlike immunoassays using disposable kits, column reagents are not expended if the sample contains no target molecules. Over 50 positive samples (>500 ng/mL TNT) have been run on one column before all the fluorescent signal molecules were depleted. For less concentrated samples, or when fewer samples are

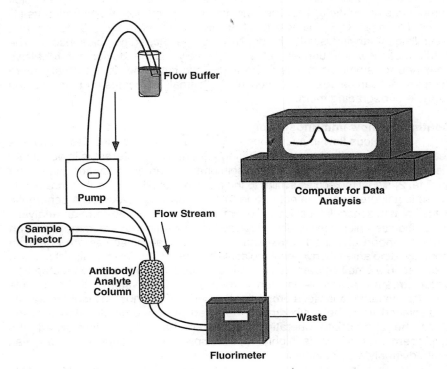

Figure 1: Schematic of the continuous flow immunosensor. The system is comprised of a peristaltic pump, a low pressure sample injector, antibody/fluorescently-labeled analyte column, fluorometer, and signal processor. A continuous aqueous flow stream is established using phosphate buffered saline containing 0.1% Triton X-100 + 12.5% ethanol.

Table 1: Compounds Detected with Flow Immunosensor

Drugs of Abuse	Explosives	Environmental Pollutants
• Cocaine & metabolites • Opiates • Amphetamines • Marijuana (THC) • Pentachlorophenol (PCP)	• Trinitrotoluene (TNT) • Dinitrotoluene (DNT) • Cyclotrimethylene-trinitramine (RDX) • Pentaerithritol tetranitrate (PETN)	• Polychlorinated biphenyls (Aroclors) • Alachlor

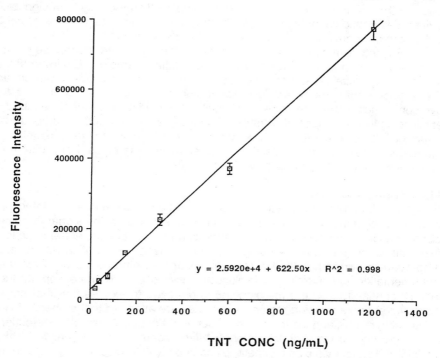

Figure 2: Standard curve for TNT detection with the continuous flow immunosensor. Fluorescence intensity is in arbitrary units. A minimum of 3 assays was performed for each concentration of TNT tested. The r^2 value for the linear region is 0.998.

tested, a single column has been used over the course of four 8-hour days without changing. Coupled with the small size of the antibody-analyte column (approximately 1/2 in x 1/4 in), this repetitive sampling capability means that the cost per test is significantly reduced from other technologies.

Finally, the laboratory prototype is small enough to fit inside two standard carrying cases. The largest components associated with the system are the personal computer used for directing the system operations and collecting the data, the fluorescence detector and the fluidics needed to pump the water through the column. A research program is currently underway to design and build a fieldable device that is approximately 4 x 3 x 7 in. The new instrument will incorporate advances in fluidics, optics and electronics.

Beyond instrumentation, other issues must be resolved before use of the continuous flow immunosensor is practical for all field samples. Environmental samples may contain oils, for example, that may significantly interfere with either the antibody columns or the fluorescent signal. We are looking at ways to overcome such matrix effects. Assay conditions, including appropriate antibody affinities, buffer conditions, flow rates, and cross-reactivity to other compounds must be determined for each analyte. We are also actively examining alternative methods for collecting and extracting samples from different environments, such as soil or oil/water mixtures.

Fiber optic biosensor

The fiber optic biosensor was developed as an ultrasensitive detection system utilizing the sensitivity and specificity of antibodies, the signal-to-noise discrimination of fluorescence measurements, and the rapid signal transduction capabilities of fiber-optic sensing. In this sensor, long optical fibers are employed to permit handling of toxic or hazardous materials away from the instrumentation and operator. The glass core is clad with silicone except for the last 12 centimeters where the cladding is removed to expose the fiber core, thereby creating a sensing region. Antibodies are attached to the fiber core in the sensing region, and the fiber is immersed in an aqueous sample. Detection of analyte occurs via a direct binding of fluorescently labeled analyte to the immobilized antibody (i.e., a competitive immunoassay between the sample and a fluorescently-labeled analyte) or a sandwich immunoassay. In the sandwich immunoassay, the antibody-coated fiber optic probe is exposed to the test sample, rinsed and then exposed to a fluorescently-labeled secondary antibody. All antibody-analyte binding with the fluorescent tag takes place on the surface of the optical fiber within the evanescent wave region (Figure 3). The evanescent wave region is generated when electromagnetic radiation from the light propagating within the core extends beyond the confines of the fiber core. Since the evanescent wave only penetrates about 150 nm into the sample, fluorescent moieties in the bulk solution have little effect on signal levels. Thus the biosensor is particularly well adapted to detection of analytes in heterogenous samples with minimal sample handling.

To facilitate application of the sample to the sensing region of the fiber probe, a chamber was constructed from a 200 µL capillary tube which can be connected to either a peristaltic pump or a syringe. The volume in this chamber can be adapted to a variety of sample sizes by changing the capillary tube. The chamber is disposable, easy to sterilize, small, and protects the fiber from breakage during shipment as well as during use. It is also possible to keep the sample in a totally enclosed system--a feature important to handling of hazardous materials.

The capability of the fiber optic biosensor for immunoassay detection of small molecules, toxins, proteins and bacteria has been demonstrated.(6) The selection of antibodies and protocol development involve many of the same types of decisions as other immunoassay procedures as far as specificity and avidity are concerned.

A competitive immunoassay was developed for the detection of TNT.(7,8) In this assay, Cy5™-labeled analyte was tested to determine the fluorescent signal associated with the absence of TNT. This labeled analyte was then removed from the immobilized antibodies with 50% ethanol in phosphate buffered saline solution. Next, the test sample containing the labeled analyte and TNT were passed over the fiber. If TNT was present in concentrations ≥ 10 ng/mL, a reduction in fluorescence signal compared to that obtained for the sample without TNT was observed (Figure 4).

One method for monitoring bacterial cells is a sandwich immunoassay. In this type of assay, a sample containing bacterial cells is incubated with an antibody-coated fiber optic probe for 5-10 minutes. Next, the probe is exposed to a fluorescently-labeled antibody which also binds to the cells of interest forming a sandwich complex. The fluorescent label is excited when it is bound to the fiber, generating an increase in signal. This assay is being optimized for *Pseudomonas cepacia*, a bacteria used to degrade trichloroethylene.

Another assay for whole cells has been developed which is unusual in that all cells in the sample are stained nonspecifically (i.e., Nile Red dye) and the capture antibody on the fiber specifically binds the cells of interest.(9) The staining procedure is simple and stained cells can be detected in 300 µL samples after 1-2 minutes. *Bacillus anthracis* concentrations as low as 3000 cells/mL were detected. The dose-response is linear for over two orders of magnitude.

To be of practical use, assays must be conducted in real samples, not simply in phosphate buffered saline, to assess the effect of the sample matrix on sensitivity and signal generation. Results similar to buffer tests were obtained in the TNT competitive assay when 90% surface water (river and harbor) or bilge water were used in place of buffer. These 'water' samples were not prefiltered or treated prior to the test and contained particulate matter. From these and other studies (10), it has been demonstrated that the fiber optic biosensor immunoassays were not adversely affected by samples that were opaque, viscous, or contain particulates.

Use of the fiber optic biosensor for analysis of environmental samples requires 1) shelf storage of the antibody-coated fiber probes for at least a year, 2) reusability of the coated fibers, and 3) automated sample handling. The first two items make the system more economically feasible. A storage stability test was performed using fibers coated with a polyclonal rabbit anti-goat IgG (specific) or goat IgG (control).(11) Compared with the activity of the

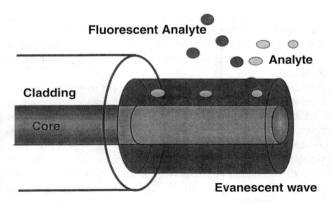

Figure 3: Evanescent wave immunosensing. Total internally reflected light travels the length of the fiber core with a small portion of the power outside the core in an area referred to as the evanescent wave. The evanescent wave in the sensing region penetrates approximately 120 nm. Only fluorescent complexes bound to the antigen/antibody complex within the evanescent wave are excited, generating a signal.

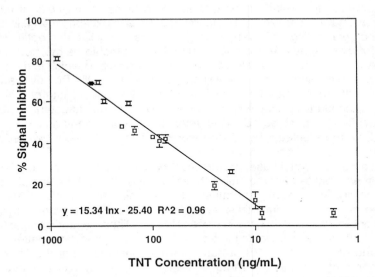

Figure 4: Standard curve for TNT detection with the fiber optic biosensor. The percent inhibition of the reference signal for various concentrations (1-1000 ng/mL) TNT are shown. A minimum of 3 assays was performed for each concentration with the exception of 200 ng/mL. The r^2 value for the linear region is 0.96.

immobilized antibody on day 0, over 80% of the activity was retained for a year, even when the fibers were stored at room temperature. Extended studies showed that activity in the antibody-coated optical fibers lyophilized in the presence of the sugar trehalose maintained over 60% of their activity when stored at room temperature and 80% of their activity when stored cold (4°C and -20°C) for over 12 months (Figure 5).

There are two issues involved when discussing reusability of the antibody-coated fiber optic probe: 1) use after negative and low analyte samples and 2) removal of analyte from immobilized antibody (regeneration). Data indicate that as long as the samples contain low levels of the analyte to be detected, assays can be performed repeatedly on the same fiber.(5) Once the fiber is saturated with analyte, however, the analyte has to be removed prior to additional assays. In addition, the competitive assays are best performed when the maximum fluorescence signal is obtained prior to the test sample but on the same fiber probe. To accomplish this, the labeled analyte must first be removed from the immobilized antibody. The method for regeneration of the immobilized antibody varies depending on the antibody-antigen affinity and chemical properties of the analyte. In the competitive TNT assay, regeneration was achieved with the probe being exposed to 50% ethanol for 1 minute. Fifteen cycles of analyte exposure/regeneration using this protocol were accomplished with < 30% loss of immobilized antibody activity. The loss of activity was accounted for by running a reference sample after every test sample and correcting the % inhibition.

The ultimate goal is to have an automated fiber optic biosensor that can be used for field screening. To reach this goal, a prototype of a portable fiber optic sensor that can monitor 4 fiber probes simultaneously has been constructed by Research International (Woodinville, WA) in collaboration with NRL. This sensor is 6 x 4 x 2 in. with all the optics, light sources and detectors for each probe on a single computer card. Studies indicated similar or slightly improved signal generation/recovery with this sensor. The improved signal may be due in part to the use of optical fibers to transmit light through the entire system as well as the QA/QC done by the manufacturer. Work is underway to automate sample and reagent delivery to the fiber probe.

At least two important problems remain to be solved before the fiber optic biosensor becomes a commercially available, widely used detection system. First, manufacturing techniques for producing the tapered fiber probes inexpensively and in large quantities must be developed. Second, procedures for minimizing fouling and protease degradation of the antibody-coated fibers must be developed for specific applications that require extended periods of use. Once these problems are solved, the fiber optic biosensor will be used for environmental and clinical monitoring.

Conclusion

The continuous flow immunosensor and the fiber optic biosensor have demonstrated an ability for sensitive detection of environmentally relevant compounds within minutes and should, in the near future, provide viable alternatives for detection of groundwater contamination at designated EPA

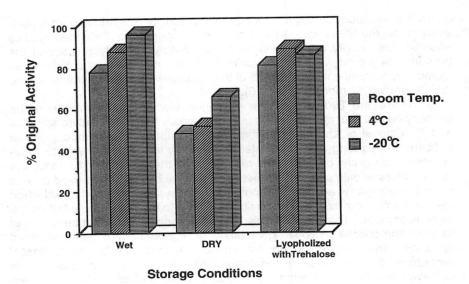

Figure 5: Immobilized antibody activity after storage. Antibodies were immobilized onto a piece of optical fiber and stored under various conditions. A portion of the fibers were stored in phosphate buffer saline (wet), others were air dried, and the last group was lyophilized in the presence of the cryoprotectant trehalose. These groups were then split into subsets for storage at three different temperatures. The percent of the original antibody activity is shown for various storage conditions after 1 year.

remediation sites. For site characterization and continuous monitoring of water effluents, the continuous flow immunosensor would be appropriate. The fiber optic biosensors could be adapted to remote monitoring of toxic agents, hazardous chemicals in storage or production facilities, and various other agents. The spectrum of possible analytes include hazards in closed environments such as engineering spaces or magazines, explosives and byproducts of explosive manufacture, pollutants, drugs or pathogenic organisms.

Literature Cited

1.   Van Emon, J.M. and Gerlach, C.L. (1995) Environ Sci Tech **29(7)**, 312A-317A.
2.   Ogert, R.A., Kusterbeck, A.W., Wemhoff, G.A., Burke, R., Bredehorst, R., and Ligler, F.S. (1992) *Anal. Letts.* **25**, 1999-2019.

3.	Whelan, J.P., Kusterbeck, A.W., Wemhoff, G.A., Bredehorst, R., and Ligler, F.S. (1993) *Anal. Chem.* **65**, 3561-3565.

4.	Wemhoff, G.A., Rabbany, S.Y., Kusterbeck, A.W., Ogert, R.A., Bredehorst, R., and Ligler, F.S. (1992) *J Immunol Methods*, **156**, 223-230.

5.	Rabbany, S.Y., Kusterbeck, A.W., and Ligler, F.S. *J Immunol. Methods*,(1994) **168**,227-234.

6.	Ligler, F.S., Golden, J.P., Shriver-Lake, L.C., Ogert, R.A., Wijesuria, D., and Anderson, G.P. *Immunomethods*, **3(2)**, 122-127, 1993.

7.	Shriver-Lake, L.C., Breslin, K.A., Golden, J.P., Judd, L.L., Choi, J.D., and Ligler, F.S. *SPIE-Optical Sensing for Environmental and Process Monitoring*, (1995) **2367**, 52-58.

8.	Shriver-Lake, L.C., Breslin, K.A., Golden, J.P., and Ligler, F.S. *Anal Chem* (1995) **34**, 2431-2435.

9.	Ligler, F.S., Shriver-Lake, L.C., and Wijesuriya, D.C. U.S. Patent # 5496700, issued March 5, 1996.

10.	Cao, L.K., Anderson, G.P., Ligler, F.S. and Ezzell, J. (1995) *J Clin. Microbiol.* **33(2)**, 336-341.

11.	Ligler, F.S., Shriver-Lake, L.C., and Ogert, R.A. (1992) *in* Proceedings of Biosensors'92 (Turner, A.P.F., Ed.) pp. 308-315, Elsevier, Oxford.

Chapter 6

Enzyme Immunoassay Analysis Coupled with Supercritical Fluid Extraction of Soil Herbicides

G. Kim Stearman, Martha J. M. Wells, Scott M. Adkisson, and Tadd E. Ridgill

Center for the Management, Utilization and Protection of Water Resources, Tennessee Technological University, Box 5033, Water Center, North Dixie Avenue, Cookeville, TN 38505

Enzyme immunoassay analysis (EIA) was coupled with supercritical fluid extraction (SFE) for the analysis of herbicides 2,4-D, simazine, atrazine and alachlor in soil. Five soils, ranging in texture from sandy loam to silty clay were fortified with 500 ng/g of herbicide, allowed to air dry, and extracted using supercritical fluid or liquid vortex extraction. Field weathered soils with incurred residues were also extracted. EIA of herbicides using a microtiter plate format were in good agreement with GC or HPLC results (mean r^2 of 0.95). SFE was performed using a Dionex model 703 extractor in the dynamic mode at 200 atm and 66°C for 3 min, followed by 340 atm extraction for 17 min. SFE recoveries with unmodified CO_2 were 7, 56, 57, and 83%, respectively for 2,4-D, simazine, atrazine and alachlor. Recoveries improved to 101, 79, 90, and 88% for 2,4-D, simazine, atrazine and alachlor, respectively, by adding an acetone:water:triethylamine modifier (90:10:1.5, v:v:v). Collection of analytes by SFE was improved by using C_{18} solid-phase traps (90% recovery) compared to liquid acetone collection (65% recovery). There were differences in extraction recoveries based on soil type.

Enzyme immunoassay analysis (EIA) has gained acceptance as a technique for the rapid determination of pesticides. EIA can be used both as a screening method and as a semiquantitative method under different conditions. EIA microtiter plate techniques are easy to use and allow many samples to be run. In many cases EIA is also less expensive than traditional GC or HPLC methods. The major problem with using the EIA technique is the cross reactivity of similar compounds. This is not a problem with soil that contains no cross reacting compounds and that is spiked and extracted shortly after spiking. However with field weathered samples, the metabolites can in some cases be more sensitive to the EIA than the parent compound.

Supercritical fluid extraction (SFE) of organics from various environmental matrices has been utilized recently to avoid using large amounts of hazardous organic solvents, commonly used in traditional extractions. SFE, when coupled with enzyme immunoassay analysis (EIA) of the extracted pesticides, requires negligible organic solvent consumption and offers an alternative, inexpensive, safe and environmentally compatible method for determining pesticides in soil samples. The purpose of this study was to develop a SFE method, and couple it with EIA for the analysis of the herbicides 2,4-D, simazine, atrazine and alachlor in soil.

CO_2 is the most commonly used supercritical fluid because it is readily available and can be converted to the supercritical state at a relatively low pressure (72 atm) and temperature (31°C). SFE extraction of pesticides from soil often requires addition of polar organic modifiers, such as acetone or methanol, to the supercritical CO_2. The purpose of the modifier can be twofold; to increase the solubility of the analyte and/or to increase the surface area of the soil, by swelling the matrix (soil) or to competitively adsorb with the analyte to the soil. Extraction temperature must be increased as the modifier percentage is increased, in order to maintain the mixture in the supercritical state. Modifiers can also be added directly to the soil in the extraction cell.

The concentration of the pesticides in the soil can be important in determining extraction recoveries, as there may be differences in recovery between pesticide spikes of 10 ppm versus 50 ppb, under identical conditions. This may be due to the fact that at lower analyte concentrations, a larger percentage of the total pesticide concentration is less accessible to the extraction solvent than at higher pesticide concentrations.

In addition to the actual extraction of the analyte from the matrix, the mode of sample collection plays an important role. Collection can be achieved either by directly eluting the sample into a liquid or by trapping on a solid phase, followed by solvent desorbtion.

In the current study, EIA is compared with GC or HPLC for the analysis of 2,4-D, simazine, atrazine and alachlor in soil. The EIA is coupled with a SFE extraction method that has been optimized with respect to modifier addition and collection of analyte.

MATERIALS AND METHODS

Soil Fortification. Properties of the herbicides used in this study are listed in Table I in order of increasing values of the soil adsorption coefficient (K_{oc}). The five soils described in Table II were fortified with 500 ng/g each of simazine and 2,4-D or with 500 ng/g each of atrazine and alachlor. Atrazine and alachlor were also added to give soil concentrations of 50 ng/g each. Soils were fortified by adding 50 mL of a herbicide-reagent grade water solution to 100 g of soil and allowing soils to air dry in a fume hood for several days.

Table I.   Herbicide properties[*]

Herbicide	Water Solubility (ppm)	Soil Adsorption Coefficient K_{oc}	Octanol-Water Partition Coefficient K_{ow}
2,4-D	900	20	443
Simazine	3	138	88
Atrazine	33	149	226
Alachlor	242	190	434

[*]SOURCE: Adapted from ref. 2

Enzyme Immunoassay Analysis. Commercial EIA 96-well microtiter plate kits (Millipore, Inc., New Bedford, MA) were used for simazine, atrazine, alachlor and 2,4-D. Eight standards including a blank were made up in the same matrix as the diluted soil extracts and were analyzed on the microtiter plate in duplicate. The procedure followed that described previously (1). Soil extracts were diluted 25:1-200:1 before pipetting into microtiter wells, dependant upon herbicide concentrations.

HPLC and GC Analysis. All confirmatory analyses by either GC or HPLC were carried out on the same extracts used in the EIA analysis. Atrazine, alachlor, and 2,4-D were analyzed by GC; simazine was analyzed by HPLC.

Atrazine and alachlor analysis was performed using a Hewlett Packard 5890 Gas Chromatograph equipped with a nitrogen-phosphorus detector (NPD) and a 30 m x 0.32 mm i.d. (0.25 μm film thickness) HP-5 capillary column. The helium carrier gas was maintained at a flow rate of 1.5 mL/minute. Helium was also used as the makeup gas at a flow rate of 10-15 mL/minute. Hydrogen and air were introduced at flow rates of 3.5 mL/min and 100-120 mL/min, respectively for operation of the N.D. The total gas flow through the detector was between 120-130 mL/min.

Analysis of 2,4-D was performed using a Hewlett-Packard 5880 GC equipped with an electron capture detector (ECD) and a 30 m x 0.53 mm i.d. (0.5 μm film thickness) SPB-5 column (Supelco, Inc., Bellefonte, PA). The analysis was carried out on the 2,4-D methyl ester. Injector and detector temperatures were maintained at 250°C and 275°C, respectively. The carrier gas was helium at a flow rate of 10 mL/minute. Nitrogen was used as the makeup gas for the ECD resulting in a total flow rate of 60 mL/minute through the detector.

Simazine analysis was performed by HPLC using a Hewlett-Packard 1090M liquid chromatograph equipped with a diode array detector, Chem Station data

Table II.   Soil physical and chemical properties *

Soil Series	Texture	Clay %	pH	Cation Exchange Capacity cmol kg^{-1}	Organic Carbon %	Surface Area m^2/g
Lindale	sandy loam	7.4	4.9	16.1	2.3	9.9
Mountview	silt loam	8.6	6.3	6.1	0.9	7.5
Baxter	silt loam	12.5	5.8	8.0	0.7	21.2
Maury	silt loam	24.0	6.3	17.5	1.3	29.6
Iberia	silty clay	49.3	6.3	40.8	1.5	46.1

(Soil Properties spans Cation Exchange Capacity, Organic Carbon, Surface Area columns)

*SOURCE: Adapted from ref. 2

processing software, a Hypersil ODS (250mm x 4 mm i.d., 5 μm) analytical column, and a Hypersil ODS (20 mm x 4 mm i.d., 5 μm) guard column (Hewlett-Packard Co., Wilmington, DE). The column was maintained at 40°C, the mobile phase flow rate was 1.5 mL/minute, and the injection volume was 25 μL. The isocratic mobile phase consisted of acetonitrile/0.1M phosphoric acid, (30:70 v:v) pH$_2$.

SFE Extraction. All SFE extractions were conducted with a Dionex Model 703 system (Dionex Corp., Sunnyvale, CA). High purity CO_2 (SFC grade with 2000 psi helium head pressure) was used throughout (Scott Specialty Corp., Plumsteadville, PA). Extractions were performed using a 3.5 mL extraction vessel. Glass wool was packed in both ends and 3 g of air-dried, ground (less than 2 mm diameter) soil was tightly packed into the extraction vessels. After several preliminary studies using various temperatures and pressures, SFE extractions were conducted at 200 atm for 3 min, followed by a 340 atm extraction for 17 min, both at 66°C. Restrictors were heated to 150°C and the collection vials were maintained at 4°C. Extractions were conducted in quadruplicate, both with and without modifiers. The modifier was acetone:water:triethylamine, 90:10:1.5, v:v:v. Analytes were collected either in 15 mL of acetone or by using solid-phase C_{18} traps (Dionex Corp., Sunnyvale, CA). The C_{18} traps were desorbed with 2 mL of acetone following SFE extraction.

Liquid Vortex Extraction. Simazine and 2,4-D were extracted from 10 g of soil with 20 mL of acetonitrile:water:acetic acid (80:20:2.5, v:v:v). Atrazine and alachlor were extracted from ten g of soil with 20 mL of acetonitrile:water (9:1, v:v). Samples were vortexed 3 times for 2 minutes each and allowed to sit overnight. They were then vortexed 4 times for 10 seconds, centrifuged, and the supernatant saved for analysis. The liquid vortex extraction achieved equal or higher recoveries of atrazine compared to the automated Soxhlet (Soxtec) extraction as described in a previous study utilizing different soils (1).

Field Incurred Samples. Soil samples containing simazine and 2,4-D were obtained from field plots located in Cookeville, Tennessee. Alachlor and atrazine containing soils were collected from field plots located in Crossville, Tennessee. The samples in both cases were collected within 1 month of herbicide application.

RESULTS AND DISCUSSION

EIA Comparison with HPLC and GC. Linear regression analysis of the EIA versus GC or HPLC results was performed. For 2,4-D, the following relationship was obtained: EIA = 1.15 GC + 22.2, r^2 = 0.81, n = 24. For simazine, the following relationship was obtained: EIA = 0.78 HPLC + 66.0, r^2 = 0.92, n = 23. For atrazine the following relationship was obtained: EIA = 1.04 GC + 22.4, r^2 = 0.98, n = 9. For alachlor, the following relationship was obtained: EIA = 0.79 GC + 35.6, r^2 = 0.96, n = 9. Thus, in all cases, the EIA results agreed closely with GC and HPLC results. The EIA results are probably more accurate at the low end of analyte concentration, because of sensitivity, while the chromatographic results may be more accurate at the high end of analyte concentration, due to increased error caused by dilution with the EIA. The EIA has a narrow standard curve range, (the standard curve range for atrazine is from 0.1-2.0 ppb), so significant dilution is necessary for extracts from soils containing high concentrations of the analyte.

Effect of Modifier on SFE Extraction. Modifier addition to SFE CO_2 improved recovery for 2,4-D (from 7 to 101%), simazine (from 56 to 79%) and atrazine (from 57 to 90%). There was no significant difference for alachlor recovery with (88%) or without modifier (78%). Alachlor is very soluble in supercritical CO_2, therefore, it is not surprising that its recovery was not improved by adding modifier. For the other herbicides both recovery and relative standard deviations (RSD) were improved by using CO_2 modified with acetone:water:triethylamine (90:10:1.5, v:v:v).

In preliminary studies no difference, in extraction recoveries between 15% and 20% acetone modifier were observed. Consequently, 15% acetone:water:triethylamine (90:10:1.5) (v:v:v) or 15% acetone:water (9:1) modifier was used. Addition of triethylamine (TEA) improved recoveries significantly, as it raised the pH, and formed a salt complex with compounds, such as 2,4-D. The use of water as a modifier has been shown to increase the surface area of clay-containing soils as a result of swelling, especially montmorrillonitic soils, such as Iberia and Maury soils. Therefore, the addition of water improves recovery of some pesticides from soils, such as the triazines in this study. Consequently, water was added to enhance the effectiveness of the acetone modifier.

Effect of Collection Method. The use of C_{18} traps gave complete recovery for sand and soil samples, spiked with herbicides in the extraction vessel and immediately extracted using SFE. Collection of the sample in liquid acetone under otherwise identical SFE conditions gave recoveries of only 55-78%. C_{18} traps improved collection because the analyte is deposited on the C_{18} trap and is not eluted into the collection vial until it is washed off the trap. This prevents formation of aerosols or volatilization of the analyte, that may have occurred with liquid collection.

Extraction recoveries and RSDs for SFE (with acetone:water:triethylamine modifier as described) using both liquid and solid phase collection are shown in Table III. Both spiked and field weathered soils are represented by this data. Herbicide recovery was improved using C_{18} traps rather than liquid acetone collection, except for simazine, where no difference was observed for either the spiked or the field weathered soils (p = 0.05, 95% confidence level). Simazine is the least water soluble herbicide (Table I) and has the lowest vapor pressure of the herbicides studied, which could explain its relatively high recovery using liquid acetone collection compared to the other herbicides.

The mean liquid vortex extraction recoveries were 78% for 2,4-D, 95% for simazine, 90% for atrazine, and 93% for alachlor. Use of the C_{18} traps resulted in acceptable recoveries (79-123%) for all of the soil herbicides. The 2,4-D recoveries were higher using SFE (101%) compared to the liquid vortex (78%) extraction, while the simazine recoveries were lower using the SFE extraction procedure (79%) compared to liquid vortex extraction (95%). No difference between simazine and

Table III.   Comparison of SFE herbicide recoveries and relative standard deviations (RSD) using collection in liquid acetone (15mL) versus trapping on C_{18} sorbent for both spiked and field weathered soils with incurred residue. SFE parameters were 200 atm at 66°C for 3 min, followed by 340 atm for 17 min using 15% (acetone:water:TEA)(90:10:1.5) (v:v:v) modifier.

	Recovery (+ RSD%)	
	Liquid Collection	C_{18} Trap
Spiked Soil		
2,4-D	55 ± 18.9	101 ± 28.8
simazine	78 ± 8.0	79 ± 14.8
atrazine	62 ± 9.2	90 ± 7.7
alachlor	66 ± 10.4	80 ± 7.2
Field-Weathered Soil		
2,4-D	66 ± 24.8	94 ± 20.0
simazine	73 ± 24.9	81 ± 13.0
atrazine	--	103 ± 7.5
alachlor	--	123 ± 9.8

*SOURCE: Adapted from ref. 2

atrazine SFE recoveries was observed when no modifier was used. However, SFE extraction of atrazine using modifier achieved more complete recovery than it did with simazine under the same conditions. No differences between SFE and liquid vortex extraction were observed for atrazine and alachlor.

SFE Extraction Versus Liquid Extraction of Field Weathered Soils . For the field weathered soils, the overnight liquid vortex extraction was assumed to give 100% recovery and all other techniques were compared to that procedure. SFE recovery for 25 field weathered 2,4-D samples was improved using solid-phase C_{18} collection (94% recovery) compared to the acetone liquid trap (66% recovery). Atrazine and alachlor extraction recoveries on 15 field weathered samples were similar for SFE with C_{18} trapping and the liquid vortex extraction. In fact, the alachlor recoveries were slightly higher using the SFE procedure (Table III).

Extraction Efficiency and Soil Properties. Soil properties, especially clay and organic matter content, determine the soil surface area and cation exchange capacity, and hence, the efficiency of extraction of the pesticides. Recoveries for SFE extraction with C_{18} trapping were compared for five soils spiked with 50 and 500 ng/g atrazine and alachlor. Alachlor recovery was lower for the Iberia silty clay soil (57%) than for the four other soils (average 94%). The Iberia soil is 50% clay which is more than double the amount of clay in any of the other soils. Also, the Iberia clay is predominately montmorillonitic clay, which readily shrinks and swells, and has a large surface area and ion binding capacity. Therefore, soil properties are important and impact the effectiveness of SFE. Achievement high recovery on one soil may not occur with another soil.

Care must also be taken when extrapolating recoveries from spiked soils to other soils. Although the Iberia soil had lower numerical recoveries than the other soils they were not significant at the 0.05 level for the 500 ng/g atrazine spike. At the 50 ng/g atrazine spike level, the Iberia (75%) and Lindale (80% recovery) soil exhibited lower herbicide recoveries than the three other soils (105% average recovery).

An extraction kinetics study for the interval from 0-20 minutes showed that the majority of herbicides were extracted in the first 5 minutes. After 15 minutes, about 80-90% of the herbicide had been extracted.

This study showed that one type of extraction protocol does not necessarily achieve high recoveries for all soil types. High extraction efficiencies on one soil were not necessarily achieved on another soil having different properties. Clay content and type, were important soil properties that seemed to influence recovery.

In this study coupling of EIA and SFE resulted in increased analytical output and lower costs. EIA compared closely with GC and HPLC results for soil extracted herbicides 2,4-D, simazine, atrazine and alachlor.

Acknowledgments

The financial support of the Cooperative State Research Service, U.S. Department of Agriculture--Water Quality Grant, under Agreement no. 93-34214-8844, and the

U.S. Geological Survey seed money grant through the University of Tennessee-Knoxville Water Resource Research Center are gratefully acknowledged.

Devon Sutherland's and Anna Bryant's assistance in herbicide analysis by GC and HPLC is gratefully acknowledged.

Literature cited

1.   Stearman, G.K., and V.D. Adams. 1992. Bull. Environ. Contam. Toxicol. 48:144-151.
2.   Stearman, G.K., M.J.M. Wells, S.M. Adkisson and T.E. Ridgill. 1995. The Analyst 120:2617-2621.

Chapter 7

Development and Application of an Enzyme-Linked Immunosorbent Assay Method for the Determination of Multiple Sulfonylurea Herbicides on the Same Microwell Plate

Johanne Strahan

DuPont Agricultural Products, Experimental Station, Wilmington, DE 19880–0402

A competitive enzyme linked immunosorbent assay (ELISA) was developed and optimized for the simultaneous analysis of multiple DuPont sulfonylurea herbicides (SUs) on the same polyantigen coated microwell plate. As many as 9 different antibodies can be assayed on this type of plate. The same excellent sensitivity, precision, and accuracy observed in the standard quantitative method is also obtained on a polyantigen coated microwell plate. Reagent optimization to increase assay sensitivity led to the development of a polyantigen coating. This ELISA format allows for a very high sample throughput and was used to screen 1500 boxes of Benlate DF fungicide (1313 discrete lots) for nine SUs at an LOD of 5 ppb in the formulated product.

Assay Format

A competitive ELISA was developed and optimized for the simultaneous analysis of more than one sulfonylurea herbicide on the same microwell plate. For the assay, multiple portions of the same sample are prepared and one specific antibody is introduced into each portion. The polyantigen coating on the microwell plate can then capture any one of nine specific antibodies which may be introduced into an aliquot (portion) of the sample.

Figure 1 is a schematic of the assay format. In step 1, the specific antibody is added to the sample. After preincubation, an aliquot of the sample is pipetted into the precoated wells on the microwell plate. During this next incubation period, the antibody that did not bind to the antigen in the sample, will now bind to the antigen immobilized on the microwell plate. The plate is then washed. In step 3, a second antibody enzyme conjugate is pipetted into the wells. The second antibody binds to the immobilized first antibody. After another wash step, the enzyme substrate is added and a color develops. The intensity of the color is inversely related to the concentration of the antigen. If there is no antigen in the sample, there is maximum color. With increasing amounts of antigen, there is decreasing color. The absorbance can be measured on a microwell plate reader and with the appropriate software, standards and controls, concentrations in samples may be calculated in a quantitative assay or estimated in a screening assay.

0097–6156/96/0646–0065$15.00/0

1. Incubate sample containing
   antigen • with specific antibody
   Y in a tube.

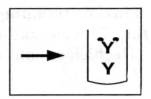

2. Add solution from Step 1. to
   microplate wells coated with
   antigen-protein conjugate.
   Incubate. Antibodies not bound
   to antigen in sample will bind to
   antigen on microplate. Wash
   microplate

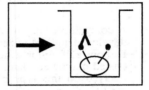

3. Add second antibody-enzyme
   conjugate 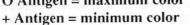 .   Incubate.
   Second antibody-enzyme binds to
   antibodies bound to microplate.
   Wash microplate.

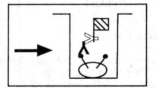

4 Add enzyme substrate.
   Incubate. Read absorbance.

   O Antigen = maximum color
   + Antigen = minimum color

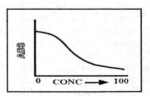

Figure 1. ELISA (Enzyme Linked Immunosorbent Assay)

Assay Components

Polyclonal antibodies specific for the sulfonylurea herbicide (SU) were produced in New Zealand white rabbits by conventional immunization procedures. The immunogen used consisted of the SU hapten conjugated to keyhole limpet hemocyanin (KLH). The resultant antibodies are not purified but left in their natural milieu, the rabbit serum.

Antibody tablets are formulated by a DuPont proprietary process. These may contain as little as 1 uL/tablet or as much as 5 uL/tablet of the antisera. They are designed so that the addition of a tablet to a certain amount of buffer results in the desired antibody titer.

Coating antigens are prepared by conjugating the SU hapten to ovalbumin using conventional conjugation protocols; i.e., activation of a carboxylic acid derivative of the antigen followed by reaction with the amines of a carrier protein (KLH) which results in a hydrolytically stable amide bond.

Second antibody alkaline phosphatase reagent is commercially available. The substrate, p-nitophenyl-phosphate (PNPP) is also commercially available.

Microwell plates are available from several vendors and the best quality plates with optimal protein-adsorptive capacity are used. Plates are coated by pipetting 200 uL/well of the coating antigen mixture in a phosphate buffer, pH 7.4 (PBS). They are coated at ambient temperature and usually left on lab bench overnight and washed in the morning as this fits in with the work flow. Plates are then stored in a zip-lock plastic bag with desiccant at 4°C.

Assay Optimization

Antibody titer is determined in a checkerboard format with different coating antigen concentrations going down the plate and different antibody dilutions going across the plate. In addition to the 0 sample to evaluate binding to the microwell plate, another sample is included to evaluate inhibition. After the optimal antibody titer is selected for 50 % inhibition in the middle of the range of sensitivity desired, then a complete standard curve is run with the optimal coating antigen concentration and antibody titer. In addition, further investigations may be made to attempt to increase sensitivity. For example, Table I shows increased sensitivity with decreasing concentrations of antibody in the assay.

Table I. Antibody Optimization

% Antibody	5 ppt	10 ppt	25 ppt	50 ppt
100	6	11	28	52
80	13	25	42	61
60	11	26	44	65
40	10	30	56	72
20	29	47	67	75

*% Inhibition = ((A-B)/A) x 100 where
A = Absorbance of Negative Control
B = Absorbance of Sample

Coating antigen is initially optimized along with the antibody in the checkerboard assay as described. The goal is to reduce the concentration of the coating antigen and hence the antibody without sacrificing the robustness of the assay or any of the assay performance specifications, i.e. precision. Table II shows the effect on the sensitivity of the assay with the optimal (minimal) antibody titer and decreased concentrations of coating antigen. The optimal concentration of coating antigen, 0.1 ug/mL, permits the use of a mixture of coating antigens without exceeding the binding capacity in the well. There are five different coating antigens/well. Four are specific for 4 different SUs and one is an analogue which captures the specific antibody from five different SUs.

Table II. Coating Antigen Optimization

[Coating Antigen]	% Inhibition*			
	5 ppt	10 ppt	25 ppt	50 ppt
0.05 μg/mL	36	53	70	78
0.1 μg/mL	33	50	68	76
0.2 μg/mL	13	21	27	30

*% Inhibition = ((A-B)/A) x 100 where
A = Absorbance of Negative Control
B = Absorbance of Sample

Incubation time is always optimized for the specific assay and specific matrix. The goal is to establish conditions that will result in a reproducible and robust assay, day to day and lab to lab. Once incubation times have been established, they are strictly adhered to. Under assay conditions, 30 minute incubation at each step is optimal. Decreasing or increasing incubation times will alter the final development time and may change the % inhibition on the curve since this is a non-equilibrium technique.

Assay development time is optimized to restore the assay time lost by using a decreased amount of antibody and coating antigen. This may be accomplished by varying the concentration of the second antibody enzyme. In Table III the effect of increasing concentration of the second antibody enzyme on assay time may be seen. This is accomplished without changing anything in the assay except development time.

Table III. Optimization: Second Antibody Enzyme (Ab-E) Conjugate

Ab-E Conjugate μg/mL	
1.2	155 minutes
2.4	110 minutes
4.8	80 minutes

Increasing 2nd Ab-E concentration decreases total assay time without affecting % inhibition.

To validate a polyantigen coated microwell plate, a comparison was made between the specific coated plate and the polyantigen coated plate for each of the nine specific SU assays. The % inhibition of each standard in each SU assay showed nearly identical results on the two plates (Table IV). Slight variations are accounted for by the controls that are included on every single plate.

Table IV. Antigen Vs. Polyantigen Microtiter Plates

| | % Inhibition* | |
SU ppt	Antigen Plate	Polyantigen Plate
15.6	20	16
31.2	34	30
62.5	60	55
125	78	75
500	91	89

*% Inhibition = ((A-B)/A) x 100 where
A = Absorbance of Negative Control
B = Absorbance of Sample

Benlate Investigation

Benlate 50 DF is a fungicide formulation which contains 50% Benomyl, the active ingredient and 50% inert ingredients. The DF stands for dry flowable.

An investigation of 1313 lots of Benlate 50 DF was done to determine if there were any SUs present in the formulated product. These were all the lots remaining in DuPont's possession and included lots manufactured between 1987 and 1991. In addition, 216 duplicate boxes were also analyzed. Nine DuPont SUs were in the screening assay. The limit of detection (LOD) was 5 parts per billion (ppb) in the dry formulation in the box.

Polyantigen coated microwell plates were ideal for this large study because the same microwell plate could be used with any antibody. Therefore, any of the nine antibodies in the study could be assayed without preparing different types of microwell plates coated with the specific coating antigen.

The most efficient plate format is custom designed to maximize sample throughput and minimize time and cost. The format chosen is shown in Figure 2 and shows 3 sets of controls and 30 samples per plate. Logistically, one antibody on a plate worked best. Alternatively, fewer samples/plate could be analyzed with all nine antibodies (Figure 3) but found this approach not as efficient with the large number of samples in this study.

The aqueous extraction protocol is shown in Table V. A final 1:100 dilution of the extract was made to minimize the interference from some of the inert ingredients in the formulation. With a limit of detection (LOD) of 5 ppb and this dilution, the actual measurement is 0.05 ppb in the assays or 50 parts per trillion (ppt) which is about 50% inhibition for most of the SU standard curves. A typical SU standard curve is shown in Figure 4. This (LOD) was optimal for a reliable and reproducible measurement given assay sensitivity and potential matrix interferences.

Table V. Extraction Protocol

1. Weigh 1 gram Benlate 50 DF into a 50 mL polypropylene centrifuge tube
2. Add 20 mL phosphate buffered saline (PBS); vortex and tumble 30 minutes
3. Centrifuge 15 minutes at 5000 RPM, 0°C
4. Decant and filter supernatant
5. Dilute supernatant a further 1:5 in PBS (final dilution = 1:100)

	1	2	3	4	5	6	7	8	9	10	11	12
A	sb	sb	sb	sb	sb	sb	sb	sb	sb	sb	sb	sb
B	Ctl 0	Ctl 5	U1	U2	U3	U4	U5	U6	U7	U8	U9	U10
C	"	"	"	"	"	"	"	"	"	"	"	"
D	Ctl 0	Ctl 5	U11	U12	U13	U14	U15	U16	U17	U18	U19	U20
E	"	"	"	"	"	"	"	"	"	"	"	"
F	Ctl 0	Ctl 5	U21	U22	U23	U24	U25	U26	U27	U28	U29	U30
G	"	"	"	"	"	"	"	"	"	"	"	"
H	nsb	nsb	nsb	nsb	nsb	nsb	nsb	nsb	nsb	nsb	nsb	nsb

Coating Antigen   All 96 wells

Control (Ctl) 0: Benlate 50 DF lot #U072390-713   Column 1
Control (Ctl) +: Benlate 50 DF lot # U072390-713   Column 2
                + 5 ppb of a sulfonylurea

Benlate 50 DF Samples (U1-U30)   Columns 3-12

Substrate Blank (sb)   Row A
Non-specific Binding Blank (nsb)   Row H

Figure 2. Microwell Plate Format Used in ELISA Screening Method for the Detection of Nine DuPont Sulfonylurea Herbicides in 1313 Lots of Benlate 50 DF Fungicide

		2	3	4	5	6	7	8	9	10	11	12
		Ab 1		Ab 2		Ab 3		Ab 4		Ab 5		
A		0	0	0	0	0	0	0	0	0	0	
B		S	S	S	S	S	S	S	S	S	S	
C		U_1	U_1	U_1	U_1	U_1	U_1	U_1	U_1	U_1	U_1	
D		U_2	U_2	U_2	U_2	U_2	U_2	U_2	U_2	U_2	U_2	
		Ab 6		Ab 7		Ab 8		Ab 9				
E		0	0	0	0	0	0	0	0			
F		S	S	S	S	S	S	S	S			
G		U_1	U_1	U_1	U_1	U_1	U_1	U_1	U_1			
H		U_2	U_2	U_2	U_2	U_2	U_2	U_2	U_2			

Microwell Plate Format
0 = negative control
S = Spiked control
U = Unknown sample

Figure 3. Microwell Plate Format Used in ELISA Screening Method for Simultaneous Analysis of Nine DuPont Sulfonylureas

Curve Fit: 4 Parameter   Corr. Coeff: 0.992
Equation: $y = (A-D)/(1 + (x/C)^B) + D$
A = 1.94   B = 1.47   C = 0.0234   D = 0.0009

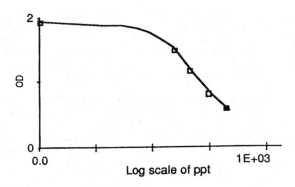

ng/mL	Mean AU (n=3)	% CV(AU)
0.00	1.964	1.4
0.025	1.522	1.6
0.050	1.204	2.5
0.100	0.868	2.1
0.200	0.603	2.0

Figure 4. Typical Standard Curve

Table VI shows typical recovery of each SU spiked into the Benlate 50 DF formulation at 5 ppb and then extracted. Recoveries were measured in quantitative assays and averaged about 90% but ranged from 80% to 100%.

**Table VI. % Recovery of 5 ppb of each SU Benlate 50 DF
(Unspiked and Spiked Measurements in ppb)**

Sulfonylurea	Unspiked	Spiked	% Recovery
Nicosulfuron	0.0	4.9	98
Metsulfuron	1.1	5.1	80
Chlorimuron	0.0	4.6	92
Tribenuron	0.0	4.2	84
Chlorsulfuron	0.0	5.0	100
Thifensulfuron	0.1	5.0	98
Bensulfuron	0.0	5.0	100
Ethametsulfuron	0.2	4.5	86
Sulfometuron	0.6	4.6	80

The controls were unspiked Benlate for the negative control and spiked Benlate at 5 ppb with each of the SUs for the positive controls. These sample were extracted and treated exactly the same as the test lots.

Assay Results

Table VII is a summary of the results. Samples from 1529 boxes, (1313 discrete lots plus an additional 216 duplicate boxes) were analyzed. Screening for 9 SUs in a box then represented 13,761 assays plus 157 X 9 controls (= 1413) or a total of 15,174 assays. Use of the polyantigen plate and the quantitative screening format significantly reduced the time and cost of this study. One technician was able to screen as many as 300 individual samples a day plus controls.

The control and sample populations shown in Table VII are clearly distinct. The control population is the Benlate 50 DF fortified at 5 ppb with each of the nine sulfonylureas compared to the unfortified Benlate 50 DF. % Inhibition is calculated vs the unfortified Benlate 50 DF. The sample populations are the 1313 discrete lots plus some duplicates. These are extracted and assayed along with the controls and % Inhibition is calculated vs the unfortified Benlate 50 DF. The criteria of three standard deviations from the mean shows no overlap except for Sulfometuron where the control and standard populations are separated by two standard deviations. There was excellent reproducibility of the fortified controls. This is a tribute to the skill of the technicians performing the assay and the robustness of the ELISA assay.

Table VII. % Inhibition (%I) of the Positive Controls and Benlate 50 DF Fungicide Lots

SU	Positive Control (% I) (5 ppb) Mean*	SD	Samples (% I) Mean**	SD
Nicosulfuron	39.5	4.6	2.2	1.9
Metsulfuron	40.9	5.6	5.1	2.5
Chlorimuron	48.7	5.8	8.2	3.3
Tribenuron	36.6	5.8	3.1	2.4
Chlorsulfuron	38.7	3.7	8.0	3.6
Thifensulfuron	40.9	4.0	1.0	1.0
Bensulfuron	41.2	5.8	1.4	1.7
Ethametsulfuron	40.9	5.1	4.0	2.6
Sulfometuron	18.8	4.3	5.1	2.6

*   n = 157 Controls (Lots U072390-713 + 5 ppb fortification of a sulfonylurea)

** 1529 Samples (1313 discrete lots and an additional 216 duplicate boxes)

% Inhibition = ((A-B)/A) x 100 where
  A = Absorbance of Negative Control
  B = Absorbance of Sample

Prototype Sulfonylurea Structure
The sulfonylurea bridge is common to all sulfonylurea herbicides.

Conclusion

Data are given here which show that the optimized ELISA is capable of low level detection and the assays are accurate, precise and reproducible. The polyantigen coated microwell plate is shown to be versatile and ideal for large studies .

Throughout this study, in all of the 1313 discrete lots of Benlate 50 DF assayed for nine DuPont SUs, there was never a single case of a Benlate sample in which any SUs were detected.

Acknowledgment

The author wishes to acknowledge the technical input and help of Jacqueline Wank in the development of the method and the further assistance of Jacqueline Wank, Catherine Valteris and Tom Gardner in the 1313 Lot analysis and the peer review and assistance writing the technical report by Dr. Christine Rankin.

Chapter 8

Immunoaffinity Extraction with On-Line Liquid Chromatography–Mass Spectrometry

Jeanette M. Van Emon[1] and Viorica Lopez-Avila[2]

[1]Characterization Research Division, National Exposure Research Laboratory, U.S. Environmental Protection Agency, P.O. Box 93478, Las Vegas, NV 89193–3478
[2]Midwest Research Institute, California Operations, 555–C Clyde Avenue, Mountain View, CA 94043–2211

Sample preparations are crucial to the success of all analytical methods. However, sample preparation is often labor-intensive and the rate limiting step of the overall analytical scheme. Certain immunochemical procedures can perform difficult separations, making subsequent analysis simpler and ensuring the specificity of the results. Immunoaffinity chromatography (IAC) is based on the ability of antibodies to extract target analytes, even from complex environmental or biological matrices. High-performance liquid chromatography-mass spectrometry (HPLC/MS) can be coupled to IAC for a complete extraction-detection system. Customized IACs for plasma, urine, water, and milk are discussed in this paper. Portable detection methods which may be suitable to site characterization are described, including capillary electrochromatography coupled with laser-induced fluorescence (CEC/LIF) detection.

Immunoaffinity chromatography (IAC) uses the specific ability of antibodies to separate an antigen (target analyte) from a complex matrix. The specific antibody is immobilized onto an inert, solid support (e.g., silica), called an immunoaffinity column and functions as a ligand. Thus, when an aqueous sample containing the target analyte is passed through the immunoaffinity column, the immobilized antibody will selectively remove the target analyte(s) from the aqueous solution. The adsorbed analyte is subsequently removed from the immunoaffinity column with a buffer of lower pH or greater ionic strength than the original solution.

This approach to sample preparation has several advantages: (a) it can be highly selective due to the antibody selectivity towards the target compound; (b) it can concentrate large amounts of sample, allowing detection of the analyte at levels that might not be possible with certain chromatographic detectors; (c) it is amenable to coupling with detection techniques such as mass spectrometry that allow a more definitive identification of the target analyte; and (d) it minimizes the use of organic solvent commonly used to extract the target analyte(s) from the aqueous solution.

Subsequent sections will describe antibody purification, antibody immobilization onto specially activated supports, packing of the immunoaffinity support, column operating conditions, elution of analyte from the immunoaffinity column, and the on-line coupling of IAC with chromatographic detectors.

Antibody Purification

The first step in the preparation of the immunoaffinity column is usually the purification of the specific antibodies. Purification of the antibodies is done routinely either by precipitation with ammonium sulfate, by ion exchange chromatography, by gel filtration, or by affinity chromatography (*1*). The procedure for precipitation with ammonium sulfate is described in Reference 1 and is summarized below. This procedure is somewhat limited to the purification of a large quantity of antiserum.

Acetate buffer (6 mM, pH 4) is added to the specific antiserum (a mixture of IgG antibodies and other proteins). The pH is adjusted to 4.6 with 0.15 M acetic acid; followed by addition of pure caprylic acid. After 30 min, the precipitate, i.e., the non-IgG protein fraction, is removed by centrifugation. The supernatant is filtered through a *5-μm* membrane filter, and mixed with a diluted phosphate buffer solution. The pH is adjusted to 7.4 with a 1 M NaOH solution. The mixture is cooled to 4°C and a solution of saturated ammonium sulfate is added. A white, colloidal IgG precipitate is formed which is removed by centrifugation. The pellet is suspended in a diluted phosphate buffer solution. This solution is dialyzed over a cellulose acetate membrane (cutoff 15,000 daltons) against several buffer changes.

The concentration of IgG antibody protein can be determined by UV measurement at 280 nm and 335 nm using equation 1:

$$\text{Concentration IgG (mg/mL)} = \frac{(E_{280}\text{ nm} - E_{335}\text{ nm})}{1.4} \quad (1)$$

in which E is the extinction at the given wavelength. The purified IgG solution is stored in a 100-fold diluted phosphate buffer solution at -20°C. The other common procedure for antibody purification is that reported by Rule and co-workers (*2*) using a protein G column. In this procedure, the antiserum is injected onto the protein G column. Weakly bound extraneous material is washed through the column with a buffer rinse. The now concentrated and purified specific antibody is desorbed with 2% acetic acid. The desorbed IgG is collected in a beaker kept in ice water and neutralized by addition of 2 M NaOH. The IgG solution is desalted and further concentrated using an ultrafiltration membrane with a molecular weight cutoff of 30,000 daltons.

Antibody Immobilization

The next step in immunoaffinity chromatography methods development is the antibody immobilization.   Factors that need to be considered in antibody immoblization include: coupling pH, coupling kinetics, amount of antibody that is immobilized onto the immunosorbent, and the activation chemistry that will be utilized for the immobilization process (3). The best strategy for the preparation of an immunoaffinity column is a combination of low antibody density and oriented coupling of the antibody via the carbohydrate moiety (3). This strategy enables the most efficient use of the immobilized specific antibody. Beads coated with protein A or G, beads with chemically activated surfaces, and activated antibodies have been used for this purpose. The most commonly used procedure is the one using beads coated with bacterial proteins A or G. These beads bind specifically to the Fc region of the specific antibodies. Thus , protein A and protein G bind to immunoglobulins in a manner that does not interfere with the specific binding to the target analyte. Protein G has the advantage of binding to a wider range of immunoglobulin species and subclasses providing a generic solid support for antibody immobilization. The activated   bead technique uses   various   immobilization   chemistries   (e.g., carbonyldiimidazole, cyanogen bromide, glutaraldehyde, N-hydroxy-succinimide, and tosyl chloride). To facilitate antibody immobilization, chemically activated supports can be used to bind the specific antibodies. These supports are available from several suppliers (e.g., BioRad, ChromatoChem, Inc.) and use several chemistries (reactive hydrazine, aldehyde functionalities).   In the third techinique, the antibody is activated first with different coupling reagents (e.g., carbodiimides, glutaraldehyde, periodate) and then coupled with the support beads. Details of antibody immobilization are given in References 3 and 4.

Column Packing and Operating Conditions

Once the antibody has been immobilized onto the solid support, the immunoaffinity support is then packed into a stainless-steel column. The technique used for packing must be gentle to prevent channeling and loss of antibody; pumped-slurry packers are preferred over the gas-activated pressure packers. Buffers such as 0.01 M phosphate and 0.1 M Tris are recommended as packing solvents. Dry packing of freeze dried materials is not recommended.

The capacity of the immunoaffinity column can be determined as follows. The column is loaded with varying amounts of target analyte from sample volumes of 5, 10, 20, and 50 mL. The sample solution is analyzed for the target analyte by immunoassay (e.g., ELISA) or by an instrumental technique, before and after passing it through the immunoaffinity column. The amount of analyte retained on the immnunoaffinity column is determined from the difference of the two measurements. Subsequently, the analyte is eluted from the immunoaffinity column, and the eluate is analyzed. Recovery is then plotted as a function of the sample volume and total amount of target analyte passed through the immunoaffinity column.

The long-term stability of the immunoaffinity column is usually established by repeatedly loading it with a known amount of target analyte (for a maximum of 50 times) and eluting it under identical conditions each time. By plotting the amount of target analyte retained by the immunoaffinity column on the y-axis, and the number of extractions on the x-axis, an indication of column stability for up to 50 analyses is obtained over a defined period of time. When the column is not in use, it is usually stored at 4°C in 0.1 M phosphate buffer (pH 7) containing 0.02 percent sodium azide.

Elution of Analyte from the Immunoaffinity Column

The elution of the target analyte(s) from the immunoaffinity column can be performed either by changing the ionic conditions of the mobile phase (e.g., lower pH buffer) or by using chaotropic buffers (5). Buffers at pH ≤ 2, such as 0.1 M glycine (pH 1.5), 0.1 M Tris-glycine (pH 1.0), 20 % formic acid (pH 1.8) and 1 M acetic acid (pH 2.0), are more effective than those at pH ≥ 10.

Chaotropic ions are the second most effective elution agents (5) and they have been used effectively in eluting high-affinity antibodies. Organic solvents such as 10% dioxane, 50% ethylene glycol (pH 11) have also been used to break the bond between the antibody and the target analyte (5). Since proteins are quite easily denatured by organic solvents, the last approach using organic solvents should be used with caution.

Detection and quantitation follows removal of the target analyte from the immunoaffinity column. Coupling IAC to HPLC or other chromatographic techniques results in an effective two dimensional separation that combines the selectivity of immunoaffinity extraction with the high resolving power of HPLC. Furthermore, the use of a specific detector such as the mass spectrometer ensures the unambiguous identification of the target analyte.

IAC On-Line with Chromatographic Detectors

To illustrate how IAC is coupled on-line with HPLC the experimental setup used by Farjam and co-workers is described (6). In a repetitive analysis mode, the first step is to flush the immunoaffinity column with water to displace the eluting solvent [e.g, methanol-water (95:5) or other solvent combination] from the previous run. The sample is then loaded followed by a water rinse to displace residual sample and any weakly-bound impurities. Simultaneously, the analytical column is preconditioned with water to remove any solvent from the previous run. Following this, the immunoaffinity column and the analytical column are switched in series and the immununoaffinity column desorbed with methanol-water (95:5) or other solvent. The eluent from the immunoaffinity column is either directed into the MS or diluted with water to reduce its methanol content, and then passed through the analytical column. Finally, the analytical column is eluted with organic solvent and the eluate is injected into the MS via the atmospheric pressure chemical ionization (APCI), electrospray, or particle beam.

The coupling of IAC with high-performance liquid chromatography-mass spectrometry (HPLC/MS) has been demonstrated (2, 7). For example, Rule and Henion (7) at Cornell University reported an on-line immunoaffinity HPLC/MS method for the extraction and detection of drugs in urine samples. A protein G column bound with specific antibody was used for the preconcentration of drugs from urine followed by HPLC or LC/MS for identification. In other work Rule et al. (2), reported the use of an aldehyde-activated silica immunoaffinity column to preconcentrate carbofuran from water and potato extract. A detection limit of 40 pg/μL in water and 2.5 ng/g in potato was achieved with an atmospheric pressure ionization quadrupole ion trap mass spectrometer. Coupling of IAC with gas chromatography (6) and liquid chromatography has also been reported for on-line sample pretreatment (8,9), analysis of human plasma (10), and determination of digoxin in serum (11). The procedure has not commonly been used for small molecules due to antibody availability, unfavorable desorption kinetics, and the emphasis on immunoassay test kits for environmental contaminants. However, there are several areas where IAC can augment current environmental analytical methods. Difficult dioxin and polychlorinated biphenyl analyses can be streamlined using immunoaffinity chromatography as a sample preparation prior to GC/MS analysis to increase sample throughput.

Application

Table I summarizes several applications of IAC. Most of the applications are for preconcentration of drugs from urine, plasma, and milk and only very few deal with preconcentration of environmental pollutants from water or aqueous extracts. IAC has been used to preconcentrate propanolol and LSD from urine samples (7), nortestosterone and related compounds from urine samples (6,8), various estrogen steroids and their glucuronide derivatives from urine samples (9,20), albuterol, digoxin, and PTH (parathyroid hormone) from plasma (10,11,12), aflatoxin M$_1$ and zearalenone from milk (15,17) and cortisol and flumethezone from urine and serum (18,19). Environmental applications of IAC were limited to carbofuran (2), atrazine (13), carbendazim (14), and chlortoluron (16). In the applications shown in Table I IAC was coupled to HPLC/UV (7,8,9,13,14,16,18 and 20), HPLC/MS (2,7,14), GC/FID (6), HPLC/radioimmunoassay (10), chemiluminescent detection (12), GC/MS (19), and ELISA (17).

We have used IAC in our laboratory and coupled it to HPLC using a diode array detector (DAD) and HPLC/MS to determine carbendazim in water. A schematic diagram of the chromatographic system used in this study is shown in Figure 1. The system is comprised of a Hewlett-Packard HP 1050 quaternary pump module with an added solvent degasser and column thermostat (Hewlett-Packard Co., Palo Alto, California), and a Suprex SFE-50 syringe pump module (Suprex Corporation, Pittsburgh, Pennsylvania). Samples of up to 250 μL were manually injected by using a Rheodyne 7126 six-port injector with an external injection loop

Table I.   Several Applications of IAC coupled with various detection systems for the analysis of biological and environmental samples

Target Analyte	Matrix	Immunoaffinity Column	Detection Technique	Reference
Aflatoxin M$_1$	Milk	Aflatest-P affinity column; aflatoxin M$_1$ is eluted with methanol.	HPLC/fluorescence	(15)
Albuterol	Human Plasma	CNBr-activated Sepharose 4B column; the analyte is desorbed with 0.1 M acetic acid containing 20% (v/v) ethanol.	HPLC/fluorescence HPLC/radioimmunoassay	(10)
Atrazine	Water	Diol-Bonded Nucleosil; atrazine is desorbed with 0.05 M phosphate buffer (pH 2.5).	HPLC/UV	(13)
Carbendazim	Water	Diol-bonded Nucleosil; carbendazim is desorbed with 0.05 M phosphate buffer (pH 2.5).	HPLC/UV HPLC/MS ELISA	(14)
Carbofuran	Surface Water Potato Extract	Aldehyde-activated silica column; the IgG is desorbed with 0.2% formic acid.	HPLC/MS	(2)
Chlortoluron	Water Plasma Urine	Aldehyde-activated porous silica; chlortoluron is desorbed with phosphate buffer-ethanol (50:50).	HPLC/UV	(16)

Continued on next page

Table I. Continued

Cortisol	Urine Serum	Aldehyde-activated porous silica; cortisol is desorbed with methanol-water (60:40).	HPLC/UV	(18)
Digoxin	Serum	Spherosil; digoxin is desorbed with 1% hydrochloric acid.	HPLC/fluo-rescence reaction detection	(11)
Estrogen steroids and their glucuronide deriatives	Urine	CNBr-activated Sepharose 4B column; the analytes are desorbed with a solution of acetonitrile-water containing excess of two cross-reacting compounds (17-β estriol and 17-β-estradiol acetate.	HPLC/UV	(9)
Estrogen steroids	Urine Plasma	CNBr-activated Sepharose 4B, the analytes are desorbed with methanol-water (95:5 v/v).	HPLC/UV	(20)
Flumethazone	Urine	Affi-gel; the analyte is desorbed with 2-propanol-water (60:40 w/w).	GC/MS	(19)
β-19-Nor-testosterone α-19-Nor-testosterone	Urine	CNBr-activated Sepharose 4B column; the analytes are desorbed with a solution containing cross-reacting steroid hormone norgestrel and transferred via a second precolumn to the analytical column.	HPLC/UV	(8)

Table I. Continued

β-19-Nor-testosterone Nor-ethindrone Norgestrol	Urine	CNBr-activated Sepharose 4B column; the analytes are desorbed with methanol-water (95:5 v/v), and subsequently reconcentrated on a reverse phase column.	GC/FID	(6)
Parathyroid hormone (PTH)	Plasma	Diol-bonded Nucleosil; the analyte is desorbed with 1% hydrochloric acid.	Chemilumi-nescence detection	(12)
Propanolol lysergic acid diethylamide (LSD)	Urine	Protein G column primed with 5 µg drug specific antibody for each injection. The IgG is desorbed from the protein G column with 2% acetic acid.	HPLC/UV HPLC/MS	(7)
Zearalenone	Milk	CNBr-activated Sepharose 4B; the analyte is desorbed with methanol.	ELISA	(17)

(Cotati, California). Larger samples of up to 40 mL, were applied directly to the immunoaffinity column using channel C of the quaternary pump.

The detectors used in our study were an HP UV/visible absorbance DAD, and a Fisons Quatro tandem mass spectrometer interfaced to the liquid chromatograph by an APCI interface (Fisons, Cheshire, UK). Data from the DAD were collected and analyzed on a HPLC[3D] ChemStation chromatography data system (Hewlett-Packard Co.). Data from the APCI-MS were collected and analyzed using MassLynx software from Fisons. The mass spectrometer was tuned for maximum sensitivity while maintaining unit resolution.

The carbendazim immunoaffinity column was prepared using Nucleosil 1000-7, which was derivatized to epoxy silica, then diol silica, and finally to aldehyde silica before immobilizing the carbendazim antibody (a gift from DuPont). Details of the procedure can be found in a recent publication (*14*).

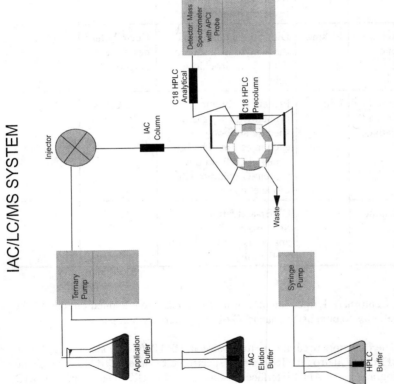

Figure 1. Experimental setup for IAC/HPLC/MS. In the repetitive analysis mode, samples are introduced through the IAC to th HPLC column, and detected by APCI-MS.

Table II summarizes the performance of the carbendazim immunoaffinity chromatography technique using HPLC-MS and HPLC-DAD and compares it to the simpler ELISA technique which does not require extraction or cleanup steps. As can be seen in Table II, the APCI-MS detector provides approximately 3 times better sensitivity for carbendazim than the DAD, and 4 times better sensitivity than reported for an ELISA. Carbendazim is well-suited to detection by APCI-MS because it is easily protonated, somewhat volatile, and undergoes little fragmentation under the soft ionization that occurs in the APCI source. For other compounds of environmental significance that have either a better chromophore or less favorable behavior in the APCI source, the DAD detector might provide greater sensitivity.

Table II. Method Performance for IAC-HPLC/MS, IAC-HPLC/DAD, and ELISA Determination of Carbendazim

	IAC-HPLC/MS	IAC-HPLC/DAD	ELISA[a]
Limit of Detection(μg/L)[b]	0.025	0.075	0.10
Linear Range (μg/L)[c]	0.025-100	0.075-100	0.25-5.0
Correlation Coefficient	0.988	0.999	0.990
Precision (%RSD)[d]	4.5	16.0	10.0

[a] ELISA method performance reported by Ohmicron Corporation.
[b] Limit of detection at S/N=3 for a sample size of 200 μL.
[c] The portion of the calibration curve showing a best-fit line with $r^2 \geq 0.998$
[d] The number of determinations was 13 for IAC-HPLC/MS and 10 for IAC-HPLC-DAD. The concentration of carbendazim in the spiked reagent water samples was 0.1 μg/L.
SOURCE: Reprinted with permission from ref. 14. Copyright 1996.

The APCI is generally considered as the universal HPLC/MS interface because of its applicability to a broad range of environmental compounds over a wide range of flow rates and mobile phase compositions. When using the APCI interface, the eluent from the HPLC system is directed through a heated pneumatic nebulizer in which the eluent is converted to a fine spray and vaporized with minimal heating. The vapor is then passed over a needle at high voltage where components are chemically ionized at atmospheric pressure and then introduced into the mass spectrometer. This straightforward technique is utilized for the analysis of compounds with diverse polarities over a wide range of solvent conditions. APCI is a soft ionization process in which only protonated molecular ions are produced with little or no fragmentation and is directly adaptable to flow rates and buffer compositions commonly used in HPLC and IAC.

The electrospray is a relatively new technique and is used to analyze molecules that bear a charge in solution. Electrospray is particularly suited to those compounds that are too polar or thermally labile to be analyzed by any other means.

In this technique the eluent from the HPLC system is injected through a hypodermic needle held at high potential causing the liquid to be converted to a spray by electric fields. As the droplets evaporate, ions are liberated by ion evaporation without the addition of heat. The electrospray technique is especially useful for large molecules, such as biomolecules, by taking advantage of multiple charging. This can effectively extend the mass range of mass spectrometers by an order of magnitude.

The particle beam is useful for providing classical electron ionization/chemical ionization (EI/CI) spectra for thermally labile, semivolatile, and nonvolatile compounds typically beyond the capabilities of GC/MS. In this technique the eluent from the HPLC system is introduced into a multistage vacuum system, in which solvent is removed, leaving a stream of dry particles that are then introduced through an orifice into a conventional EI/CI ion source.   Each of the MS techniques described here is compatible with IAC. An objective of IAC research is to provide efficient sample preparations for MS methods.

In collaboration with Sandia National Laboratories in Livermore, CA, applications of IAC coupled with capillary electrochromatography and laser-induced fluorescence detection (CEC/LIF) are in development. In CEC, an electric field is applied across a capillary tube (packed with a chromatographic support) to generate an electroosmotic flow that can be used for chromatographic separations. The separation efficiency that can be achieved under these conditions exceeds that obtained with pressure-driven flow, such as in HPLC, and can even approach that obtained by capillary GC if submicron-sized particles are used. In addition, because no HPLC pump is required, analytical methods based on this phenomenon may be adaptable to portable or remote sensing applications.

Although in comparison to HPLC/MS, CEC promises (a) superior resolution, (b) decreased solvent consumption, and (c) short run times for a broad class of analytes, CEC suffers from the lack of a sensitive universal detector. For certain compounds, LIF detection provides a route to ultrasensitive analytical methods; however, only relatively few compounds exhibit native fluorescence. Furthermore, the small injection volumes used in CEC usually lead to only a moderate concentration sensitivity. Many of the advantages offered by CEC cannot be fully realized if traditional sample extraction and cleanup procedures are used. One way to augment the advantages of CEC/LIF is by coupling it to IAC. A demonstration of CEC/LIF interfaced to IAC using polynuclear aromatic hydrocarbons (PAHs) as target analytes is in progress. See Figure 2 for the experimental setup.

This technique involves four steps. First, the sample is electrokinetically injected (i.e. injection with the aid of a voltage) onto the IAC column packed with IgG PAH antibodies bound to silica beads. Nonspecifically bound material is removed by washing the IAC column with buffer. Second, a dissociating buffer desorbs the bound PAHs from the IAC column and preconcentrates them on the HPLC analytical column. Third, the PAHs are eluted and separated with gradient CEC and detected by LIF. Finally, a regenerating buffer is applied to the IAC

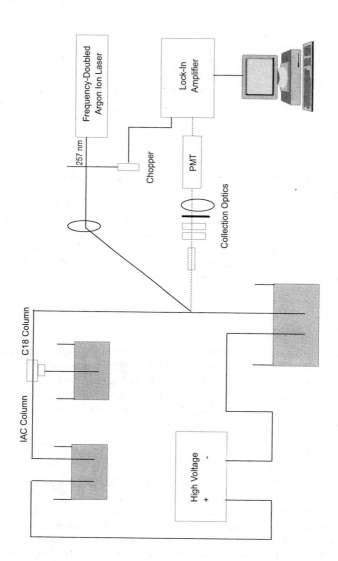

Figure 2. Experimental setup for IAC/CEC/LIF. Following IAC extraction, an electric field across a packed capillary tube generates an electroosmotic flow for chromatographic separation. Note that no HPLC pump is required, making the method suitable for field work and remote sensing applications.

column to prepare it for the next injection.   By using a 100-um id x 30-cm length fused-silica capillary column packed with C18-bonded silica, 7 target PAHs can be separated with efficiencies greater than 100,000 theoretical plates per meter in less than 30 minutes (*21*).   An intracavity-doubled argon-ion laser operating at 257 nm is used to detect the PAHs by LIF. Results of this study were presented at the 1996 Pittsburgh Conference in Chicago, Illinois (*21*).   For analytes that do not natively fluoresce, a chomatographic competitive binding immunoassay using fluorescent labels can be employed.   Analyte from the sample and a known amount of analyte tagged with a fluorescent label compete for the limited number of antibody binding sites on the immunoaffinity column.   Either the fraction of bound label, or the unbound fraction that passes through the column, can be used to determine the amount of analyte in the sample.

Conclusion

Analytical chemists can profit from the incorporation of immunochemical methods in their laboratories. Extraction and cleanup using immunoaffinity columns is an easy first step into antibody-based methods.   By coupling the separation strength of immunoaffinity with the quantitation capabilities of HPLC/MS, CEC/LIF, and other detection systems, analysts can maximize quality of results and minimize reliance on organic solvent sample preparations.   Leveraging immunochemistry can improve sample throughput for instrumental analysis, enabling indepth detection systems to be used for large-scale environmental monitoring and human exposure assessment studies.

Acknowledgments

The U.S. Environmental Protection Agency (EPA), through its Office of Research and Development (ORD), funded and collaborated in the research described here. It has been subjected to the Agency's peer review and has been approved as an EPA publication. Neither the EPA nor ORD endorses or recommends any trade name or commercial products mentioned in this article; they are noted solely for the purpose of description and clarification.

Literature Cited

1.   Katz, S.E.; M.S. Brady, "High-Performance Immunoaffinity Chromatography for Drug Residue Analysis," *J. Assoc. Off. Anal. Chem.* **1990**, 73, 557-560.
2.   Rule, G.S.; A.V. Mordehai; J. Henion, "Determination of Carbofuran by On-line Immunoaffinity Chromatography with Coupled-Column Liquid Chromatography/Mass Spectrometry," *Anal. Chem* **1994**, 66, 230-235.
3.   Matsow, R.S.; M.C. Little, "Strategy for the Immobilization of Monoclonal Antibodies on Solid-Phase Supports," *J. Chromatogr.* **1988**, 458, 67-77.
4.   Hoffman, W.L.; D.J. O'Shannessy, " Site-Specific Immobilzation of Antibodies by their Oligosaccliaride Moieties to New Hydrazide Derivatized Solid Supports," *J. Immunological Methods* **1988**, 112,113-120.

5. Phillips, T.M., "High Performance Immunoaffinity Chromatography-An Introduction," *LC.* **1985,** 3, 962-972.

6. Farjam, A.; J.J. Vreuls; W.I.G.M. Cuppen; U.A.Th. Brinkman; G.J. deJong, "Direct Introduction of Large-Volume Urine Samples into an On-Line Immunoaffinity Sample Pretreatment-Capillary Gas Chromatography System," *Anal. Chem.* **1991,** 63, 2481-2497.

7. Rule, G.S.; J. Henion, "Determination of drugs from urine by on-line immunoaffinity chromatography high-performance liquid chromatography-mass spectrometry," *J. Chromatogr.* **1992,** 582, 103-112.

8. Farjam, A.; G.J. de Jong; R.W. Frei; U.A.Th. Brinkman; W. Haasnoot; A.R. M. Hamers; R. Schilt; F.A. Hug, "Immunoaffinity Pre-Column for Selective On-line Sample Pretreatment in High-Performance Liquid Chromatography Determination of 19-Nortestosterone," *J. Chromatogr.* **1988,** 452, 419-433.

9. Farjam, A.; A.E. Brugman; A. Soldaat; P. Tiommerman; H. Lingeman; G.J. de Jong; R.W. Frei; U.A.Th. Brinkman, " Immunoaffinity Precolumn for Selective Sample Pretreatment in Column Liquid Chromatography: Immunoselective Detection," *Chromatography.* **1991,** 31,469-477.

10. Ong, H.; A. Adam; S. Perreault; S. Niarleau; M. Bellemare;P. Du Souich, "Analysis of Albuterol in Human Plasma Based on Immunoaffinity Chromatographic Cleanup Combined with High-Performance Liquid Chromatography with Fluorimetric Detection," *J. Chromatogr.* **1989,** 497, 213-221.

11. Reh, E., "Determination of Digoxin in Serum by On-Line Immunoadsorptive Cleanup High Performance Liquid Chromatographic Separation and Fluorescence-Reaction Detection," *J. Chromatogr.* **1988,**443,119-130.

12. Hage, D.S.; P.C. Kao, "High Performance Immunoaffinity Chromatography and Chemiluminescent Detection in the Automation of a Parathyroid Hormone Sandwich Immunoassay," *Anal. Chem.* **1991,** 63, 586-595.

13. Thomas, D.H.; M. Beck-Westermeyer, D.S. Hage, " Determination of Atrazine in Water Using Tandem High Performance Immunoaffinity Chromatography and Reversed-Phase Liquid Chromatorgraphy," *Anal. Chem.* **1991,** 66, 3823-3829.

14. Thomas, D.H.; V. Lopez-Avila; L.D. Betowski; J. Van Emon, "Determination of Carbendazim in Water by High-Performance Immunoaffinity Chromatography with Diode-Array or Mass Spectrometric Detection," *J. Chromatogr.* **1996,** 724, 207-217.

15. Ioannov-Kakouri, B.; M. Christodaoulido; E. Christou; E. Constantinidou, " Immunoaffinity Column/HPLC Determination of Aflatoxin M_1 in Milk," *Food & Agricultural Immunology,* **1995,** 7,131-137.

16. Shahtaheri, S.J.; M.F., Ketmeh; P. Kwasowski; D. Stevenson, " Development and Optimization of an Immunoaffinity-Based Solid-Phase Extraction for Chlortoluron," *J. Chromatogr. A.* **1995,** 697,131-136.

17. Azcona, J.I.; M.M. Abouzied; J.J., Pestka, "Detection of Zearlenone by Tandem Immunoaffinity-Enzyme Linked Immunosorbent Assay and Its Application to Milk," *J. Food Prot.* **1990,** 53(7), 577-80,627.

18. Nilsson, B., "Extraction and Quantification of Cortisol by Use of High-Performance Liquid Affinity Chromatography," *J. Chromatogr.* **1983**, 276, 413-417.

19. Stanley, S.M.R.; B.S. Wilhelm; J.P. Rodgers; H. Bertschinger, "Immunoaffinity Chromatography Coupled with Gas Chromatography-Negative Ion Chemical Ionization Mass Spectrometry for the Confirmation of Flumethazone in the Equine," *J. Chromatogr.* **1993,** 614, 77-86.

20. Farjam, A.; A. E.Brugman; H. Lingeman; U.A.Th. Brinkman, "On-line Immunoaffinity Sample Pre-treatment for Column Liquid Chromatography: Evaluation of Desorption Techniques and Operating Conditions Using an Anti-estrogen Immuno-precolumn as a Model System," *Analyst,* **1991,** 116, 891-896.

21. Thomas, D.; V. Lopez-Avila; C. Yan; D.S. Anex; J.M.Van Emon, "Immunoaffinity Chromatography/Capillary Electrokinetic Chromatography with Laser-Induced Fluorescence Detection," *Book of Abstracts, Pittcon'96* **1996**, 789.

Chapter 9

Sensitive Analyte Detection and Quantitation Using the Threshold Immunoassay System

Kilian Dill

Molecular Devices Corporation, 1311 Orleans Drive, Sunnyvale, CA 94089

A sensitive detection system has been developed to quantitatively determine levels of various solution analytes. The Threshold Immunoassay System is commercially available, widely adaptable, and employs a unique detection system based upon a silicon chip. In this chapter, the mechanism of the Threshold Immunoassay System will be explained and examples of applications will be provided. The advantages of this system include reduction in distortions due to solid-phase/liquid-phase interactions, large dynamic range, and high sensitivity. An example is provided that illustrates the usefulness of the Threshold Immunoassay System for the detection of the herbicide, atrazine.

The Threshold Immunoassay System has been available for about 7 years (1,2). It is an analytical system used for quantitative determination of various solution analytes (3) using a silicon chip for detection (4,5). In this chapter we will discuss the theory and operation of the Threshold Immunoassay System using the light-addressable potentiometric sensor (LAPS). Furthermore, we will present data for a variety of assays which will indicate speed, sensitivity, and versatility of the system.

Threshold is one of the most sensitive and reliable immunoassay systems on the market and has gained broad acceptance in commercial, government, and academic laboratories. Most notably, the system is currently being used by the pharmaceutical industry for the detection of impurities in the production of biopharmaceuticals by companies seeking FDA approval for the release of their products (1, 6-9). Contaminants include DNA and host cell proteins which are often present in genetically engineered products. Contaminants in monoclonal antibody products include protein G (or A) as a column material used in the antibody purification process (1, 9, 10). Threshold can also be used to detect large molecules by using sandwich immunoassay formats and may be used to detect small molecules by using competitive immunoassay formats (11). Thus, the system is exceedingly versatile and adaptable. Furthermore, it is clearly one of the most sensitive detection systems available.

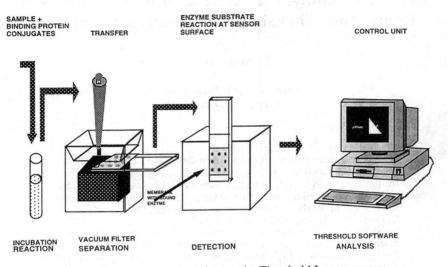

Fig. 1. Immunoassay detection on the Threshold Immunoassay
System.

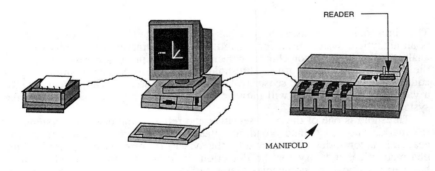

Fig. 2. Threshold Immunoassay System.

Sandwich immunoassay formats are accomplished on the Threshold System via a four step process (12). [1] The first step involves the incubation of the analyte with the labeled binding proteins required to form a sandwich in solution phase. This allows all molecular associations to be accomplished in solution thus avoiding artifacts caused by a solid phase. [2] Analytes are specifically captured and concentrated via vacuum filtration. This is accomplished when the immunocomplex is bound to the 100 μ thick biotin-coated membrane as the solution is passed through it. [3] Detection of urease-labeled antibody conjugates is accomplished by the enzymatic hydrolysis of urea in the microvolume adjacent to a silicon chip. [4] Data are analyzed with Molecular Devices' proprietary software program.

Figure 1 shows a flow diagram of the assay protocol. First, the sample is incubated in a test tube with the various assay components present, such as the biotin and fluorescein labeled anti-analyte antibodies and the biotin binding protein, streptavidin. After the end of the incubation period, the sample is filtered through a biotin-coated nitrocellulose membrane and the immunocomplex is concentrated and captured on the membrane, localized to a 3 mm spot. Assay volumes can vary, but in most cases 100 μL to 2 mL of liquid are used. In a second step, a solution containing the urease conjugate of an anti-fluorescein antibody is passed through the membrane and bound to the immunocomplex; this urease-conjugate together with the enzyme substrate provides the signal generator for the system. Eight individual and independent assay sites are present on each immunocapture membrane (supported on a plastic stick) which can be used to assay different concentrations of a given analyte or detect entirely different analytes. Detection occurs via a light addressable potentiometric sensor (LAPS) when the membrane is placed into the detector where it is pressed against a LAPS chip. The detector (or Threshold reader) is filled with the enzyme substrate in a buffered solution. All eight assay sites on each stick are detected simultaneously. The enzyme is urease and the substrate is urea; the signal is the change of pH detected by the LAPS chip. The data are then analyzed by the Threshold software.

Figure 2 shows the actual Threshold Immunoassay System, including the computer, printer and workstation. The workstation includes both a vacuum manifold for immunofiltration and the reader with the LAPS sensor. The reader can be removed and conveniently disassembled, for cleaning, if protein fouling occurs. Figure 2 shows a workstation with one immunofiltration manifold. This manifold can accommodate four sticks for a total of 32 assay sites. Two additional auxiliary manifolds can be added for a total of 96 assays sites which can be vacuum filtered simultaneously. Use of the appropriate standards and controls on the sticks reduces the number of assay sites available for sample testing to 66.

The mainstay of the assay system is the LAPS sensor. Figure 3 shows the basics of the sensor. It is a silicon chip with 8 identically etched sites. The membrane containing the various captured chemistries is aligned and pressed against the chip with a plunger to form tightly sealed 0.6 μL microvolumes. As mentioned earlier, the signal generator in this system is an enzyme, which, upon acting on the substrate, produces an increase in pH within this microvolume. It is this pH change that is monitored by LAPS.

The exact mechanism by which the pH change is monitored on the chips by LAPS is as follows: A bias potential is applied to the external circuit that connects the chip to the controlling electrode. The bias potential, together with the pH-dependent surface potential at the pH-sensing sites, determines the magnitude and direction of the electric field in each respective pH-sensing region of the semiconductor adjacent to the dielectric. Illumination of the semiconductor with light emitting diodes (LEDs) near a pH sensing site produces a photopotential resulting in a transient photocurrent to flow in a circuit external to the semiconductor chip. Changes of pH on the chip surface adjacent to any particular pH-sensing region alters the bias voltage required to produce a given photocurrent when a LED is used to illuminate the semiconductor near that region. In operation, as the pH is altered, the inflection point of the photocurrent/applied voltage curve shifts on the potential axis (see Figure 4). It is this rate in the shift (μV/s) that is monitored and is directly related to the change in pH.

What makes the Threshold Immunoassay System such a good analytical system? There are several attributes. First, antigen-antibody binding takes place in solution phase rather than on a solid phase. This greatly speeds up the process and enhances antibody and antigen stability (12). Secondly, the detection and quantitation occurs in a very small volume adjacent to each pH-sensing region on the LAPS chip. The small volume is created by a movable piston which compresses the immunocapture membrane against the LAPS chip. The sensitivity, per unit of assay time, is often 10-100 times greater than that observed for colorimetric or enzyme-linked immunosorbent assays (ELISA). Thirdly, the immunocapture membrane greatly facilities capture and analysis of particulate antigens such as biological cells and spores. The detection system, however, does not require integral contact of the enzyme label on the surface (as does other detection systems). In the Threshold System, protons diffuse rapidly from the labeled enzyme to the sensor surface. Additionally, the Threshold System can also be used to detect large and small molecules; sizes may vary from a few hundred atomic mass units (cAMP, saxitoxin, atrazine) to molecular weights in the millions (DNA).

The assay format need not be limited to typical immunosandwich assay formats. For small molecules, detection is made possible by an indirect detection method (11). In this case, the competitive inhibition is utilized in several different formats. Simply, the unknown quantity of sample analyte is allowed to compete with a known quantity of labeled analyte for the antibody binding site. A decrease in the expected signal is then related to the detection of the (non-labeled) analyte in the sample being tested. Alternatively, an indirect detection assay format may be used where the analyte to be used in the competitive inhibition studies is covalently bound to streptavidin (11). More complicated binding inhibition studies may be performed in a complex system involving receptors. In this case, an immunosandwich is formed as the agonists and antagonists compete for the receptor binding site.

Physical parameters may also be extracted using the Threshold System. Using some of the binding schemes described above, binding parameters (absolute and relative) such as K_a and K_d have been derived for antibody/antigen, lectin/carbohydrate, and agonist/antagonist/receptor binding pairs (11-17). We have found that very accurate K_d values can be derived that can be corroborated by literature values or by other physical techniques. Kinetic rate constants for binding (association and dissociation) may also be derived (18). Association rate constants (k_{on}) may be measured for binding processes, such as antigen-antibody, that occur at a moderate rate (18). The off-rate (k_{off}) may be then obtained from the values of k_{on} and K_a.

To show the sensitivity of the Threshold Immunoassay System, the mass sensitivities for selected molecules are presented in Table I. The molecules in this table represent large and small molecules and detection based upon use of monoclonal and polyclonal antibodies. As one can see, limits of detection occur over a wide range of analyte concentrations (about 1-10 pg per assay). Two cases represent the detection of small molecules using monoclonal antibodies with low

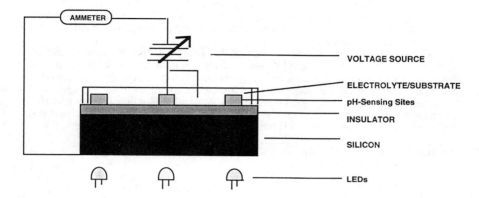

Fig. 3. The Light Addressable Potentiometric Sensor (LAPS).

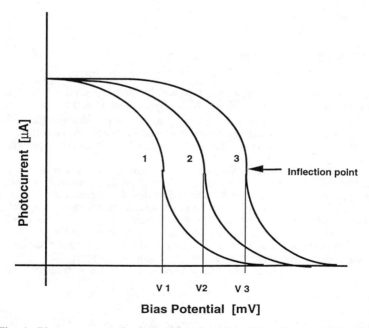

Fig. 4. Photocurrent-voltage curves in LAPS and the effects of pH. V1
represents a low pH value and V3 represents a higher pH value.

affinity constants ($K_a \sim 10^7$ - 10^8 M^{-1}). Results for both cases show that use of these low affinity antibodies results in analyte limits of detection which are considerably higher. An attempt to achieve lower limits of detection by increasing the quantity of the low affinity antibody used in the assay (to increase assay response) will only result in additional background signal which reduces the specific signal even further.

The reason for the high sensitivity of the Threshold System lies in several factors. Firstly, we utilize liquid phase binding, eliminating any steric problems associated with solid phase binding which reduces the quantity of bound analyte. Secondly, the Threshold System uses the filtration-capture method. This method captures, specifically, only the bound immunocomplexes while the nonspecific material is simply filtered through the membrane. Because the immunocomplexes are captured on a small area, the sample is concentrated. This allows solution

TABLE I. Selected Molecules Tested Using the Threshold
          Immunoassay System

Molecule	Limit of Detection
DNA	2 pg
IL-2	50 pg/mL
Neuropeptide	10 pg/mL
Transferrin	20 pg/mL
Ricin	3 pg
*Saxitoxin	1 ng
*cyclic AMP	3 ng

* Low affinity monoclonal antibody was used

binding to occur in 2 mL volumes or larger, but the captured material will be detected in a microvolume (0.64 µL). Thirdly, for our signal generator, we use the enzyme urease which has an exceedingly high turnover number [20 $\times 10^5$ mol urea/min/mol urease (19)] and the enzymic reaction is measured in volumes of 0.6 µL. Lastly, the detector associated with the LAPS system produces very low electronic noise.

Except for DNA, the immunosandwich assays are performed in the Immuno-Ligand Assay (ILA) format. It is a generic format for larger molecules containing multiple epitopes, which can bind two antibodies simultaneously.   For small molecules, where only a single epitope is present, a competitive assay format is performed utilizing a polyclonal or monoclonal antibody. The biotin and fluorescein labeling reagents are sold commercially (Molecular Devices holds the patent on biotin-DNP-NHS labeling reagent) and the assay can be set up in less than a day; the assay scheme is shown in Figure 5. Molecular Devices also provides a kit containing labeling reagents, the capture reagent, enzyme-conjugate, substrate, biotin-coated sticks, and buffers. The antibody is first labeled with fluorescein or biotin. The capture reagent is present as streptavidin, a protein with an affinity constant of 10^{15} M^{-1} for biotin and has four binding sites (20). The signal generator is urease conjugated to an anti-fluorescein antibody.

When the analyte and labeled antibodies are mixed together we have the reaction or binding stages. This is done in the liquid phase. The separation stage is actually the filtration-capture stage. The reaction mixture is filtered under vacuum through a 0.45 micron biotin-coated nitrocellulose membrane. Only a

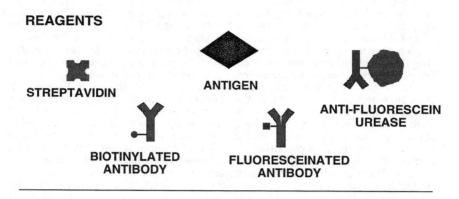

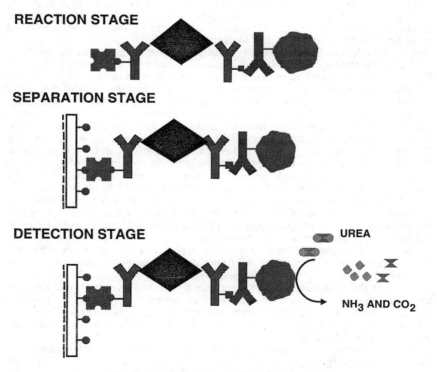

Fig. 5. Immuno-Ligand Assay (ILA) stages.

complete immunocomplex (sandwich) which is captured will provide a signal. Although biotinylated antibody may be captured without the analyte and fluoresceinated antibody being present, no signal will be generated. The enzyme reagent (urease conjugated to an anti-fluorescein antibody) then attaches to the bound immunocomplex. The detection stage occurs when the fully formed immunocomplexes are then placed in buffer (pH 6.5) containing the substrate urea (Figure 5). Under microvolume conditions, a pH change is readily observed when urea is hydrolyzed by urease into ammonia and carbon dioxide.

A typical result from this type of assay system is shown in Figure 6, a standard detection curve for the ILA for protein G. Note the straight line response for the analyte. Furthermore, the detection range is over 2 log units and the lower limit of detection is 2 pg. This is a typical example for a highly sensitive assay for a larger molecular weight analyte. The inset to Figure 6 shows the lower levels of analyte concentration. The background signal, resulting from nonspecific binding of proteins to the membrane, limits the level of detection.

The detection of an analyte more relevant to this audience is that of the herbicide atrazine, which is used throughout the agricultural industry (21). The structures of atrazine and its biotin-DNP derivative are given in Figure 7. Atrazine is one of many related herbicides containing the triazine ring structure which block plant photosynthesis. It is only sparingly water soluble, which makes it difficult to work with. In order to determine if the Threshold system can be used to detect a molecule this small, some assay modifications were made (indirect detection method). Monoclonal anti-atrazine antibody was obtained from Dr. Thomas Giersch of the University of Munich (22). For the indirect detection mode, labeled atrazine was required in order to sequester the antibody on the membrane (stick) and obtain a signal. The basic assay diagram is shown in Figure 8. In this system, the atrazine has been labeled with biotin (23) tethered by 1,3 diaminopropane and the antibody binds to this species. Using this assay format, the atrazine is bound to streptavidin and pre-coated onto the membrane. In the solution phase, fluorescein labeled anti-atrazine antibody is mixed with a sample being tested that may contain the analyte. In the absence of atrazine in the sample solution, the antibody should be bound to the membrane and give the largest response (signal) possible. When a large quantity of atrazine is present in the sample, the antibody binds predominantly to the atrazine in solution and is subsequently washed through the membrane without being captured. Thus, the presence of a large quantity of atrazine in solution would result in a very low signal. Not shown in the diagram is the urease conjugate of the anti-fluorescein antibody. The addition of the enzyme reagent (not shown in Figure 8) to the immunocomplex is subsequently accomplished by the addition of the reagent to the immunocomplex bound to the membrane.

Assay development followed a typical competitive assay approach. Once the biotinylated atrazine (b-atrazine) and fluoresceinated antibody were produced, an antibody loading study was performed. The amount of antibody that can be captured by the membrane must be determined. This will give the upper limits of the signal that can be generated when no free atrazine is present. The protocol was as follows. A b-atrazine solution was mixed with capture reagent such that the biotin/streptavidin was about 1/5. Six micrograms of this material was then loaded onto the membrane.

This was followed by increasing quantities of antibody and a fixed quantity of the urease-antibody conjugate (2 μg/test). The sample was washed with wash buffer and the stick was placed into the reader filled with a buffer containing the substrate, urea. The reader output can then be plotted to obtain the graph shown in Figure 9.

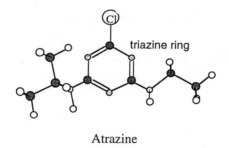

Fig. 6. Recombinant protein G standard curve.

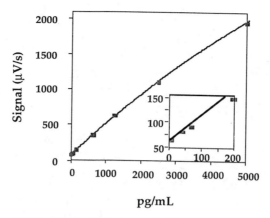

Atrazine

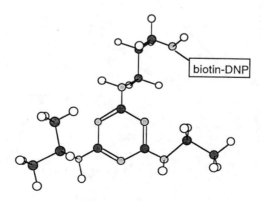

Biotin-DNP Atrazine

Fig. 7. Structures of atrazine and biotin-DNP derivative of atrazine.

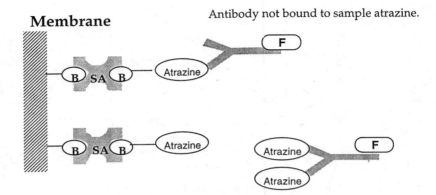

Fig. 8. Atrazine assay format.

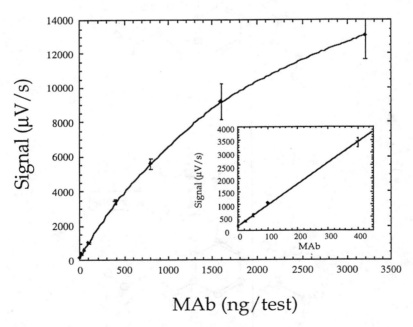

Fig. 9. Anti-atrazine MAb loading study.

As can be seen in Figure 9, up to 3.2 μg of antibody can be loaded onto the membrane. However, the upper end of the loading curve is non-linear. This can be related to a number of items, including the fact that enzyme is inhibited by the high pH. The lower end of the antibody loading curve is quite linear. To avoid any artifacts, such as overloading the membrane, we chose to use antibody at the 400 ng/test range for our experiments. Typically, 150 ng/test of each labeled antibody are used in an ILA assay format. Note in Figure 9 that the concentration of antibody used in our assays will limit our signal output and the limit of detection. Use of additional antibody in the experiments produces unwanted nonspecific binding resulting in a larger background signal.

Figure 10 shows the results of our atrazine assay using the format described above. The protocol is slightly different than for the MAb loading study described above as we have analyte to be tested in this case. The antibody (400 ng/test) was mixed with a fixed quantity of a solution containing varying amounts of atrazine. The solution was allowed to incubate for one hour at room temperature. Following this, the membrane was pre-wet with buffer and the streptavidin/biotin-atrazine mixture was loaded onto the membrane. The membrane was washed to remove unbound, labeled atrazine and then the antibody-containing sample was loaded onto the membrane and vacuum filtered. Following this, 200 μL of the anti-fluorescein urease conjugate was added (1.5 μg) and also filtered, followed by another 0.5 mL wash of the membrane. Sticks were then read. Enough solution was made such that each sample afforded three assays for that concentration of atrazine.

These are very preliminary results, based upon a few experiments. Figure 10 shows that as the quantity of atrazine increased in solution, the signal response of the assay decreased. The curve is not quite sigmoidal in shape as one would expect, but this could result from several factors. One is that the monoclonal antibody is not truly a pure monoclonal antibody but rather a mixture containing predominantly one population. Another possibility is that linking biotin to atrazine altered some of the material (not a single product) and this causes non-uniform capture of the antibody. Never-the-less, this preliminary experiment does show that the assay is feasible and can be further optimized. The midpoint of this monoclonal indicates that the value of K_d is about 1×10^{-7} M. This value indicates that the affinity of this antibody for atrazine is not optimal. The limit of atrazine detection in our system is about 200 pg/mL.

As mentioned, this is a very early trial experiment and many improvements can be made to increase the level of detection. Firstly, a higher affinity antibody would definitely improve the limit of detection. The antibody concentrations that were used were well below the K_d value for this system, limiting our detection of atrazine. Secondly, the assay format can be changed to improve detection. By mixing the b-atrazine/solution atrazine, streptavidin, and antibody all in one mixture, a true competitive assay can be achieved and, more importantly, the capture efficiency of this antibody would be improved. Another enhancement is to use a different means of linking biotin to the atrazine. For instance, attachment of the biotin label to one of the existing N-alkyl substituents. This could be used instead of the present attachment of biotin via displacement of the chlorine atom on atrazine. As it turns out, the chlorine atom on atrazine is crucial for the effective binding to the herbicide because the antibody that we obtained appears to lose affinity for atrazine when the chlorine atom is missing. All the above mentioned items could drastically lower the level of atrazine detection.

To show that this system works for other small molecules, we obtained data for the detection of the red tide toxin, saxitoxin. The format was very similar to that described for atrazine. For competitive binding, saxitoxin (STX) was covalently linked to streptavidin. In this way, the streptavidin-STX conjugate

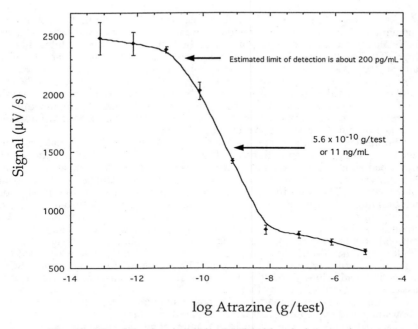

Fig. 10. Titration of anti-atrazine MAb with solution atrazine.

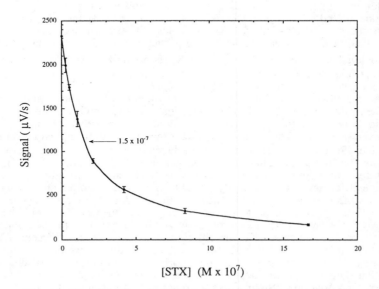

Fig. 11. Titration of anti-saxitoxin MAb with solution saxitoxin.
Reprinted with permission from reference 11. Academic Press, Inc. 1994.

competed with free STX in solution (competitive inhibition) for the binding sites of an anti-STX antibody labeled with fluorescein. The monoclonal antibody was kindly provided by Dr. Tan Chan from the Southwest Research Institute (24). Unfortunately, this monoclonal had a low affinity for the toxin ($7 \times 10^6 M^{-1}$). It is for this reason that the limit of detection for STX is high using this assay scheme (see Table I). The titration data are presented in Fig. 11. Binding constant data are obtained by inverting the concentration of STX used at the midpoint. The result once again indicates that small molecules can be detected, and that the assay protocols can be modified to meet the needs of the user.

We have shown that the Threshold Immunoassay System is at the forefront of sensors currently available for the detection of small molecular weight analytes and larger biomolecules . It is versatile and can also be used to detect cells and spores. All events occur in solution phase and the assay is versatile, thus enabling many formats. Furthermore, the assay can be modified to include almost any assay volume, ranging from 100 μL to 2 mL. The greatest attribute of the system is sensitivity. Clearly, we can routinely detect pg quantities for most analytes.

Acknowledgment
The author thanks Jayne Blomdahl in her tireless effort in the proofreading of the manuscript.

References
1. McNabb, S.; Rupp, R.; Tedesco, J. L. *Bio/Technology* **1989**, 7, 343-347.
2. Libby, J. M.; Wadda, H. G. *J. Clin Microbiol.* **1989**, 27, 1456-1459.
3. Merrick, H.; Hawlitschek, G. *Biotech Forum Europe* **1992**, 6, 398-403.
4. Hafeman, D. G.; Parce, J. W.; McConnell, H. M. *Science* **1988**, 240, 1182-1185.
5. Owicki, J. C.; Bousse, L. J.; Hafeman, D. G.; Kirk, G. L.; Olson, J. D.; Wada, H. G.; Parce, J. W. *Annu. Rev. Biophys. Biomol. Struct.* **1994**, 23, 87-113.
6. Robinette, R. S. R.; Herbert W. K. *J. Immunol. Methods* **1993**, 159, 229-234.
7. Per, S. R.; Aversa, C. R.; Johnston, P. D.; Kopasz, T. L.; Sito, A. F. *BioPharm.* **1993**, Nov.-Dec., 34-40.
8. Workman, W. E. *Pharmacopeial Forum* **1995**, 21, 479-484.
9. Eaton, L. C. *J. Chromatogr. Biomed Appl.* **1995**, in press.
10. Briggs, J.; Panfili, P. R. *Anal. Chem.* **1991**, 63, 850-859.
11. Dill, K.; Lin, M.; Poteras, C.; Fraser, C.; Hafeman, D. G.; Owicki, J. C.; Olson, J. D. *Anal. Biochem.* **1994**, 217, 128-138.
12. Panfili, P. R.; Dill, K.; Olson, J. D. *Current Opinion Biotech.* **1994**, 5, 60-64.
13. Rogers, K. R.; Fernando, J. C.; Thompson, R. G.; Valdes, J. J.; Eldefrawi, M. E. *Anal. Biochem.* **1992**, 292, 111-116.
14. Fernando, J. C.; Rogers, K. R.; Anis, N. A.; Valdes, J. J.; Thompson, R. G.; Eldefrawi, A. T.; Eldefrawi, M. E. *J. Agric. Food Chem.* **1993**, 41, 511-516.
15. Dill, K.; Blomdahl, J. A.; Lydon, S. R.; Olson, J. D. In Proceedings of the U. S. Army Edgewood Research, Development, and Engineering Center 1993 Conference on Chemical Defense Research, Aberdeen Proving Ground, Maryland.
16. Dill, K.; Olson, J. D. *Glycoconj. J.* **1995**, 12, 660-663.
17. Dill, K.; Fraser, C.; Blomdahl, J. A.; Olson, J. D. *J. Biochem. Biophys. Methods* , in press.
18. Hafeman, D. G.; Dill, K., manuscript in preparation.
19. Blakelely, R. L.; Zerner, B. *J. Molec. Cat.* **1984**, 23, 263-292.

20. Green, N. M. *Methods Enzymol.* **1990**, 51-67.
21. Merck Index, An Encyclopedia of Chemicals and Drugs. The 9th Edition, **1976**, Rahway, NJ.
22. Giersch, T. *J. Agric. Food Chem.* **1993**, 41, 1006-1011.
23. Tom-Moy, M.; Baer, R. L.; Spira-Solomon, D.; Doherty, T. P. *Anal. Chem.* **1995**, 67, 1510-1516.
24. Huot, R. I.; Armstrong, D. L.; Chanh, T. C. *J. Toxicol. Environ. Health* **1989**, 27, 3810-393.

Chapter 10

Optical Sensing Technology for Environmental Immunoassays

Stephen L. Coulter and Stanley M. Klainer

FCI Environmental, Inc., 1181 Grier Drive, Building B, Las Vegas, NV 89119

Immunoassay kits have been developed for field use in environmental monitoring. While they satisfy most of the requirements for an effective field method, they still suffer from the complexity of their use, and the difficulty in obtaining a reproducible endpoint. The marriage of competitive immunoassay to optical sensors results in a simple, solid-state system which provides a single step method for environmental monitoring.

Needs for environmental monitoring are constantly increasing. Effective environmental monitoring requires the use of field analytical instrumentation. Environmental monitoring is typically a transitioning of laboratory analyses to field analyses. Most analytical methods used today are not suitable for the field. They were primarily designed for laboratory operation and lack the critical elements necessary for field environmental analyzers:

1) Superior performance characteristics;
2) Operational methods which maximize the performance of the analyzer (ease of use);
3) In situ measurements (no reliance upon sampling and preparation);
4) Real-time responses;
5) No sampling artifact;
6) Low cost for evaluation of multiple samples (encourages broader testing schemes);
7) Portability.

While some of the existing monitoring equipment has characteristics which make them useful for field use, a major barrier is still the ability to perform analyses in situ. Fiber optic sensors using either chemical or biological sensing mechanisms are inherently well-suited for in situ methods. The development of solid-state, fiber optic instrumentation based upon fluoroimmunoassay techniques provides not only a higher level of performance characteristics (sensitivity, selectivity) over other environmental, field instrumentation, but also provides a field methodology which is extremely simple to use.

0097–6156/96/0646–0103$15.00/0

FOCS® SENSORS

FCI Environmental has designed products based upon a fiber optic chemical sensor (FOCS®) technology. The FOCS® sensors utilize optical waveguides (e.g., fiber optics) coated with proprietary sensing chemistries which are designed to interact with specific environmental contaminants. The sensing package comprises a light source which provides the output through the waveguide, the sensing waveguide and the appropriate detector. The light transmission through the waveguide is based upon the phenomenon of total internal reflectance. The transmitted light interacts with the sensing chemistries on the surface of the waveguide. This interaction directly affects the signal transmitted to the photodetectors. The FOCS® design can accommodate a number of optical sensing mechanisms and a number of detection methods. The sensing mechanism can be based upon organic, inorganic or biologic responses. The method of detection can be based upon any of the common optical detection methods including those listed below.

SENSING MECHANISMS	
Organics	Antibodies/Antigens
Dyes	Enzymes
Polymers	Biologics
Inorganics	Organometallics

⇓

DETECTION METHODS	
Fluorescence	Refraction
Absorbance	Chemiluminescence

The current PetroSense® instruments are based upon the FOCS® technology. Detection is based upon refractive index changes as hydrocarbons interact with the fiber optic. The PetroSense® sensors detect total petroleum hydrocarbons (TPH) in the parts per million range. They meet the requisite performance criteria for field analyzers. They have the necessary sensitivity and selectivity for field screening instruments.

The primary advantage that the PetroSense® analyzers have is their mode of operation. The PetroSense® sensors operate in vapor, in water and at the interface between vapor and water. They provide real-time, in situ, analytical measurements of environmental contaminants at low cost. By taking in situ measurements, the need for sampling, storage and transportation is eliminated. The greatest source of error in typical field monitoring methods comes form this sampling and handling (1). The ability to have the data in real time enables more accurate mapping of a contaminated site by allowing the field investigator to know when additional data is needed at the time of evaluation. If these additional analyses are delayed and data/samples are collected at different times, artifact is introduced into the comparisons due to the temporal changes in the contamination site. The combination of low cost per analysis, ease of use and real-time results encourages a more thorough site analysis.

COMPETITIVE IMMUNOASSAY OPTICAL SENSOR

Whereas the current PetroSense® sensors meet all the criteria for effective field screening instruments, the goal of FCI Environmental is to improve both the selectivity and the sensitivity of these sensors in order to provide analytical sensors. The current sensors detect a composite TPH concentration in the parts per million range (0.1 to 40,000 ppm of TPH.) The goal is to be able to detect parts per billion levels of selected contaminants and to be able to selectively detect specific contaminants or classes of contaminants. Immunoassay methodologies provide the selectivity desired. By utilizing a fluorescent detection scheme, the sensitivity of the sensor is improved to less than a part per billion.

Immunoassay kits have been commercialized to take advantage of the specificity of the method. Table I provides some examples (not an exhaustive compilation) of commercial kits used for field screening:

Table I. Commercial Environmental Immunoassay Kits

Target Analyte	Companies Involved in Environmental Immunoassays	
BTEX	D-TECH EnSys Millipore	Quantix SDI
PAHs	D-TECH EnSys	Millipore Quantix
Chlorophenols	EnSys Millipore	Ohmicron
PCBs	D-TECH EnSys	Millipore Ohmicron
TNT	D-TECH EnSys	Millipore Quantix
Dioxins	EnSys	SDI
Pesticides	Millipore Ohmicron	Quantix

The problem is that most field screening kits require multiple steps (typically 6 to 10 steps), have significant time delays (often 20 minutes) and are difficult to use in terms

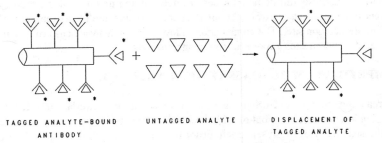

Figure 1. FCI Environmental Competitive Immunoassay Scheme

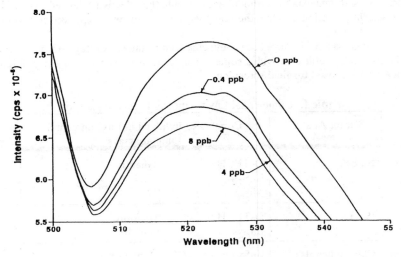

Figure 2. Competitive Immunoassay Response

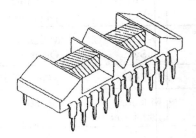

Figure 3. Chip Level Waveguide Sensor

of obtaining an accurate, reproducible endpoint. The strategy of FCI Environmental is to develop a solid-state, single step immunoassay sensor for environmental analyses which requires less than 5 minutes per assay. The selectivity of the immunoassay, the sensitivity of fluorescence detection and the ease of use with a single step, solid-state configuration may provide the next major breakthrough for environmental field screening instrumentation.

The immunoassay techniques employed at FCI Environmental are based upon competitive immunoassay methods. Some of the considerations in the development of a competitive immunoassay sensor include the lack of sensitivity due to high background noise, inadequate antibody loading on the substrate and inefficient exchange of the environmental analytes (antigens) for the "tagged" analytes on the sensor. In a competitive immunoassay optical sensor, the desired antibody is immobilized to the optical waveguide. Tagged antigens or analytes are bound to the antibody, with a "tag" typically being a fluorophore. When the environmental analyte comes into contact with the sensor, it displaces the tagged analyte, causing a change in the fluorescent signal intensity.

The company's first demonstration of the technique was for a "drugs of abuse" application (2). The antibody for cocaine was immobilized to the sensor surface, followed by the complexation of the antibody with fluorescently tagged cocaine. When the sensor was exposed to cocaine in a sample (as in Figure 1), the competitive reaction resulted in a loss of the fluorescent signal. The detection limits of this sensor were determined to be less than 400 parts per trillion (see Figure 2.)

The fabrication of the solid-state immunoassay optical sensor utilizes monoclonal, polyclonal or pooled monoclonal antibodies to establish the selectivity necessary for a given application. The selection of the tag for the analyte controls the efficiency of exchange between the analyte in the sample and the initially bound analyte on the sensor. Standard coupling methods were applied to the system to generate membrane-based sensors and FOCS® sensors. The efficiency of these coupling methods was maximized for the highest analyte loading.

"IMMUNOASSAY ON A CHIP"

To maximize the efficiency of the solid-state immunoassay sensor, a chip level waveguide sensor platform has been developed (3). This chip sensor incorporates the light source, optical waveguide and photodetectors into a single sensing element. The chip is designed to be "dual-armed" with a sensing arm and a reference arm. It can also be fabricated with a single arm, or even multiple arms for a variety of sensors on a single platform. There are several advantages of the chip design in the areas of size, performance, manufacturability and cost. The smaller design allows for the reduction in the size of the overall probe. The performance of the sensor is improved due to an increase in sensitivity and a reduction of artifact (noise). Signal intensity has been improved by reducing the number of interfaces in the design and through the configuration of the sensing element which minimizes the amount of lost light. The incorporation of a referencing channel with balancing of the incident light signal which allows for the subtraction of any humidity and temperature effects, minimizes the level of artifacts (noise). Additional noise can come from undesired association of the tagged analyte with the sensor. Efforts were made to increase the yield in the binding of the tagged analyte and then to block the remaining sites with bovine serum albumin. This served to minimize the unused active binding sites, a significant source of background noise. The design characteristics of the optical chip configuration are designed to address the level of background noise which is the other barrier to improved performance mentioned earlier.

The manufacturability of the chip design is a major benefit. The complexity of the assembly process is reduced based upon the incorporation of several components into the single chip. Issues involved with reproducibility of the alignment of the components are greatly reduced with automated fabrication of the chips. The yield of the process is increased, the inspection of the goods is simplified and the overall efficiency is improved. The obvious result of these improvements in sensor performance and manufacturability is a cost reduction for the sensor.

The current focus at FCI Environmental is the adaptation of the FOCS® immunoassay configuration to the chip. Two products have currently been developed on the chip for petroleum hydrocarbons and for indoor air quality monitoring. The use of immunoassays will allow the development of a large number of environmental sensors based upon simply changing the antibodies on the chip. With a common platform, the sensors become interchangeable for applications to areas such as indoor air quality, industrial hygiene or personal exposure monitors.

Literature Cited

1.   *Field Measurements, Dependable Data When You Need It;* United States Environmental Protection Agency, Solid Waste and Emergency Response; EPA/530/UST-90-003: September, 1990.
2.   Li, H.; Goswami, K.; Klainer, S. *The Use of Fiber Optic Chemical Sensors (FOCS) For the Detection of the Effluvia from Illicit Drugs and the Materials Used to Prepare These; Final Report for Phase I;* Advanced Research Projects Agency Contract No. DAA D05-92-C-0022: December, 1993.
3.   Saini, D.P. *Chip Level Waveguide Sensor;* U. S. Patent No. 5,439,647; August 8, 1995.

Applications and Evaluations

Chapter 11

Recent Developments in Immunoassays and Related Methods for the Detection of Xenobiotics

Ingrid Wengatz, Adam S. Harris, S. Douglass Gilman,
Monika Wortberg, Horacio Kido, Ferenc Szurdoki,
Marvin H. Goodrow, Lynn L. Jaeger, Donald W. Stoutamire,
James R. Sanborn, Shirley J. Gee, and Bruce D. Hammock

Departments of Entomology and Environmental Toxicology,
University of California, Davis, CA 95616

Very few rapid, cost-effective methods for the analysis of hazardous substances in humans and the environment are available. Immunoassays are among these methods and are becoming established for measurement of toxic materials in the environment. In addition immunoassays are suitable for monitoring human exposure to xenobiotics. Advantages of immunoassays also include sensitivity, specificity, applicability to a wide variety of compounds and adaptability to laboratory or field situations. The use of these assays facilitates development of good models for human exposure, movement of groundwater contaminants, and research on remediation systems.

An important objective of our research is to develop assays to assess human exposure to xenobiotics. Metabolites of these xenobiotics may serve as biomarkers in toxicity and exposure assessment studies. Immunoassays for biomarkers include triazine mercapturates, nitrophenols, and pyrethroid metabolites. Traditionally immunoassays have been used as single-analyte methods, but now class-selective and multi-analyte assays for environmentally relevant compounds have been successfully demonstrated. A further step in simplifying and improving sensitivity of assays, as well as the development of field portable devices, is the implementation of near-infrared fluorescence detection. Also developed in this laboratory are assays to detect heavy metals using chelators instead of antibodies.

The most recent work in this laboratory encompasses a variety of goals. One goal is to develop immunoassays that can be employed for monitoring programs to detect toxic chemicals in environmental samples. Immunoassay techniques can be very advantageous for this task. Advantages of immunoassays have been

0097–6156/96/0646–0110$15.00/0

reviewed numerous times (*1-3*), they include achievement of high analyte sensitivity and selectivity with minimal sample preparation, high throughput of samples and therefore cost effectiveness for monitoring programs.

Besides detecting parent compounds or environmental degradates in environmental samples, it is of importance to have tools to monitor human exposure to xenobiotics, e.g. pesticides. In this respect immunoassays for pesticide metabolites can be very helpful. Some of the analytes we are currently developing assays for do not fit into the category of easy target analytes. Usually if the target analyte is fairly large, hydrophilic, stable, nonvolatile, and foreign to the host animal (*4*) it is fairly easy to develop an enzyme immunoassay (EIA). Yet our group has been able to develop immunoassays for relatively small molecules (less than 400 molecular weight) such as 4-nitrophenol (*5*), 1-naphthol (*6*), and monuron (*7,8*), hydrophobic chemicals such as pyrethroids (*9*), hydrolytically unstable chemicals such as carbaryl (*10*), and volatile chemicals such as thiocarbamates (*11*). Table II gives an overview of immunoassays developed in our laboratory.

Another focus is to improve existing immunoassay formats, by increasing the speed of the assay or sensitivity, or simplylifying handling procedures (e.g. by using NIR fluorescent labels or using multianalyte formats). Lastly we are exploring assays to detect metal ions based on technology that is similar to immunoassay.

Pesticide Immunoassays

1) *s*-Triazine Herbicides.   The *s*-triazines are used extensively worldwide as herbicides and are prone to cause environmental contamination problems, particularly in groundwater. Often they can be used as markers of agrochemical pollution. It is most likely true that if *s*-triazines are detected in an environmental sample, other agrochemicals may also be detected. Our long-term project with *s*-triazines is aimed at several of the most common *s*-triazines and their environmental and mammalian metabolites (Figure 1). High affinity polyclonal and monoclonal antibodies were obtained from a series of logically designed *s*-triazine haptens. With these antibodies, useful assays have been developed for several parent *s*-triazines and some important metabolites (*12-17*). The lowest limit of detection that has been achieved for one of the atrazine assays is 30 pg/ml (*17*). High tolerance for matrices like urine and organic solvents was shown with some of the assays. Triazines and/or metabolites have been analyzed with these assays in samples of soils, natural waters, foods and urine.

1a) Parent Compounds.   Presently our efforts are concerned with the quantitative determination of simazine in the presence of atrazine. A new immunizing antigen was based on a hapten design whereby one of the larger alkylamino groups of the triazine ring was replaced with a methylamino substituent (Figure 1, where R^1 = Cl, R^2 = CH_3, R^3 = $(CH_2)_5COOH$). Presumably antibodies would have a geometry which would accommodate a methylamino moiety (and perhaps an ethylamino) but not the bulkier isopropylamino group of atrazine. Indeed some success was achieved with the generation of an assay more selective for simazine than atrazine (*18*). Although not the perfect hapten for a totally selective simazine assay, this assay does provide better antibodies for simazine recognition and will be utilized in our multianalyte methods development.

We have now embarked on the preparation of an immunogen using a triazine hapten with two smaller alkylamino substituents on the triazine ring, namely the mercaptopropanoic acid derivative of a bis-methylamino triazine (Figure 1, $R^1 =$ $SCH_2 CH_2COOH$, $R^2 = R^3 = CH_3$). We anticipate that antibodies to this hapten would accommodate both ethylamino groups of simazine, while theoretically less likely to form an antibody-atrazine complex because of the larger isopropylamino appendage of atrazine.

1b) Multianalyte Methods. The structural similarities of the triazine herbicides present opportunities to develop multianalyte immunoassays. Immunoassay traditionally has been used as a single-analyte method, however, it is frequently observed that "specific" antibodies bind to a number of structurally similar compounds, rather than being monospecific for one analyte. This phenomenon, which occurs with both mono- and polyclonal antibodies, is named cross-reactivity. Especially in the case of small analytes such as pesticides, many cross reacting compounds may exist for a given antibody.

If the presence of cross-reacting compounds in a sample is unknown, false data will be obtained during immunoanalysis when assuming a single analyte. On the other hand, cross-reactivity enables the use of antibodies as a screening tool for multiple analytes or for a whole class of analytes. Usually this approach yields a sum signal which is not weighted for a specific compound, but rather indicates whether a certain class of analytes is present or absent. It is possible to carefully design immunizing haptens that will give assays that are specific for a single molecule or will detect a range of related structures. Class-selective assays are now in use in some laboratories and on the market for a small number of environmentally relevant compounds.

The idea of simultaneous immunochemical detection of multiple analytes which do not cross-react has previously been demonstrated. One approach is to use dual labels for two analyte analysis, thereby performing two independent assays on the same solid phase. It can be based on the use of two different enzymes (*19*), radioactive markers (*20*), fluorophores (*21,22*), metal-labels (*23*) or others. If only a single label is used but no cross-reactivity occurs, spatial resolution of capture antibodies allows more complex multianalyte analysis as was described for the multispot immunoassay based on fluorescence detection (*24-26*). The dual label approach as well as the spatial resolution approach, however, depend on the use of "monospecific" antibodies or on the presence of analytes that do not interfere with each other at their actual concentration levels. Especially in clinical chemistry even minute cross-reactivities can be undesirable when these interfering compounds are present at a much higher level than the analyte(s) of interest.

Recently, several groups have taken advantage of cross-reactivity of antibodies to perform multianalyte immunoassays within a class of compounds. Muldoon et al. (*27*) quantitated ternary mixtures of the triazine herbicides atrazine, simazine and cyanazine in pesticide rinsate. The mathematical approach used a linear extension of the four-parameter curve fit and was based on the assumption that the log standard curves of different analytes are parallel. The limit of detection for triazines in this assay system was 200 ppb, which is feasible for rinsate analysis but not for trace analysis in drinking water. In their multianalyte ELISA (MELISA) Jones et al. (*28*) extended the four parameter log-logistic curve to mixtures where the log plots were not assumed to be parallel. Wortberg et al. (*29*) applied the MELISA methodology for analysis of ternary and quaternary mixtures of triazines at low to sub-ppb levels, using mono- and polyclonal antibodies. The principle was to use a set of data produced by an array of triazine antibodies which exhibit different cross-reactivity patterns.

Multivariate statistical analysis as a means of identifying (and sometimes quantifying) analytes is a well-established method. An example for the quantitation of a single but previously unidentified triazine herbicide in a sample by pattern recognition has been described by Cheung et al. (*30*). The underlying assumption is that the characteristic pattern generated by each analyte is consistent over a certain range of its concentration. The authors' approach comprised principal component analysis, minimum estimates of variance and K nearest neighbors cluster analysis. Karu et al. (*31*) investigated four alternative methods of multivariate analysis: discriminant analysis, maximum likelihood analysis, classification and regression trees and computational neuronal networks. The all-monoclonal assay system was applied to triazine herbicides, phenyl urea herbicides and avermectins.

The simultaneous quantification of four analytes seems to be the limit, both in terms of accuracy and cost due to the number of antibodies and calibration curves involved. It would be advantageous to have a method that allows categorizing analytes into certain subgroups and thus narrows down the number of antibodies needed for MELISA. A subgroup would comprise analytes with similar substitution patterns or the same functional group, resulting in highly cross reacting analytes. To choose a suitable subset of antibodies capable of categorizing analytes from a larger pool Wortberg et al. (*32*) used cluster analysis.

As the cross-reactivity of antibodies among compound classes or groups of compounds becomes less, the power of multianalyte techniques increases. The long term goal remains to develop general mathematical approaches which will facilitate solving future multianalyte problems based on cross-reacting antibodies. The power of this technique is enhanced by the availability of antibody libraries with a variety of cross-reactivities.

2) Organophosphorus Insecticide Metabolites.

Enzyme immunoassays have been used only recently for the assessment of human and wildlife exposure to toxic chemicals, such as pesticides (*33*). For assessing exposure to pesticides and other chemicals, immunoassays can be used for the detection of parent compounds and/or metabolites in biological matrices, especially urine and blood. The organophosphorus (OP) insecticides remain widely used today and are responsible for adverse effects in humans and in some wildlife species. When metabolized, many OPs release stable aryloxy leaving groups. The presence in urine of these leaving groups, free or conjugated, is considered to be a reliable indicator of OP exposure. We are developing and validating EIAs, based on rabbit and sheep polyclonal antibodies, for selected metabolites of particular OP insecticides.

The target analytes for this project are metabolites of some common OP insecticides: the 4-nitrophenol and monosubstituted 4-nitrophenol metabolites of five different OPs (methyl and ethyl parathion, EPN, dicapthon, and fenitrothion) and the 3,5,6-trichloro-2-pyridinol metabolite of chlorpyrifos (Figure 2). By directing these immunoassays against the hydrolysis products of the parent OPs, the assays can be used for detecting both the parent OPs (following enzymatic or chemical hydrolysis of samples) and the metabolites in biological and environmental monitoring studies.

2a) 4-Nitrophenols. One of our immunoassays for 4-nitrophenols has a limit of detection as low as 0.5 ng/mL for 4-nitrophenol, depending on the matrix (5). It has been applied for the analysis of 4-nitrophenol and parathion (following a hydrolysis step) in soils (34). Using the approach outlined in Figure 3, this assay is being applied for the analysis of 4-nitrophenol in urine samples from humans accidentally exposed to methyl parathion.

2b) 3,5,6-Trichloro-2-pyridinol (TCP). TCP is another OP metabolite target of assay development. The first immunizing hapten used resulted in an immunoassay selective for chlorpyrifos, but not for TCP. Efforts to synthesis a new TCP immunizing hapten for the development of this assay continue.

3) Pyrethroid Insecticides and Metabolites. Fenvalerate and fenpropathrin (Figure 4) belong to the present generation of photostable pyrethroids, which were developed by a series of modifications of the natural pyrethrins during the last few decades. They represent highly potent insecticides with relatively low mammalian toxicity. Both pyrethroids are predominantly used in crop protection. Conventional methods for the detection of pyrethroids like gas chromatography (GC) and high pressure liquid chromatography (HPLC) involve multistep sample cleanup procedures. A simple immunoassay for the detection of the pyrethroid permethrin has already been reported (35).

Our objective is the development of ELISAs (9) and NIRDIAs (near infrared fluorescence detection immunoassays) for the analysis of fenvalerate and fenpropathrin in environmental samples, as well as their human metabolites such as the example in (36) (Figure 5). Metabolites can often be used as biomarkers to monitor human exposure to xenobiotics (37). Table I lists the most desirable characteristics for a biomarker.

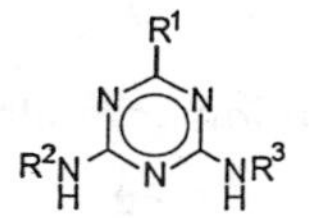

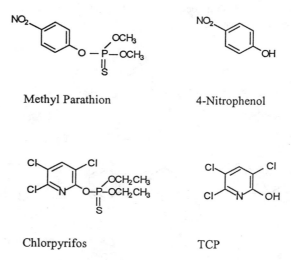

Compound	R^1	R^2	R^3
Atrazine	Cl	CH_2CH_3	$CH(CH_3)_2$
Cyanazine	Cl	CH_2CH_3	$CNC(CH_3)_2$
Simazine	Cl	CH_2CH_3	CH_2CH_3
Propazine	Cl	$CH(CH_3)_2$	$CH(CH_3)_2$
Hydroxyatrazine	OH	CH_2CH_3	$CH(CH_3)_2$
Hydroxysimazine	OH	CH_2CH_3	CH_2CH_3
Desethylatrazine	Cl	H	$CH(CH_3)_2$
Desisopropyl-atrazine	Cl	CH_2CH_3	H
Prometon	OCH_3	$CH(CH_3)_2$	$CH(CH_3)_2$
Prometryn	SCH_3	$CH(CH_3)_2$	$CH(CH_3)_2$
Terbutryn	SCH_3	CH_2CH_3	$C(CH_3)_3$

Figure 1. Structures of common *s*-triazines and some of their metabolites

Methyl Parathion

4-Nitrophenol

Chlorpyrifos

TCP

Figure 2. 4-nitrophenol metabolite of methyl parathion and 3,5,6-trichloro-2-pyridinol (TCP) metabolite of chlorpyrifos

parathion or fenitrothion

Conjugated 4-nitrophenols in urine

**Enzymatic hydrolysis
of urine samples**

Free 4-nitrophenols in sample

A. No sample prep   **C. SPE columns**

B. Liquid-liquid Extraction

**Analysis by 4-Nitrophenol
Enzyme Immunoassay**

Figure 3. Scheme for the detection of 4-nitrophenols in human urine samples as a biomarker of exposure.

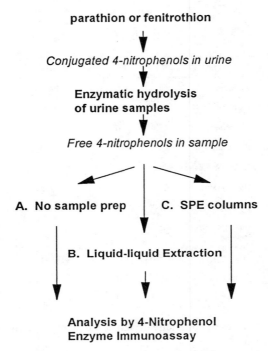

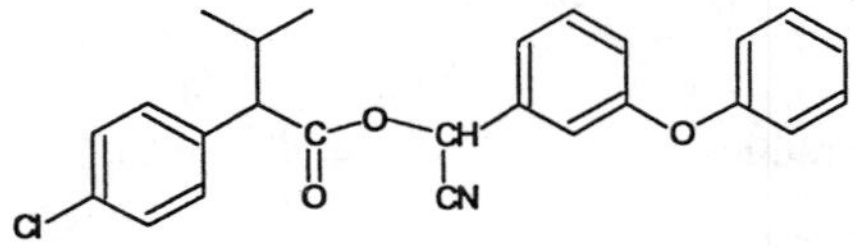

Fenvalerate

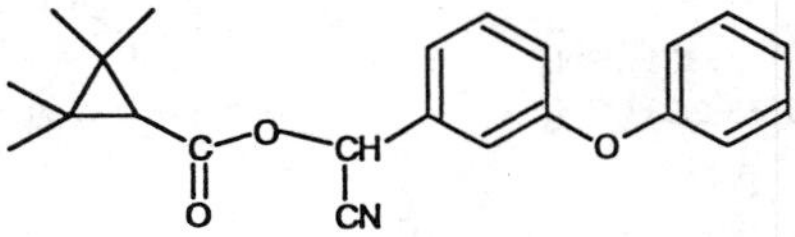

Fenpropathrin

Figure 4. Structures of the pyrethroids fenvalerate and fenpropathrin

Table I. Criteria for Selecting a Diagnostic Metabolite for Human Monitoring

Major metabolite

Consistent proportion of total metabolism

Availability of highly sensitive immunoassay

Ease of clean up and/or concentration

Low cost, high speed of analysis

NIR Fluorescence Detection

Near infrared (NIR) fluorescence has recently gained importance because inexpensive semiconductor lasers and photodiodes have become commercially available in the past decade. These technical advances as well as the development of NIR fluorogenic labels allow the detection of NIR-tagged small molecules, antigens, and antibodies with high sensitivity. Interferences due to autofluorescence and light scattering are significantly diminished in the far visible and NIR regions (ca. 600-1000 nm) compared to the ultraviolet-visible (UV-VIS) area of the spectrum. Fluorescent chromophores with high molar absorbance values (e_M) and quantum yields (Q) are suitable as NIR-labels. This label can be a synthetic compound as in Fig 6 (e_M NN382: ca. 1-2 x 10^5, Q: 0.1-0.8)(*38,39*) with extensive conjugation in the molecule. Semiconductor laser-induced fluorescence detection of NIR-labeled biomolecules at concentrations as low as 4 x10^{-14} M was reported (*40*). The same group also demonstrated that a covalently linked NIR-fluorescent tag can serve as a useful reporter system for immunoassays. An ELISA uses an enzyme label to yield an amplified sensitive signal, whereas in the NIRDIA an analyte labeled with a NIR fluorescent dye provides the very sensitive signal. To simplify the purification of the fluorogenic tracer, NIR dye (Figure 6) and analyte were coupled using biopolymers as carriers. NIRDIAs for the detection of insecticidal pyrethroids and the herbicide bromacil are under development.

Chelate Assay for Heavy Metals

Several toxic heavy metals pose very serious hazards on the environment and human health. An example of the heavy metal hazards to wildlife is the situation at Clear Lake, California, where the Sulphur Bank mercury mine is being investigated as a Superfund site. Some typical analytical methods for the highly selective detection of toxic heavy metals at ppb/ppt levels are flameless atomic absorption spectrometry (AAS), atomic fluorescence spectroscopy, inductively coupled plasma mass spectrometry, and electrochemical techniques. These methods have some drawbacks that are typically encountered when using instrumental analysis. They require very expensive and sophisticated equipment, highly qualified personnel, the sample throughput is limited and they are not suitable for on-site analysis. Recently developed immunoassays present an interesting alternative for the detection of certain metal ions (*41,42*). These novel

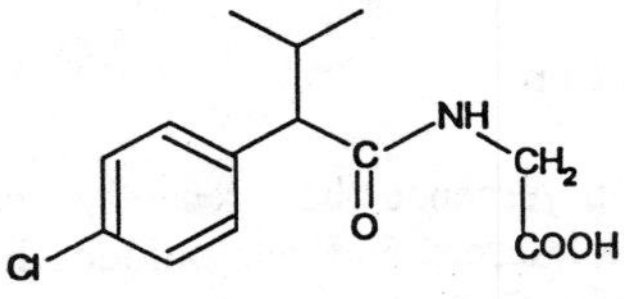

Metabolite

Figure 5. Structure of the analyte, a fenvalerate metabolite (glycine conjugate of p-chlorophenylisovaleric acid)

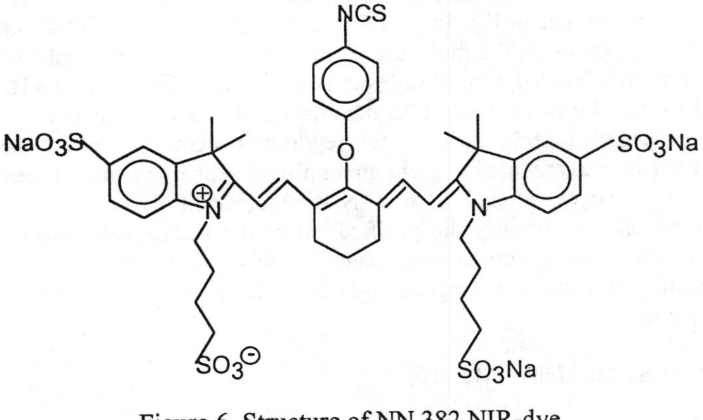

Figure 6. Structure of NN 382 NIR-dye

techniques involve raising antibodies against metal chelates. Reardan (*43*) reported the selective recognition of an EDTA-type indium chelate by monoclonal antibodies. In our group a simple analytical method for the detection of mercuric ions at low ppb levels has been developed (*44*). It combines the inexpensive ELISA methodology with the selective, high affinity recognition of Hg^{2+} ions by dithiocarbamate chelators, without using antibodies. A schematic presentation of the chelate assay principle is given in Figure 7.

Polychlorinated Dibenzo-p-Dioxins

The chemical and toxicological properties of polychlorinated dibenzo-*p*-dioxins (PCDD's) along with the apparent pervasiveness of these chemicals in the environment, has created a demand for new analytical methods for these compounds. These new methods ideally would be extremely selective, would exhibit very low detection limits, and would be applicable to very large numbers and wide varieties of samples in a cost-effective manner. Immunoassays are well-suited to address many of the extreme analytical challenges presented by polychlorinated dibenzo-*p*-dioxins (*45,46*).

Our efforts to develop immunoassays for polychlorinated dioxins are directed toward three main goals. The first is to improve the performance of the ELISA based on the existing monoclonal antibodies to 2,3,7,8-tetrachloro dibenzo-*p*-dioxin, TCDD (*47*). By altering the structure of the competing hapten, changes in sensitivity and selectivity can be realized (*48*). The second goal is to develop new antibodies with greater sensitivity and selectivity by improved hapten / immunogen design. Design considerations include improved mimicry of the target analyte and addressing concerns about the lipophilicity of the molecule by using rigid handles. Because of the toxicity of TCDD and associated regulation, the third goal is to develop less toxic surrogate standards for routine analysis of TCDD. Collaborators at the USDA in College Station, Texas and in the Department of Environmental Toxicology at UC Davis are making substantial contributions to this research project. Molecular modeling techniques are being used to help design improved TCDD haptens and surrogate standards(*46*). Toxicities of surrogate standards developed in this project are being assessed using a cell-based bioassay which uses induced expression of a luciferase gene to indicate aromatic hydrocarbon receptor-mediated toxicity of TCDD analogs (*49*).

Figure 8 shows a representative set of haptens and surrogate standards that have been synthesized for this project. Polyclonal antibodies have been generated to seven haptens and are being evaluated for use in immunoassays. Monoclonal antibodies are being developed to some of the haptens. These same haptens are being used as coating antigens to attempt to improve the performance of previously developed monoclonal antibodies as mentioned above. Preliminary results indicate that equivalent or improved sensitivity can be achieved using these coating antigens. Several of the surrogate standards are being evaluated for both their response in the ELISAs and for their toxicity. Initial results with the trichloromethyl analog of TCDD indicate that it behaves nearly identically to authentic TCDD in the ELISA and it is likely to be considerably less toxic than the parent compound.

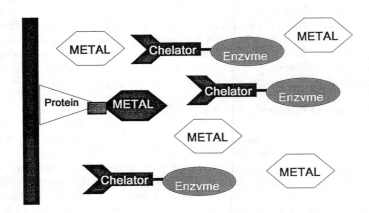

Figure 7. Schematic of chelate assay principle

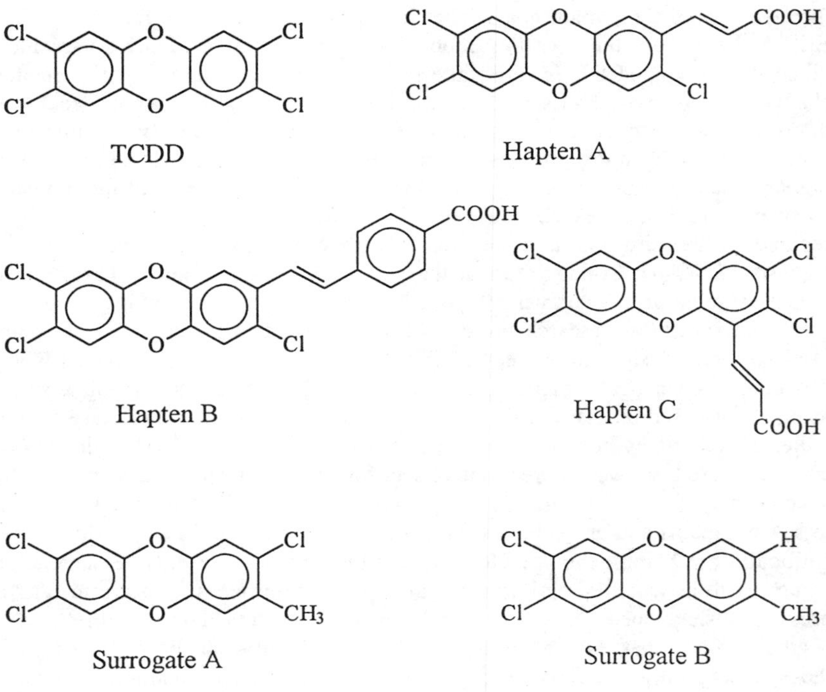

TCDD

Hapten A

Hapten B

Hapten C

Surrogate A

Surrogate B

Figure 8. TCDD and haptens

Conclusions

In our laboratory many enzyme immunoassays have been developed to detect pesticides and other toxic compounds. Table II gives an overview of antibodies developed and assays used in this laboratory. Much experience has been gained in the development of herbicide EIAs, particularly assays for triazines. Recent EIA developments for metabolites of triazines, organophosphates and pyrethroids can benefit from this accumulated knowledge about the influence of the hapten structure, resulting in logical hapten synthesis strategies and conjugation methods, as well as more know-how in immunization procedures, antibody screening, coating antigen structure and assay development. In some cases it is more desirable to detect the metabolites rather than the parent compounds if, for instance, the parent compounds decompose or metabolize in the environment or organism rapidly, or if the metabolite or break-down product is toxicologically more relevant. In this case a selective assay is needed. Metabolites can also provide a basis for the development of group specific antibodies, for example 3-phenoxybenzoic acid, a pyrethroid moiety can serve as a common hapten a whole group of pyrethroids. Our current work mainly focuses on meeting the challenges encountered in immunoassay development: cross-reactivity, better detection methods and handling lipophilic analytes.

Since many triazines show very similar structures, like atrazine and simazine, it is generally a problem to generate a selective antibody, which does not show any cross-reactivity. In this laboratory a more selective assay for simazine was developed. In another approach the different cross-reactivities of antibodies for triazines were used as an advantage in developing a multi-analyte method.

The use of new fluorogenic labels enables the development of new assay formats that lead to or use novel instrumentation that is even more suitable for environmental analysis. The NIR dyes as a new tracer show some advantages compared with enzyme tracers. They provide at least the same sensitivity, yet save time, because the assay does not need a step for substrate turnover by an enzyme. NIR dyes allow the use of inexpensive laser diode technology, which also facilitates the design of small geometric instrumentation.

Our efforts using lipophilic haptens such as TCDD and pyrethroid mimics have already shown some success and will therefore help to find new ways to adapt immunoassays as an appropriate analytical method for a wider range of compounds. This work also demonstrates that new promising applications of antibodies and novel reporter molecules are still emerging. There is a continuing need for new approaches in hapten design and coupling chemistry. These and other approaches will serve to make immunochemical methods more valuable in the environmental sciences. They also will facilitate the development of biosensors and probes which can be integrated with other analytical procedures.

Table II: Immunoassays Developed in this Laboratory

Compound	Reference	Compound	Reference
Amitrole	(50)	Pyrethroids and metabolites	(51,52)
Bentazon	(53)	Tetrachlorodi-benzodioxin	
Benzoylphenyl ureas	(54-56)	Thiocarbamates	(11,57)
Bromacil	(58,59)	Triazines and metabolites	(12,13,15, 16)
Carbaryl	(10)	Trichlopyr	[a]
Chlorpyrifos	[a]	2,4,5-Trichlorophenoxy-acetic acid	[a]
Fenoxycarb	[b]	Triton series X and N detergents	(60)
Glyphosate	[b]	Urea herbicides	(7,8)
Metals (Hg and others)	(44,61)	*Bacillus thuringiensis* β-exotoxin	(62)
Naphthalene and metabolites	(6,63,64)	*Bacillus thuringiensis* δ-endotoxin	(65)
Nitrophenols and other nitroaromats	(5)	*Alternaria* and *Fusarium* toxins	[a]
Paraquat	(66-68)		

[a] Those lacking citations are currently under investigation.
[b] Antibodies have been made for these compounds, but not optimized.

Acknowledgments

The research was funded in part by National Institute of Environmental Health Sciences Superfund Research Program, 5 P42 ES04699-08 and Environmental Protection Agency Center for Ecological Health Research, CR819658, U.S.E.P.A. Cooperative Agreement CR819047, Lawrence Livermore National Laboratory Center for Accelerator Mass Spectrometry Agreement B291419 under Prime Contract No. W-7405-ENG-48 and the USDA Forest Service NAPIAP - R827. Although the work reported here was sponsored in part by the U.S.E.P.A, no endorsement by the Agency is implied.

Literature Cited

1. Hammock, B. D.; Mumma, R. O. In *Pesticide Analytical Methodology*; Harvey, J., Jr. ; Zweig, G., Eds.; ACS Publications: Washington, D. C., **1980**; p 321-352.
2. Cheung, P. Y. K.; Gee, S. J.; Hammock, B. D. In *The Impact of Chemistry on Biotechnology: Multidisciplinary Discussions*; Phillips, M. P.; Shoemaker, S. P.; Middlekauff, R. D. ; Ottenbrite, R. M., Eds.; American Chemical Society: Washington DC, **1988**; p 217-229.
3. Meulenberg, E.; Mulder, W.; Stoks, P. *Environ. Sci. Technol.* **1995**, *29*, 553.
4. Hammock, B. D.; Gee, S. J.; Harrison, R. O.; Jung, F.; Goodrow, M. H.; Li, Q. X.; Lucas, A. D.; Székács, A.; Sundaram, S. In *Immunochemical Methods for Environmental Analysis*; Van Emon, J. M. ; Mumma, R. O., Eds.; American Chemical Society: Washington, D.C., **1990**; p 112-139.
5. Li, Q.-X.; Zhao, M.-S.; Gee, S. J.; Kurth, M.; Seiber, J. N.; Hammock, B. D. *J. Agric. Food Chem.* **1991**, *39*, 1685-1692.
6. Krämer, P. M.; Marco, M.-P.; Hammock, B. D. *J. Agric. Food Chem.* **1994**, *42*, 934-943.
7. Karu, A. E.; Goodrow, M. H.; Schmidt, D. J.; Hammock, B. D.; Bigelow, M. W. *J. Agric. Food Chem.* **1994**, *42*, 301-309.
8. Schneider, P.; Goodrow, M. H.; Gee, S. J.; Hammock, B. D. *J. Agric. Food Chem.* **1994**, *42*, 413-422.
9. Wengatz, I.; Stoutamire, D. W.; Gee, S. J.; Hammock, B. D. *J. Agric. Food Chem.* , in preparation.
10. Marco, M.-P.; Gee, S. J.; Cheng, H. M.; Liang, Z. Y.; Hammock, B. D. *J. Agric. Food Chem.*. **1993**, *41*, 423-430.
11. Gee, S. J.; Miyamoto, T.; Goodrow, M. H.; Buster, D.; Hammock, B. D. *J. Agric. Food Chem.* **1988**, *36*, 863-870.
12. Goodrow, M. H.; Harrison, R. O.; Hammock, B. D. *J. Agric. Food Chem.*. **1990**, *38*, 990-996.
13. Harrison, R. O.; Goodrow, M. H.; Hammock, B. D. *J. Agric. Food Chem.* **1991**, *39*, 122-128.
14. Lucas, A. D.; Schneider, P.; Harrison, R. O.; Seiber, J. N.; Hammock, B. D.; Biggar, J. W.; Rolston, D. E. *Food Agric. Immunol.* **1991**, *3*, 155-167.
15. Lucas, A. D.; Jones, A. D.; Goodrow, M. H.; Saiz, S. G.; Blewett, C.; Seiber, J. N.; Hammock, B. D. *Chem. Res. Toxicol.* **1993**, *6*, 107-116.
16. Lucas, A. D.; Bekheit, H. K. M.; Goodrow, M. H.; Jones, A. D.; Kullman, S.; Matsumura, F.; Woodrow, J. E.; Seiber, J. N.; Hammock, B. D. *J. Agric. Food Chem.* **1993**, *41*, 1523-1529.
17. Schneider, P.; Hammock, B. D. *J. Agric. Food Chem.* **1992**, *40*, 525-530.
18. Wortberg, M.; Goodrow, M. H.; Gee, S. J.; Hammock, B. D. *J. Agric. Food Chem.* , accepted.
19. Blake, C.; Al-Bassan, M. N.; Gould, B. J.; Marks, V.; Bridges, J. W.; Riley, C. *Clin. Chem.* **1982**, *28*, 1469.
20. Wians, F.; Dev, J.; Powell, M.; Heald, J. *Clin. Chem.* **1986**, *32*, 887.

21.   Hemmila, I.; Holttinen, S.; Pettersson, K.; Lovgren, T. *Clin. Chem.* **1987**, *33*, 2281.
22.   Vuori, J.; Rasi, S.; Takala, T.; Vaananen, K. *Clin. Chem.* **1991**, *37*, 2087.
23.   Hayes, F. J.; Halshall, H. B.; Heinemann, W. R. *Anal. Chem.* **1994**, *66*, 1860.
24.   Ekins, R.; Chu, F.; Biggart, E. *Analytica Chimica Acta*. **1989**, *227*, 73-96.
25.   Kakabakos, S. E.; Christopulos, T. K.; Diamandis, E. P. *Clin. Chem.* **1992**, *38*, 338.
26.   Parsons, R. G.; Kowal, R.; LeBlond, D.; Yue, V. T.; Neagarder, L.; Bond, L.; Garcia, D.; Slater, D.; Rogers, P. *Clin. Chem.* **1993**, *39*, 1899.
27.   Muldoon, M. T.; Fries, G. F.; Nelson, J. O. *J. Agric. Food Chem.* **1993**, *41*, 322-328.
28.   Jones, G.; Wortberg, M.; Kreissig, S. B.; Bunch, D. S.; Gee, S. J.; Hammock, B. D.; Rocke, D. M. *J. Immunol. Meth.* **1994**, *177*, 1-7.
29.   Wortberg, M.; Kreissig, S. B.; Jones, G.; Rocke, D. M.; Hammock, B. D. *Anal. Chim. Acta.* **1995**, *304*, 339-352.
30.   Cheung, P. Y. K.; Kauvar, L. M.; Engqvist-Goldstein, A. E.; Ambler, S. M.; Karu, A. E.; Ramos, L. S. *Anal. Chim. Acta*. **1993**, *282*, 181.
31.   Karu, A. E.; Lin, T. H.; Breiman, L.; Muldoon, M. T.; Hsu, J. *Food Agric. Immunol.*. **1994**, *6*, 371.
32.   Wortberg, M.; Jones, G.; Kreissig, S. B.; Rocke, D. M.; Gee, S. J.; Hammock, B. D. *Anal. Chim. Acta.* **1996**, *319*, 291-303
33.   Harris, A. S.; Lucas, A. D.; Krämer, P. M.; Marco, M.-P.; Gee, S. J.; Hammock, B. D. In *New Frontiers in Agrochemical immunoassay;* Kurtz, D. A.; Skerritt, J. H. ; Stanker, L., Eds.; AOAC Intl., Arlington, VA, **1995**, 217-235.
34.   Wong, J. M.; Li, Q. X.; Hammock, B. D.; Seiber, J. N. *J. Agric. Food Chem.* **1991**, *39*, 1802-1807.
35.   Stanker, L. H.; Bigbee, C.; Emon, J. V.; Watkins, B.; Jense, R. H.; Morris, C.; Vanderlaan, M. *J. Agric. Food Chem.* **1989**, *37*, 834-839.
36.   Wengatz, I.; Stoutamire, D. W.; Hammock, B. D. **1996**, in preparation.
37.   Harris, A. S.; Wengatz, I.; Wortberg, M.; Kreissig, S. B.; Gee, S. J.; Hammock, B. D. In *Effects of Multiple Impacts on Ecosystems* ; Cech, J. J.; Wilson, B. W. ; Crosby, D. G., Eds.; Lewis Publishers: Chelsea, Michigan, in press.
38.   Patonay, G.; Antoine, M. D. *Anal. Chem.* **1991**, *63*, 321-327.
39.   Shealy, D. B.; Lipowska, M.; Lipowska, J.; Narayanan, N.; Sutter, S.; Strekowski, L.; Patonay, G. *Anal. Chem.* **1995**, *67*, 247-251.
40.   Strekowski, L.; Lipowska, M.; Patonay, G. *J. Org. Chem.* **1992**, *57*, 4578-4580.
41.   Wylie, D. E.; Carlson, L. D.; Carlson, R.; Wagner, F. W.; Schuster, S. M. *Analytical Biochemistry*. **1991**, *194*, 381-387.
42.   Wylie, D. E.; Lu, E.; Carlson, L. D.; Carlson, R.; Babacan, K. F.; Schuster, S. M.; Wagner, F. W. *Proc. Natl. Acad. Sci. USA*. **1992**, *39*, 4104-4108.

43. Reardan, D. T.; Meares, C. F.; Goodwin, D. A.; McTigue, M.; David, G. S.; Stone, M. R.; Leung, J. P.; Bartholomew, R. M.; Frincke, J. M. *Nature* . **1985**, *316*, 265-268.
44. Szurdoki, F.; Kido, H.; Hammock, B. D. In *Immunoanalysis of Agrochemicals: Emerging Technologies*; Nelson, J. O.; Karu, A. E. ; Wong, R. B., Eds.; American Chemical Society: Washington, D.C., **1995**; p 248-264.
45. Sherry, J. P. In *Immunoanalysis of Agrochemicals: Emerging Technologies*; Nelson, J. O.; Karu, A. E. ; Wong, R. B., Eds.; American Chemical Society: Washington, D.C., **1995**; p 335-353.
46. Stanker, L. H.; Recinos, A., III; Linthicum, D. S. In *Immunoanalysis of Agrochemicals: Emerging Technologies*; Nelson, J. O.; Karu, A. E. ; Wong, R. B., Eds.; American Chemical Society: Washington, D.C., **1995**; p 72-88.
47. Stanker, L. H.; Watkins, B.; Rogers, N.; Vanderlaan, M. *Toxicology* **1987**, *45*, 229-243.
48. Goodrow, M. H.; Sanborn, J. R.; Stoutamire, D. W.; Gee, S. J.; Hammock, B. D. In *Immunoanalysis of Agrochemicals: Emerging Technologies*; Nelson, J. O.; Karu, A. E. ; Wong, R. B., Eds.; American Chemical Society: Washington, D.C., **1995**; p 119-139.
49. El-Fouly, M. H.; Richter, C.; Giesy, J. P.; Denison, M. S. *Env. Toxicol. Chem.* **1994**, *13*, 1581.
50. Jung, F.; Székács, A.; Li, Q.; Hammock, B. D. *J. Agric. Food Chem.* **1991**, *39*, 129-136.
51. Wing, K. D.; Hammock, B. D. *Experientia* **1979**, *35*, 1619-1620.
52. Wing, K. D.; Hammock, B. D.; Wustner, D. A. *J. Agric. Food Chem.* **1978**, *26*, 1328-1333.
53. Li, Q. X.; Hammock, B. D.; Seiber, J. N. *J. Agric. Food Chem.* **1991**, *39*, 1537-1544.
54. Wie, S. I.; Sylwester, A. P.; Wing, K. D.; Hammock, B. D. *J. Agric. Food Chem.* **1982**, *30*, 943-948.
55. Wie, S. I.; Hammock, B. D. *J. Agric. Food Chem.* **1982**, *30,* 949-957.
56. Wie, S. I.; Hammock, B. D. *J. Agric. Food Chem.*. **1984**, *32*, 1294-1301.
57. Gee, S. J.; Harrison, R. O.; Goodrow, M. H.; Braun, A. L.; Hammock, B. D. In *Immunoassays for Trace Chemical Analysis: Monitoring Toxic Chemicals in Humans, Food, and the Environment.;* Vanderlaan, M.; Stanker, L. H.; Watkins, B. E. ; Roberts, D. W., Eds.; American Chemical Society: Washington, D.C., **1991**; p 100-107.
58. Szurdoki, F.; Bekheit, H. K. M.; Marco, M.-P.; Goodrow, M. H.; Hammock, B. D. *J. Agric. Food Chem.* **1992**, *40*, 1459-1465.
59. Bekheit, H. K. M.; Lucas, A. D.; Szurdoki, F.; Gee, S. J.; Hammock, B. D. *J. Agric. Food Chem.*. **1993**, *41*, 2220-2227.
60. Wie, S. I.; Hammock, B. D. *Anal. Biochem.* **1982**, *125*, 168-176.
61. Szurdoki, F.; Kido, H.; Hammock, B. D. *Bioconjugate Chem.* **1995**, *6*, 145-149.

62. Bekheit, H. K. M.; Lucas, A. D.; Gee, S. J.; Harrison, R. O.; Hammock, B. D. *J. Agric. Food Chem.* **1993**, *41*, 1530-1536.

63. Marco, M.-P.; Nasiri, M.; Kurth, M. J.; Hammock, B. D. *Chem. Res. Toxicol.* **1993**, *6*, 284-293.

64. Marco, M.-P.; Hammock, B. D.; Kurth, M. J. *J. Org. Chem.* **1993**, *58*, 7548-7556.

65. Cheung, P. Y. K.; Hammock, B. D. In *Biotechnology for Crop Protection*; Hedin, P. A.; Menn, J. J. ; Hollingworth, R. M., Eds.; American Chemical Society: Washington, DC, **1988**; p 298-305.

66. Van Emon, J. M.; Seiber, J. N.; Hammock, B. D. In *Bioregulators for Pest Control*; Hedin, P. A., Ed.; ACS Publications: Washington D.C., **1985**; p 307-316.

67. Van Emon, J. M.; Hammock, B. D.; Seiber, J. N. *Anal. Chem.* **1986**, *58*, 1866-1873.

68. Van Emon, J. M.; Seiber, J.; Hammock, B. D. *Bull. Environ. Contam. Toxicol.* **1987**, *39*, 490-497.

Chapter 12

A Status Report on Electroanalytical Techniques for Immunological Detection

Omowunmi A. Sadik and Jeanette M. Van Emon

Characterization Research Division, National Exposure Research Laboratory, U.S. Environmental Protection Agency, P.O. Box 93478, Las Vegas, NV 89193–3478

Immunoassays are commonly used for the detection of low levels of a specific target analyte or group of analytes. The search for new nonradioactive detection methods has resulted in a plethora of immunoassay techniques utilizing different types of labels, the most common being enzymes. Electrochemical immunoassays are based on modifications of enzyme immunoassays with the enzyme activities being determined potentiometrically or amperometrically. Other nonradioactive assays have been designed which are not adapted from spectrophotometric detection methods. Each of these assays, including electrochemical enzyme immunoassays, is discussed briefly in this chapter. In addition, various electrochemical immunosensors reported recently and focuses on the use of electropolymerized conducting polymers (CPs) in amperometric immunochemical sensors are discussed. An electrochemical immunosensor is described for the analysis of polychlorinated biphenyls using a CP-based immunosensor. The sensor produced adequate linear response characteristics and sensitivities that are comparable with results obtained using the enzyme-linked immunosorbent assay (ELISA) technique.

Immunochemical techniques are based on the interaction of antibodies (Ab) with antigens (Ag). An antibody is a protein that recognizes a target analyte (i.e., antigen) or group of analytes and then reacts specifically with it to form a complex. An immunogen is a compound that induces the formation of specific antibodies. Various labels can be used to produce a range of assays, including radioimmunoassays (RIA), fluorescence immunoassays (FLIA), enzyme immunoassays (EIA) and chemiluminescence immunoassays (CLIA). The wide dynamic range and low detection limits of electroanalytical techniques are helpful in the development of electrochemical immunoassay as an alternative approach to spectrophotometric and radiometric detection procedures.

0097–6156/96/0646–0127$15.25/0

The use of electrochemical methods for immunoassay procedures follows a tradition of technical expansion. Breyer and Radcliff first demonstrated the polarographic detection of azo-labeled antigen in a homogeneous immunoassay (1). In the 1950s, Berson and Yalow developed the area of radioimmunoassays (2). Since the first electrochemical immunoassay experiments, significant progress has been made in modern electrochemical instrumentation, microelectronics, and the development of new classes of electrode materials. These developments have further stimulated interest in combining immunoassay procedures with electrochemical measurements. The principles have been used to achieve very low detection limits when coupled to chemical amplification systems, such as enzymes. Electrochemical detection techniques are generally inexpensive, fast, and amenable to automation.

Electrochemical monitoring of immunological reactions holds great promise as a practical alternative to assays involving radioactive labels. Several electrochemical immunoassay approaches have been proposed. These include: multianalyte immunoassays involving anodic stripping voltammetric detection of different metal labels (3), capillary electrochemical enzyme immunoassay coupled to flow injection analysis systems for digoxin and atrazine (4,5), the use of interdigitated array electrodes for small-volume voltammetric enzyme immunoassay (6), and homogeneous amperometric immunoassay for theophylline (7,8). The successful implementation of electrochemiluminescence labels for immunometric assays of hormones, cancer markers and nucleic-acid hybridization assays was reported by Yang et al., (9). A separation-free enzyme immunoassay utilizing microporous gold electrodes with self-assembled monolayer antibody electrodes was also reported (10).

A new and highly promising approach for the detection and amplification of Ab-Ag interactions involves the incorporation of immobilized antibodies into conducting polymer films or membranes (11,12). Conducting polymer membranes (CPMs) are particularly attractive for biosensor applications where a range of different immobilized analytes can easily be assessed. The interaction with the analytes can also be controlled electrochemically through the application of electrical potentials. Such sensors have been used successfully with pulsed amperometric detection in a flow injection system (11- 14). CPM-based biosensors have been proposed for direct and continuous detection of low concentrations of biological and organic analytes in process streams, environmental samples and biological fluids (15-17). The use of bilayer lipid membranes for electrochemical transduction of immunological reactions has also been reported (18). A review that discusses the principles of electrochemical immunoassay protocols based on the measurement of a faradaic current was recently reported (19). Other recent work discusses the concept of light-addressable potentiometric immunosensors (20).

Our aim in this chapter is to give a status report on electroanalytical techniques employed for immunological detection, and also to present some results obtained in the successful application of conducting polymer-based immunosensors for environmental applications.

Electrochemical Immunoassay

Immunoassays are commonly categorized as "heterogeneous", (in which Ab-bound Ag is separated from free Ag), or "homogeneous", (in which no separation steps are involved). Features of electrochemical immunoassays are shown in Table I. Direct electrochemical immunoassays monitor changes in electrical properties of the Ab-Ag binding events. In this case, the sensitivity is directly proportional to the amount of Ab present. However, in some competitive assays, labeled Ags compete with unlabeled Ags for a limited number of Ab binding sites. The Ab-bound Ags (labeled and unlabeled) are separated from the free Ags and the signal produced by the Ab-bound labeled antigens is then measured. The signal intensity of the bound phase is inversely proportional to the concentration of the unlabeled Ag. Competitive sandwich assays involve the incubation of excess primary Ab with an Ag. The Ab-Ag complex is then incubated with a labeled secondary Ab which binds to the first Ab. The unbound labeled Ab is rinsed away and the bound labeled Ab is measured. The signal intensity is related to the amount of primary Ab present, which is inversely related to the amount of Ag (analyte) in the sample.

A class of heterogeneous enzyme immunoassays with electrochemical detection has emerged and is known as electrochemical enzyme immunoassay (ECIA). ECIA is the result of the modifications of enzyme immunoassays with the enzyme activity being determined electrochemically (21,22). ECIA is based on the labeling of specific Ab or target analyte with an enzyme that catalyzes the production of an electroactive product. Different formats of ECIA have been reported that depend on the label type, assay format, and the electrochemical techniques employed. They are discussed below.

Competitive ECIA: Microwell plates are first prepared by attaching specific Ab to the inside walls through passive adsorption or covalent bonding. An equilibrium is established between the bound Ab, the analyte, and enzyme-labeled Ag. This may take several minutes or hours depending on the analyte and the configuration of the orientating reagents. After the incubation step, the unbound reagents are washed away and the substrate is added. At a fixed time, the sample is withdrawn and analyzed for electroactive products. The general procedure in ECIA for the determination of analyte is represented in Figure 1. Alkaline phosphatase is used in voltammetric immunoassays with phenyl phosphate as the substrate and it catalyzes the hydrolysis of the p-nitrophenyl phosphate ester to yield phenol and phosphoric acid. The enzyme-generated phenol is easily detectable by either a flow injection analysis with electrochemical detection (FIA-EC), or by liquid chromatography with electrochemical detection (LC-EC). A typical quantitation curve shows current vs. concentration of analyte standard. The electrochemical signal decreases with increasing analyte concentration. ECIA of this type has been demonstrated for IgG, α-feroprotein, digoxin and morphine using unmodified electrodes (21-23).

Table I.   Features of Electrochemical Immunoassay Techniques Reported

Target Substance	Measurement Technique	Assay Format	Flow Mode	LOD	Ref.
Thyroxin(T4)	Capacitive	Direct	Stirred	10^{-9}M	18
IgG	Potentiomentric	Direct	No	ppb	34
HSA	Amperometric	Direct	No	ppb	36,11,12
Thaumatin	PAD	Direct	Yes	ppb	13
P-Cresol	PAD	Direct	Yes	ppb	14
hCG	Potentiometric	Direct	No	5μM	49
2,4-D	Potentiometric	Direct	No	1μg/L	50
Atrazine	Amperometric	Comp[a]	Yes	0.10μg/L	5
Glucose	Amperometric	Comp[a]	No	ppb	17
Digoxin	Amperometric	Comp[a]	Yes	3.8×10^{-12}M	4,22
Theophyline	Amperometric	Comp[a]	Yes	<25ng/mL	8
Atrazine	Conductimetric	Comp[a]	No	<0.025μg/L	37
hCG	Amperometric	Sw[b]	Yes	2.5units/L	25
CEA	ECL (DET)	Comp[a]	Yes	200fmol/L	31

LOD=limit of detection, PAD=pulsed amperometric detection, ppb=parts per billion, Comp[a]=competitive Sw[b]=Sandwich, CEA=carcinoembryonic antigen, ECL= electrochemiluminescence reaction, DET= direct electron transfer.

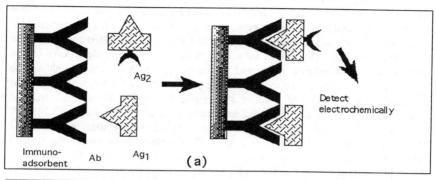

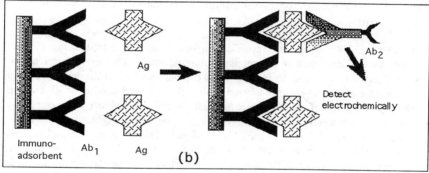

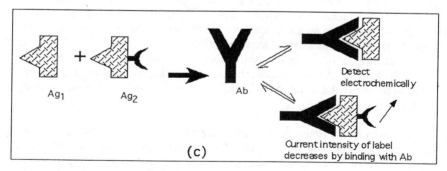

Figure 1: General procedures employed in electrochemical immunoassays: (a) competitive enzyme immunoassay, (b) sandwich immunoassay, (c) homogeneous immunoassay.

Sandwich ECIA: Sandwich immunoassay is another ECIA format, particularly useful for larger molecules that can accommodate binding by two Abs simultaneously. The procedure is depicted in Figure 1. An Ab specific for the analyte is first adsorbed on the microwell plate. Then a sample containing the analyte and enzyme-labeled specific Ab is added to the plate, forming a sandwich complex with the immobilized Ab. After incubation, the unbound enzyme-labeled Ab is rinsed away and the substrate is added. The rate of formation of the products is proportional to the amount of enzyme-labeled Ab bound to the walls, which is proportional to the concentration of analyte in the sample. In one case, a sandwich assay was performed in which aminophenyl was used as a substrate instead of nitrophenyl, and the aminophenol product was detected anodically with an FIA system (22,24). A separation-free sandwich enzyme immunoassay was recently demonstrated for hCG using microporous gold electrodes and self-assembled monolayer/immobilized capture antibodies (25). The assay performed effectively in both buffer and whole human blood with a detection limit of 2.5 units/L hCG in blood. This assay type is comparable to most heterogeneous enzyme immunoassays that require multiple washing steps.

Capillary ECIA: Another modification of enzyme immunoassays with electrochemical detection is the capillary immunoassay. This assay involves the covalent attachment of the Ab to a modified capillary surface. Competition occurs between the analyte and enzyme-labeled analyte for a limited number of Ab binding sites. The detection of the enzymatic product, *p*-aminophenol, is carried out using amperometric measurement in a flow injection analysis set-up. Quantitation is made by measuring the peak height of the FIA-EC signal. This assay procedure has been demonstrated for digoxin with detection limits of 3.8×10^{-12} M (4), and for atrazine with detection limits of 0.10 µg/L (5). Capillary enzyme immunoassay with electrochemical detection introduces several advantages in comparison with conventional immunoassay methodology. The covalent attachment of Ab to a modified capillary can be easily controlled so that the Ab binding sites are located distal to the point column attachment. This decreases the possibility of losing Ab activity as a result of the immobilization process. The small sample size of the capillary (22µL) significantly reduces the amount of reagents required, and the time required for molecules to reach the surface in a narrow capillary is short, resulting in a faster assay (5).

Homogeneous ECIA: Homogeneous electrochemical enzyme immunoassay involves the competitive binding of a labeled electroactive species to the Ab. The electrochemistry of the label is significantly changed such that the separation of bound and free Abs is unnecessary. The quantitation is based on changes in the intensity of the signal that occurs when the enzyme-labeled Ag binds with the Ab to form an Ab-Ag* complex (Figure 1). After equilibration is achieved the ability to distinguish between free Ag* and Ab-Ag* can be distinguished, enabling the immunoassay to be carried out without a separation step. When an enzyme label is used, the immunoassay relies on a reduction in the rate of enzyme catalysis occurring as the Ab binds to the enzyme-labeled Ag, enabling the free Ag* to be identified electrochemically from Ab-Ag*.   The earliest examples of these assays are homogeneous competitive immunoassays for oestriol with mercuric acetate as a label, and the binding of ferrocene-labeled morphine to its antibody (23).

Non-Enzymatic Immunoassays: These are based on the use of non-enzymatic labels. Examples of these are listed in Table II. The Ag is labeled with an electroactive group. The labeled Ag is reducible or oxidizable in a potential range over which the Ag is electroinactive, thus enabling the labeled Ag to be distinguished from the unlabeled Ag (21-23)

Table II. Examples of Enzyme Labels and Electroactive Functional Groups Employed in Electrochemical Immunoassays

Alkaline phosphatase
Glucose Oxidase
Horseradish peroxidase
Urease
Glucose-6-phosphate dehydrogenase
Mecuric (III)
Ferrocenyl group
Dinitro group
Azo Group
2,4-Dinitrophenyl group

Immunoaffinity Chromatography with Electrochemical Detection

Electrochemical immunoassays are based on the combination of high performance immunoaffinity chromatography (HPIC) with electrochemical detection of the column effluent (22).   The Ab is covalently attached to a chromatographic column so that the binding sites are active and accessible.   The analyte (Ag) and the enzyme-labeled Ag mixtures are injected into the immunoaffinity column.   Competitive binding occurs with the immobilized Ab, followed by the injection of enzyme substrate.   Next, the electroactive product is oxidized or reduced, at a thin-layer detector, and the peak area is quantitated.   Acidic buffers are then passed through the column to displace the bound and enzyme-labeled Ag.   The amount of electroactive product formed is proportional to the amount of enzyme-labeled Ag bound to the column, which is inversely proportional to the amount of Ag in the sample. Generally in immunoaffinity chromatography with electrochemical detection, the injected sample flows through a thin layer electrochemical cell where the enzymatic product (e.g., phenol) is oxidized or reduced, at a solid working electrode (Figure 2).   One advantage of this procedure is its low detection limit which is typically in the low picogram-per-milliliter range.   The enzyme-amplified electrochemical signal

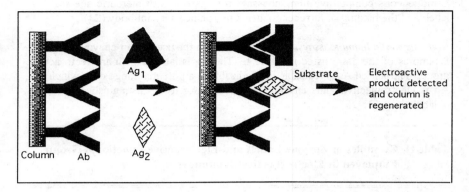

Figure 2: Major steps involved in immunoaffinity chromatography with electrochemical detection.

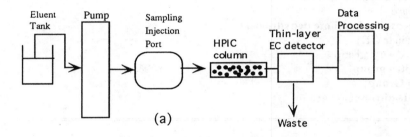

(a)

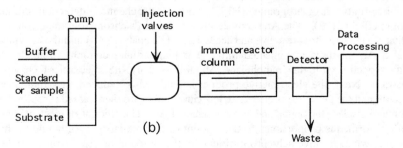

(b)

Figure 3: Schematic diagrams of instrumentation for (a) immunoaffinity chromatography with electrochemical IC-EC detection, (b) flow injection immunoassay (FIIA) methods.

(Figure 3a) is a function of the Ab-Ag binding rather than the ability to detect the product, and this can result in even lower detection limits.

Flow Injection Immunoassay

The combination of immunoassays and FIA systems offers the advantage of high sample throughput. This enhances the development of automated, field-portable and miniaturized laboratory systems. Flow injection immunoassay (FIIA) methods, including immunosensors discussed below, may have potential for on-line monitoring of waste stream effluents and groundwater. FIIA involves the immobilization of Abs on solid particles packed onto a small flow-through reactor (Figure 3b). The analyte incorporated in an eluent enters the reactor chamber where the immunochemical reaction occurs. FIIA methods based on different detection principles have also been developed (26). An example of the electrochemical detection principle is a competitive potentiometric FIIA for the determination of theophyline (27).

An assay involves the competition between an enzyme-labeled Ag and an analyte (unlabeled Ag) for a limited amount of primary Ab binding sites. The mixture is introduced via a flow-injection system into the secondary Ab reactor. The reactor-bound enzyme activity, measured by flowing an appropriate substrate solution through the reactor, is inversely proportional to the concentration of the free analyte in the sample. A single assay takes about 15 minutes, including the time for regeneration of the reactor. The advantage to using a liposome-encapsulated dye with an electroactive species rather than an enzyme and a chromogenic substrate has been shown in an automated FIIA system (28). In comparison with automated microwell plate assays, automated FIIA systems are generally simpler, faster, and reusable. Similar to microwell plate assays, FIIA systems can be calibrated (28).

Electrochemiluminescence Immunoassay

Electrochemiluminescence (ECL) is a nonradioactive detection method, also based on electrochemical principles. ECL involves the initiation of a chemiluminescence reaction at an electrode through the application of a selected potential. In ECL, a critical alignment of three constituents (Figure 4a) is required for the electrochemical reaction to occur. A precursor molecule (PRM) diffuses to an electrode surface to be activated. This association with the electrode results in a very fast electron transfer reaction. The transfer reaction initiates the excitation of a reporter molecule $(Ru(bpy)_3^{2+})$ in close association with the electrode. This ultimately results in the emission of a photon of light at a specific wavelength (29,30).

In a typical immunoassay, specific Abs (Ab_1) are bound to magnetic beads. A second specific Ab (Ab_2) capable of recognizing a different epitope on the same target is made into a reporter molecule through the attachment of an ECL label (Figure 4b). The specific molecules are then incubated with both antibodies, while the two Abs are "sandwiched" to the Ag at different sites. This Ab-Ag sandwich is passed through a flow-cell with a photomultiplier tube (PMT) where it mixes with

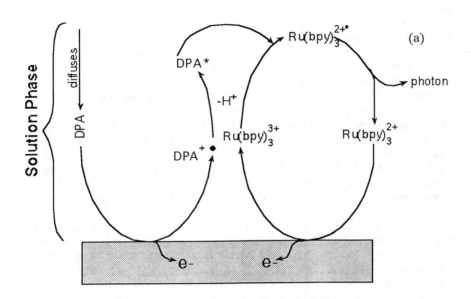

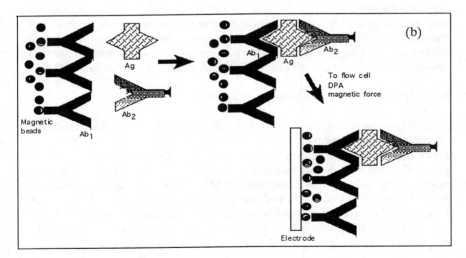

Figure 4: Scheme of electrogenerated chemiluminescence immunoassay: (a) electrochemiluminescence reaction, DPA = polyaromatic hydrocarbon 9,10-diphenyl anthracene, $Ru(bpy)_3^{2+}$ = metal complex [bpy=2,2'-bipyridine], (b) Steps involved in ECL immunoassay procedure.

the PRM in a buffer solution. A magnetic force is applied to capture the magnetic beads on the electrode surface. This stabilizes the target molecule and its attached reporter for maximum detection by the PMT. A continued flow of buffer solution removes all unbound reagents.

The electrochemical measurement is performed by the application of an electrical potential to the immunoelectrode. Some clinical applications developed on the basis of this approach include cancer markers such as CEA (carcinoembryonic antigen), AFP (α-fetoprotein), and PSA (prostate-specific antigen), hormones such as digoxin, and infectious disease markers such as hepatitis B surface antigen (31). The ECL has a wide detection range. The lower limits of detection are in the 200 fmol/L range and quantitation extends across six orders of magnitude (31).

Immunosensors

A considerable amount of attention is now being given to the development of analytical biosensors based on immunochemical principles. Biosensors can provide continuous, rapid, *in-situ* measurement of for clinical diagnosis, measurement of food freshness and contamination, fermentation process control and environmental monitoring (16,17). In general, a biosensor consists of a biospecific sensing element that is responsive to a given property or substance. This biospecific element is in contact with or integrated into a suitable transducer. The transducer is required to detect the interaction of the biospecific sensing element and the analyte. Immunosensors can be classified into four types depending on the transduction principles employed.   These are electrochemical, piezoelectric, optical and thermometric. Immunological biosensors may utilize direct, competitive or sandwich detection methods. This allows for high analyte specificity and sensitivity in the parts-per-billion (ppb) or parts-per-trillion (ppt) range.

Irrespective of the transduction principles employed, a practical immunosensor should be specific, provide fast response time, be reversible, and be capable of direct detection of a specific immunoreaction with minimal sequential addition of immunochemical reagents. In addition, other desirable characteristics of a practical immunosensor include the possibility of continuous flow measurements, and multianalyte and multimatrix sample measurements.   Ideally, there would be no need for sample preparation steps when analyzing complex samples. Also the immunosensor system should be small, and capable of convenient signal processing and should be suitable for integration into other devices that allow remedial actions to be taken after sensor measurements. Direct and indirect immunosensors based on the principles of enzyme and fluorescence immunoassays have been widely reported (17, 32). The following section discusses electrochemical immunosensors and the principles employed.

Electrochemical Immunosensors

Electrochemical detection methods are generally inexpensive. They provide a wide dynamic range, have low detection limits and are suitable for automation. A true immunosensor should directly detect immunological reactions by measuring the changes in surface potentials. However, this has not provided sufficient sensitivity and usually some type of chemical amplification (e.g,. an enzyme) is required. The features of electrochemical immunosensors reported in the recent literature are shown in Table I. Of the numerous electrochemical techniques for the detection of immunological reactions reported, the most promising ones are presented below.

Potentiometric Immunosensors: Potentiometric immunosensors utilize changes in potential that occur when the Ab (or Ag) on an electrode binds with its specific corresponding binding partner. The assays are based on the assumption that proteins in aqueous solution are polyelectrolytes, and that an Ab protein possesses a net electrical charge. This electrical charge of the Ab will be affected on binding to an Ag with the exception of the isoelectric point of the Ab. The change in electrical charge resulting from the binding of an Ab to an Ag is measurable from the net electrical charge on the Ab alone. Thus, the inherent signal measured is derived from the potential difference between the Ab-immobilized electrode and a reference electrode and it will depend on the concentration of the free Ag. Direct potentiometric immunoassay is often accompanied by only small changes in measured potential. This necessitates signal amplification. Devices that use field-effect transistors (FETs) are examples of the application of a voltage between the source and drain electrodes, controlled by the strength of the electrical field generated by the gate (Figure 5). This procedure helps to provide amplification, and the conductivity of the channel region is then measured. This technique has been applied for several analytes and is detailed in a review (33).

Amperometric Immunosensors: In amperometric assays, the current recorded when electroactive species are either oxidized or reduced is measured at a solid electrode such as platinum, gold, or glassy carbon. The current obtained has a linear relationship with the concentration of the electroactive species. There are many ways of measuring immunological reactions with amperometric immunosensors. Some detection techniques rely on direct formats, such as the pulsed amperometric immunoassay (12,13); others are the indirect immunoassay formats which use labeled species and washing steps (21). Interesting applications have been reported for IgG, HSA and thaumatin proteins (12,13,34). Immunoassays that are based on amperometric detection principles provide more sensitivity due to the linear relationship between the current and the concentration, whereas potentiometric detection assays measure the logarithmic relationship between the potential and concentration. Amperometric immunosensors can be coupled to other laboratory instrumentation such as liquid chromatography (LC-EC) and flow injection analysis (FIAEC).

Capacitive Immunosensors. Capacitive detection is based on the fact that the capacitance of an electrolytic capacitor depends on the thickness and behavior of the dielectric layer placed on the surface of the metal plate. The resultant changes in the dielectric constant from Ab-Ag interactions are then measured. The dielectric constant is a measure of the ability of a substance to store energy in response to the application of electric fields. Any variation in the surface potential results in a shift in the capacitance versus voltage curve. A commercial system based on this principle has been developed for the determination of glucose using concanavalin A as the immobilized reagent and glucosamine catalase conjugate as the competitor (35). Other systems based on the use of conducting polymer membranes (CPMs) for the measurement of Ab-Ag interactions have been reported (36,14).

Conductimetric Immunosensors: This system measures changes in conductivity of a testing medium. As with capacitive detection, the conductivity changes are due to the alteration of the double layer at the immunoelectrodes. This conductivity is measured for the monitoring of immunological reactions. Two kinds of conductimetric devices are available. One utilizes a bilayer lipid membrane to mimic a natural membrane cell containing the Ab (18). In this case, the selective interaction with the analyte results in a measurable transient current. The other available system is the conductimetric detection that uses chemiresistors. This is based on the measurement of changes in electrical conductivity of CPs which contain the appropriate dopants. Conducting electroactive polymers can serve two simultaneous purposes: (a) as a solid phase to support the immobilized Abs, and (b) as the measuring device through conductivity modulation by a dopant, such as iodine. This has been employed in the detection of glucose. The product of glucose oxidation is used by a second enzyme to reduce I^- to I_3^- which is incorporated into the CPM to produce a change in conductivity (37).

Conducting Electroactive Polymers

Over the last two decades, a range of polymers have been studied with the view of using them as sensing materials. Many of these polymers have been covered in some excellent reviews (15,16,38). They have been used mainly to immobilize functional groups which are capable of initiating some chemical and electrochemical activity at the electrode surface. In some cases, they are used for electrocatalytic purposes: to capture analytes via ion exchange, as enzyme hosts, for precipitation, complexation, or for biospecific Ab-Ag or enzyme reactions (15). CEPs such as polythiophene, polypyrrole, polyfuran and polyaniline, some of which are represented Figure 6, are a new class of electrode materials characterized by the ability to be reversibly oxidized or reduced through the application of electrical potentials. The main structural feature of all conducting polymer materials is the extended *p*-conjugated system, a sequence of alternating single and double bonds in the polymer chain.

Conducting polymer-based immunochemical sensors will be illustrated in this chapter by the use of polypyrrole (PPy). This is the most commonly used polymer because of its ease of deposition, low cost of monomer and chemical stability. The electrochemical polymerization of pyrrole occurs via an oxidative coupling in a

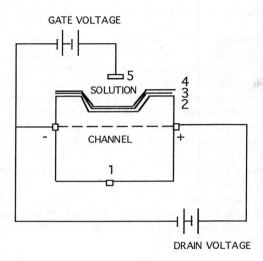

Figure 5: Schematic diagram of a potentiometric field effect transistor (FET) electrode: (1) substrate, (2) SiO_2 insulator, (3) Si_3N_4 passivation, (4) biochemically selective membrane, (5) gate electrode.

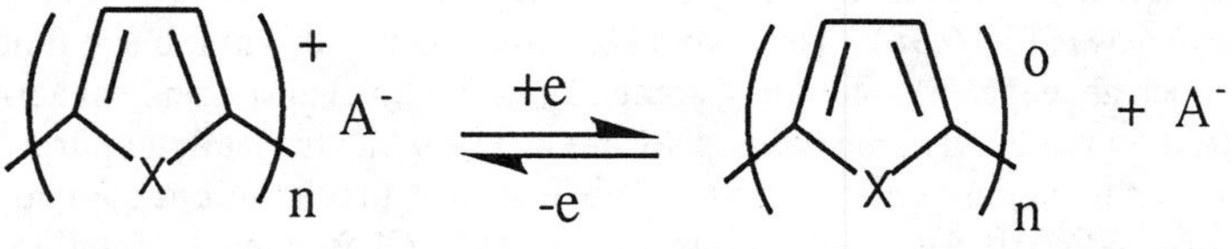

Figure 6: Changes in the properties of CEPs through the application of electrical potentials. **X** can be S, N or O and **A** is the counterion incorporated during synthesis ; **n** is the monomer/counterion ratio.

solution of pyrrole monomer and a suitable supporting electrolyte (or counterion). PPy is reversibly oxidized in a single process with the corresponding switching from the insulating state to the conducting state. Analytes are capable of direct and indirect electron transfer at PPy surfaces, thus the polymer is a suitable candidate for the immobilization of species such as enzymes, antibodies, polyelectrolytes, etc. These properties, and their dependence on type and concentration of the analyte, enable the utilization of PPy for sensing purposes.

Sensors based on conducting polymers utilize various reagents such as enzymes, antibodies, polyanions, metaloporphyrins, catalytic systems and different functional groups (15). These sensors exhibit improved sensitivities and faster response time compared to conventional electroanalytical methods of analysis. Also, surface modifications using CEP materials provide a more powerful analytical tool due to the ease of deposition, while the conductivity of the deposit allows easy charge transfer from the outer surface to the electrode surface by both the direct and mediated electron transfer protocols (38).

Although the use of antibodies for sensing purposes enables the development of extremely sensitive immunosensors, some practical difficulties exist due to the generation of sensitive, reproducible analytical signals from the Ab-Ag reaction. These difficulties are attributed to the lack of a faradaic (or electron transfer) signal and to the essentially irreversible nature of the Ab-Ag process (11). Moreover, unlike enzyme-based sensors, Ab-based biosensors are more prone to these difficulties because there is no catalytic event. Hence, other systems are needed to complete transduction and amplification of the binding event. Several attempts at overcoming these limitations have been addressed. These include the use of potential measurements (23,29), indirect amperometric immunoassay (23,40), and direct measurements that allow changes in sensor surface to be determined after the Ab-Ag interaction (36). In all these cases, the procedures are time-consuming because many steps are required and the equilibration time is lengthy. The procedures also suffer from the need for chemical regeneration of the Ab-Ag interactions.

To overcome the problem of time-consuming multistep procedures needed for the detection of Ag, some workers reported application of applied electric oscillation having a slit cell consisting of two electrodes and a glass slide for the detection of *candida Albicans* (41).They recorded a reaction time of about 5 minutes as against the usual 1-2 days normally required to complete the interaction. The traditional approach was extremely tedious, requiring constant visual observation and manual calculations of the agglutination rates.

Recently, in our laboratory, we demonstrated the use of antibody-containing conducting polymer electrodes with pulsed amperometric detection in an FIA mode (11,12). The performance of this system with respect to electrical signal generation, reproducibility, reusability, time and selectivity was investigated. The system was shown to overcome many of the limitations of electrochemical immunosensors addressed earlier. These problems are discussed below.

Improvement of Electrochemical Immunosensors Performance

In overcoming the difficulties associated with the generation of rapid, sensitive and reversible Ab-Ag interactions, it was shown that the use of Ab immobilized onto CEP matrices together with a pulsed potential waveform enabled selective molecular recognition (11, 12). This detection technique was used for human serum albumin (HSA) as a test case. The case showed the possibility of controlling Ab-Ag interactions while reversible analytical signals were obtained. The CEP was used as the sensing electrode in this novel signal generation technology. This sensing electrode was coupled to a periodic, or other transient, pulsed voltage waveform. The voltage waveforms induced changes in the CEP such that a detectable interaction with a target analyte was obtained in a reversible manner.

As an extension of these studies, the incorporation of other Abs and analytes into conducting polymers to introduce sophisticated molecular recognition capabilities has been investigated (13,14). These polymer materials can then be used as the basis of electrochemical sensors provided an appropriate electrical response can be derived from the interaction of the bioactive sensing surface with the molecule of interest. Until recently, this has proven to be particularly challenging with electro-immunological sensors, since there is no obvious sensitive signal generation mechanism for reversible molecular interactions. These studies were also designed to gain further insight into the general applicability of this new immunological sensing technology, and to investigate the possibility of developing a practical immunological sensing system.

It is believed that the attainable sensitive and reversible nature of this detection technique should allow the CEP to detect analytes such as anions, cations, organic acids, amines, metal complexing groups, antigens, antibodies, enzymes, DNA, organochlorine pesticides and others. The selective detection obtained using this technique also provides stable and reproducible signals without electrode fouling or hysteresis effects. This technology was patented in the United States and has been filed as an international patent application (42,43).

This new technology has now ushered in the development of assays for proteins and other analytes (e.g., HSA, thaumatin [a natural sweet protein similar to aspartame and saccharin], para-cresol and other phenolics, as well as the selective detection of ions such as phosphates and nitrates (11- 14, 44-47)). In each case, the sensor performance gave a detection limit in the low ppb range, wide dynamic range, rapid response time (minutes) and a reproducibility of <5% RSD over 10 consecutive injections. We are now at the forefront of developing this research for monitoring environmental contaminants for various applications. So far, we have recorded improvements in Ab-immobilization procedures and further optimization in the analytical detection principles and hope to apply these principles to environmental measurements. The following summary gives the results obtained for the development of conducting polymer-based immunosensor for PCBs.

PCB Immunosensor: Currently in our laboratories, research continues on the use of CEP-based immunosensors for direct measurement of target environmental analytes. In this work, the transduction is via a novel electrochemical sensing, enabling measurements in liquids. In this sensing system, a low-cost, portable electronic module sends an alternating potential wave across the flowing sample and the reference electrode which are sensitive to potential changes at the surface of the device. The sensitive and selective recognition of the analyte is achieved without the use of labels. This is accomplished by the use of selective Abs which have specificity and affinity for the target analytes.

When an assay is performed for analyte, an Ab is electrochemically immobilized within the conducting, electroactive polymer deposited on a substrate. A known concentration of the target analyte interacts with the immobilized Ab via a flowing stream. A potential waveform is simultaneously applied to initiate the immunological reaction. The transient event of the binding reaction produces electrostatic alterations at the CEP surface giving rise to rapid response signals. The CEP, as the basis of transduction of the Ab-Ag interactions, produces signals with no hysteresis and the sensor response correlates quantitatively with the analyte concentration.

An electropolymerized, anti-PCB Ab immobilized CEP electrode was used for the detection of various aroclors. In these experiments, the analytical signal was generated by applying a pulsed waveform (with pulse frequency of 120 and 480 ms) between 0.40 and -0.60V.

The oscillating potential reversibly drives the Ab-Ag binding process. At an applied positive potential, the binding reaction is encouraged, whereas at negative potential the binding reaction is discouraged, and loss of signal results. The current arising from this process was monitored in real time.

Figure 7 shows a typical current versus time read-out obtained using the electrochemical immunosensor. This shows that well-defined signals (with no hysteresis) due to the Ab-Ag interaction were obtained. The detection of other aroclor standards was carried out using the sensor. The results also showed that the linear dynamic range of the immunosensor was 0.3 to100 ng/mL with a correlation coefficient of 0.998 for Aroclor 1242. The method detection limits for Aroclors 1242, 1248, 1254 and 1013 were 3.3, 1.56, 0.39 and 1.66 ng/mL respectively. Cross reactivity experiments showed that the immunosensor exhibited high selectivity for PCBs. The highest cross reactivity measured for the chlorinated phenolic compounds relative to Aroclor 1242 was less than 2%. This should allow the sensor to be used in the presence of potential interferences such as chlorinated anisoles, benzenes and phenols. Figure 8 shows the typical calibration curve obtained for the detection of Aroclor 1242 in which the polymer was used as the immunosensing

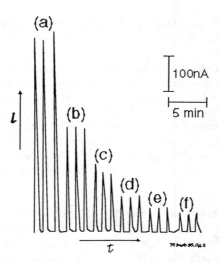

Figure 7: Typical FIA signals obtained for the detection of Aroclor 1242 using polypyrrole-anti PCB (PP/APCB) immunosensor. (a) 100, (b) 50, (c) 25, (d) 12.5, (e) 1.56, (f) 0.39 ng/ml. Conditions: $E_{initial}$ = +0.40 V (120 ms), E_{final}= -0.6V (480 ms), eluent = 0.05 M phosphate buffer (pH 7.4), flow rate = 0.5 ml/min. PP/APCB electrode (0.2 M pyrrole in 1000 fold dilution of antibody) coated on silver substrate in a thin layer cell. Stainless steel auxiliary electrode, Ag/AgCl (3 M KCl) reference electrode.

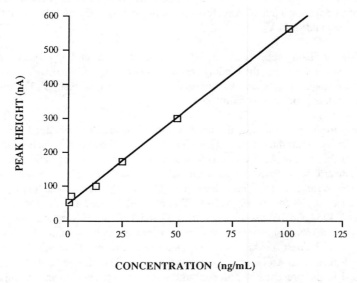

Figure 8: Calibration curve for the detection of Aroclor 1242 using PP/APCB immunosensor. Conditions as in Figure 5.

electrode. This result shows that the response was linear over the range investigated (up to 100 μg/L) and the correlation coefficient (r) was 0.998.

This assay procedure, as demonstrated for PCBs, showed that an electrochemical immunosensor with linear response characteristics and sensitivity is appropriate for sensing PCBs. The results are comparable to data reported for PCBs using an enzyme-linked immunosorbent assay (ELISA) technique (48). The obvious advantages of this technique are; rapid detection time, elimination of separation steps, suitability for continuous operation, and minimal operator attendance. It is expected that this method will be widely applicable for the detection of contaminants in groundwater and effluent monitoring.

Conclusions

Electrochemical immunoassays, including immunosensors, offer all the analytical advantages of conventional immunoassay methodology, such as high sensitivity and selectivity. Additional benefits of electrochemical immunoassays include increased control over the rate and course of immunological reactions, ability to amplify and generate analytical signals *in-situ*, and the option of electrochemically modifying the electrode surface to prepare tailor-made biochemically selective layers for analytes of interests. Several approaches to electrochemical immunoassays have been described. They include electrochemical enzyme immunoassays, electrochemical detection based on immunoaffinity chromatography and flow injection analysis, electrochemiluminescence immunoassays and electrochemical immunosensors. Detection by electrochemical sensors can be inexpensive. And, when coupled to electrochemically active functional layers such as conducting polymers, they are suitable for rapid and sensitive detection and quantitation of analytes in complex matrices.

Examples of conducting polymer-based immunosensors for the detection of PCBs were given. We have also shown that the data obtained for the sensitivity and selectivity of PCB sensing are comparable to those reported for an ELISA technique, yet a more rapid detection was obtained. This method of analysis could be employed for the analysis of other organochlorine compounds provided the antibodies are readily available. This class of immunosensors holds great promise due to the possibility of miniaturization on a chip and integration into a complete automated system. Conducting polymer-based sensors promise a more sensitive, cost-effective, practical and less-time consuming approach to direct electrochemical immunoassays. The potential application of this family of sensors in the fabrication of sensor arrays for use in clinical and environmental analyses is therefore envisaged. Even though immunoselective electrodes have been extensively used in medical research for patient diagnosis and making macromolecular investigations, increased awareness will further their use in the screening of environmental contaminants. Industrial and process controls will provide other markets for these valuable analytical tools.

Acknowledgments

The research in this document has been funded by the U.S. Environmental Protection Agency through a National Research Council Associateship. It may not necessarily reflect the views of the Agency and no official endorsement should be inferred.

Literature Cited

1.   Breyer, B.; Radcliff, F.J.; *Nature*, 1951, 167, 79.
2.   Berson, S.A.; Yalow, R.S.; Bauman, A.; Rothschild M.A.; Newerly K.; J. *Clin. Invest*, 1956, 35, 170.
3.   Hayes F.; Halsall B.; Heineman W.R.; *Anal. Chem.*, 1994, 66, 1860.
4.   Kaneki, N.; Xu, Y.; Kumari, A.; Halsall, B.; Heineman, W.R.; *Anal. Chim. Acta*, 1994, 287, 253.
5.   Jiang, T.; Halsall, H.B.; Heineman, H.B.; *J. Agric. Food Chem.*, 1995, 43, 1098.
6.   Niwa, O.; Xu, Y.; Halsall, B.; Heineman, W.R.; *Anal. Chem.*, 1993, 65, 1559.
7 . Athey, D.; McNeil, C.; Bailey, W.; Hager, H., Mullen, W.; Russel, L.; Meyerhoff, M.E.; *Anal. Chem.*, 1994, 66, 1369.
8 . Palmer, D.; Edmonds, T.; Seare, N. *Anal. Chem.*, 1993, 26,1425.
9.   Yang, H.; Leland J.K.; Yost D.; Massey R.J.; *Biotech.* 1994, 12, 193.
10. Wilner, I.; Blonder, R.; Dagan, A. *J. Am. Chem. Soc.*, 1994, 116, 9365.
11. Sadik, 0. A.; Wallace, G. G.; *Anal. Chim. Acta.*, 1993, 297, 209.
12. Sadik, 0. A.; Wallace, G. G.; *Proc. 207th ACS National Meeting*, March 13 -18, 1994, San Diego, CA., USA., 70, 178.
13. Sadik, O.A.; John, M.J.; Wallace G.G.; Bamett D.; Clarke, C.; Laing, D.G.; *Analyst*, 1994, 119, 1997.
14. Barnett, D.; Laing, D.G.; Skopec, S.; Sadik, O.A.; Wallace, G.G.; *Anal. Lett.*, 1994, 27, (No. 13), 2417.
15. Bidan, G.; *Sensors & Actuators*, 1992, B6, 45.
16. Sadik, Omowunmi; *J. Analytical Methods & Instrumentation, (In* Press).
17. Biosensor Technology Fundamentals & Applications, (Eds: Buck, R.P.; Hatfield, W.E.; Umana, M.; and Bowden, E.F.;) Marcel Dekker, New York, 1990, 219.
18. Nikolelis, D.; Tzanelis, M.; Krull, U,; *Anal. Chim. Acta* , 1993, 282, 257.
19. Cardosi M.F.; Birch S.W.; Higgins I.J.; Methods in Immunological Analysis, Masseryeff, R.; Albert W.; Staines, N.; (Eds.; VCH: Weinheim) Germany, 1993, 1, 359.
20. Panfili, P.; Dill, K.; Olson, *J. Curr. Opin. Biotechnol.*, 1994, 5, 60.
21. Ngo, T. T.; Ed. "Electrochemical Sensors in Immunological Analysis". N.Y. Plenum Press, 1987.
22. Xu, Y.; Halsall, H.B.; & Heineman, W.R.; In Immunological Assays & Biosensor Technology for the 1990s, Nakamura, R.M.; Kasahara, Y.;RechnitzG.A. (eds.) Washington, DC, 1992, 291.
23. Green, M.J.; *Philos Trans. R.. Soc. London*, Ser. B. 1987, B316, 135.

24. Gil, E.P.; Tang, H.T.; Halsall, W.R.; Heineman, W.R.; Misiego, A.S.; *Clin. Chem.*, 1990,36,6625.
25. Duan C.; Meyerhoff, M.E.; Anal. Chem., 1994, 66, 1369.
26. Puchades, R.; Maquieira, A.; Atienza, J.; and Montoya, A.; Crit. Rev. Anal. Chem, 1992, 23, 301.
27. Palmer, D.A.; Edwards, T.E.; Seare, N.J.; *Anal. Lett.*, 1993, 26(7), 1425.
28. Durst, R.A.; Siebert, S.T.A.; Roberts, Larsson-Kovach I.M.; Reeves S.G.; 2nd *Bioelectroanalytical Symposium,* Matrafured, Budapest, 1992.
29. Bard, A.J.; & Faulkner, L.R.; *Electrochemical Methods,* New York, 1980, 621.
30. Knight, A.W.; & Greenway, G.M.; Analyst, 1994, 119, 879.
31. Yang, H.; Leland, J.K.; Yost, D.; & Massey, R.J.- *Biotechnology,* 1994, 12, 193.
32. Place, J.F.; Sutherland, R. M.; Dahne, C.; *Biosensors,* 1985, 1, 321.
33. Pranitis, D.M., Telting-Diaz M., and Meyerhoff, *Crit. Rev. Anal. Chem.,* 1992, 23, 163.
34. Taniquchi, I.; Fujiyasu, T.; Eguchi, H.; Yasukouchi, K.; Tsuji, I.; Unoki, M.; *Anal. Sci.;* 1986, 2, 587.
35. Bresler, H.S.; Lenkevich, M.J.; Murdock , J.N.; Roblin, R.O.; *ACS Symp.Ser,* 1992, 511, 89.
36. John, R.; Spencer, M.J., Smyth M.R.; Wallace, G.G.-, *Anal. Chim. Acta,* 1991, 249, 381.
37. Sandberg, R.G.; Van Houten, L.J.; Schwartz, R.P.; Dallas, S.M.; Michael, J.C.; & Narayanswamy, V.; *ACS Symp. Ser. 1992,* 511, 81.
38. Zotti, G. *Synthetic Metals,* 1992, 51, 373.
39. Bush, D. L.; Rechnitz, G.A.; *Anal. Lett.,* 1987,20, 178.1
40. Monroe, D.; *Crit. Rev. Clin. Sci.,* 1990,*28, 1.*
41. Karube, I., Gotoh, M.; In: Analytical Uses of Immobilized Biological Compound for Detection, Medical and Industrial Uses (Guilbault, G. G., Mascini M., eds), *Reidel Publ. Co.,* 1988, 267.
42. Sadik et. al., *U.S. Patent* No. U.S.5,403,451 of March 4, 1994
43. Sadik et al, *International Patent Application No.* PCT WO94-20841, Sept 15, 1994
44. Sadik, O.A.; Wallace, G.G.; *Electroanalysis,* 1993, *5, 555.*
45. Sadik, O.A.; Wallace, G.G.; *Electroanalysis,* 1994, 6, 860.
46. Ongarato, D. O.; Sadik, O.A.; Wallace, G.G.; Riviello, J.M.; *J.Electroanal. Chem.* (In Press)
47. Sadik, O.A.; Talaie, A.; Wallace, G.G.; *Intel. Mat. Syst. and Stritc.,* 1993, 4, 123.
48. Johnson J.C.; Van Emon J.M. EPA/600/R-94/112, June, 1994.
49. Blackburn, G.F.; Talley, D.B.; Booth, P.M.; Durfor, C.N.; Martin, M.T.; Napper,
    A.D.; Rees, A.R.; *Anal. Chem.,* 1990, 62, 221 1.
50. Alvarez-Icaza, M.; Bilitewski, U.; *Anal. Chem.,* 1993, 65, 525.

Chapter 13

A First Application of Enzyme-Linked Immunosorbent Assay for Screening Cyclodiene Insecticides in Ground Water

Tonya R. Dombrowski[1], E. M. Thurman[1], and Greg B. Mohrman[2]

[1]U.S. Geological Survey, 4821 Quail Crest Place, Lawrence, KS 66049
[2]Rocky Mountain Arsenal, 72nd and Quebec Street, Building 111, Commerce City, CO 80022

A commercially available enzyme-linked immunosorbent assay (ELISA) plate kit for screening of cyclodiene insecticides (aldrin, chlordane, dieldrin, endosulfan, endrin, and heptachlor) was evaluated for sensitivity, cross reactivity, and overall performance using ground-water samples from a contaminated site. Ground-water contaminants included several pesticide compounds and their manufacturing by-products, as well as many other organic and inorganic compounds. Cross-reactivity studies were carried out for the cyclodiene compounds, and results were compared to those listed by the manufacturer. Data obtained were used to evaluate the sensitivity of the ELISA kit to the cyclodiene compounds in ground water samples with a contaminated matrix. The method quantitation limit for the ELISA kit was 15 µg/L (as chlordane). Of the 56 ground-water samples analyzed using the ELISA plate kits, more than 85% showed cyclodiene insecticide contamination. The ELISA kit showed excellent potential as a screening tool for sites with suspected ground-water contamination by insecticides.

The Rocky Mountain Arsenal (RMA), a 70 km² tract of land 14.5 km northeast of Denver, Colorado, has been the site of several intensive chemical research and production projects. From the 1940s to the 1960s chemical agents, rocket fuels, and weapons were manufactured at the site by the U.S. Army. In the mid-1940s, production facilities at the Arsenal were leased to private industry, and agricultural pesticides were manufactured by what is now a division of Shell Chemical Company.
Waste and other industrial by-products of the manufacturing processes from both the U.S. Army and Shell Chemical Company were disposed of according to the accepted protocols of that time period. Since the first contaminant problems were reported in the mid-1950s, innovative containment and clean-up measures have been undertaken on a massive scale by all organizations associated with RMA property (1). Target analytes and "fingerprint" compounds (materials known to be unique to the RMA) have been identified. Aggressive monitoring policies have resulted in the establishment of a network of over 1200 ground

water wells, extending throughout the RMA property and into the surrounding areas. The number of analyses required to support these monitoring programs combined with sampling constraints due to well volume, hazard and accessibility, and the complex matrix present in many areas of the RMA present a major challenge in the areas of time, expense, and overall analysis complexity.

Because a method for rapidly identifying areas of significant interest or for reliably screening a large number of samples (many potentially in the field) would be beneficial when dealing with such a large and complex site, the cyclodiene enzyme-linked immunosorbent assay (ELISA) plate kit (Millipore Corporation[1], Bedford MA) seemed applicable to the RMA as it would provide specific, rapid results with minimal sample preparation and volume. ELISA utilizes the highly specific binding sites of antibodies, which recognize a single compound or class of compounds, to provide information on the presence and concentration of those compounds in a sample. Because of the specific nature of this technique, many other contaminants present in the same sample can be effectively "screened out," and the analyte(s) of interest reproducibly determined. The specificity of the binding site, however, is based on chemical structure, and needs to be adequately characterized to determine the type or configuration of compounds to which the kit will respond, particularly when dealing with a complex and highly uncharacterized sample matrix (*2,3*). Specific compounds of interest to this study conducted by the U.S. Geological Survey in cooperation with the U.S. Department of the Army from March 1994 to October 1995 were the cyclodiene insecticides: aldrin, chlordane, dieldrin, endosulfan, endrin, and heptachlor (see Figure 1). The purpose of this paper is to discuss the results obtained from this study, in which the cyclodiene ELISA kit was evaluated as a screening tool for insecticide contamination in highly complex ground-water matrices. The cyclodiene ELISA kit used in this study was rated by the manufacturer as having a linear concentration range of 5 to 100 µg/L (as chlordane). Historical gas chromatography (GC) data from the RMA wells showed cyclodiene concentrations that ranged from non-detectable levels to about 100 µg/L (composite values from the four cyclodienes, aldrin, chlordane, dieldrin, and endrin, were evaluated separately) with the majority of the wells showing levels below 10 µg/L. Given this information, and the linear concentration range specified for the cyclodiene ELISA, this kit appeared well suited as a screening tool for ground-water samples at the RMA. In several areas of the Arsenal, however, ground water containing both insecticides and herbicides in combination with many other organic and inorganic compounds was present; therefore, it was possible that compounds with structural similarities to the cyclodiene insecticides were present. As the ELISA kit had not been used previously with samples in such a complex ground-water matrix, the sensitivity, and cross reactivity of the kit, as well as the possibility of matrix interferences, had to be fully evaluated.

Experimental Methods

All sample analyses were performed using the ELISA plate kit for cyclodienes according to manufacturer's instructions (Millipore Corporation, Bedford MA) using reagents included with the plate kits. Cross-reactivity and sensitivity studies were carried out using standards obtained through the Army Standard Analytical Reference Materials (SARM) repository. Separate standard solutions for aldrin, chlordane, dieldrin, endosulfan, endrin, and heptachlor were prepared

---

[1] The use of brand names in this paper is for identification purposes only and does not constitute endorsement by the U.S. Geological Survey.

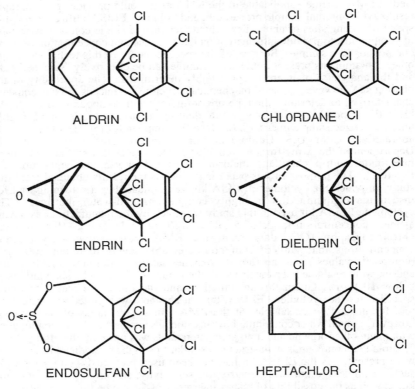

Figure 1. Chemical structures of the cyclodiene insecticides. Heavy lines indicate bridge position above the plane of the paper. Dashed lines indicate bridge position below the plane of the paper.

from neat or solid standard material. Appropriate amounts of each of these materials were weighed out and dissolved in hexane (Fisher Scientific, Pittsburgh PA). These standard solutions were then diluted to 1,000.0, 100.0, 10.0, 1.0, and 0.1 μg/L concentrations using distilled water as the final diluent. Stock standard concentrations were 2,000,000 μg/L to ensure that dilute standards would contain less than 0.5% organic solvent. Two blank solutions were evaluated in the cyclodiene kit study, distilled water produced in-house, and the negative control included with the kit reagents. The analyses of ground water samples were performed using a chlordane standard at concentrations of 0.0, 5.0, 25.0, and 100.0 μg/L.

The linear concentration range of the ELISA kit was determined from a plot of B/B_0 versus the log of the corresponding standard concentration. B/B_0 values represent the optical density of the sample solutions divided by the optical density of the negative control solution. Sensitivity information on each compound was obtained by using the IC_{50} (concentration of the compound required to give a B/B_0 value of 50%) and LDD (least-detectable dose, defined as B/B_0 of 0.90) values. These were calculated from the standard curve of B/B_0 versus the log of the corresponding standard concentration.

All 56 ground-water wells involved in this study were sampled from May 1994 thru June 1994. All wells sampled were in the unconfined aquifer system and represented a geographical distribution that included a range of contaminant concentrations from relatively uncontaminated to very contaminated water. Sampling was carried out by personnel from the U.S. Geological Survey. Duplicate samples were taken at a frequency of 1 in 10, (duplicates equaled 10% of the total samples). Sample and trip blanks were collected with the same frequency. All pertinent geological information (well depth, screened interval, pumping method, and water level) was noted at the time of sample collection. Samples were collected in clean, baked, amber glass jars with Teflon-lined lids, stored at or below 4^0 C, shipped within 48 hours of sampling, and analyzed on receipt.

Results and Discussion

The data obtained from the cross-reactivity studies are listed in Table I. The study concentrations show the same overall trends as noted in product information supplied by the manufacturer. The highest sensitivity is exhibited toward endosulfan, while the lowest is toward aldrin. A tentative hypothesis that has been proposed to explain this difference in sensitivity centers on the chlorinated bridge structure spanning the primary chlorinated ring as seen in the chemical structures of the compounds of interest (See Figure 1). It is believed that this bridge (common to all compounds investigated) may be the antibody binding site. The additional bridge on the secondary ring in aldrin, dieldrin, and endrin also may affect binding as may the presence of electronegative atoms or structural components. The presence of an oxygen atom near the chlorinated bridge (as in the ring oxygens in endosulfan and the oxygenated "bridge" structure of endrin) then would serve to stabilize the binding site and enhance binding capability. The oxygenated bridge in dieldrin is structurally removed from the binding site a considerable distance as this molecule is in a "trans" conformation, and therefore the added binding enhancement is not as readily apparent as in the "cis" conformation compound (endrin) and the sterically unhindered heptachlor. The presence of a chlorine atom or a double bond near the chlorinated bridge seems to destabilize the binding site, resulting in decreased binding efficiency and decreased sensitivity for aldrin, chlordane, and heptachlor. The double bond present in aldrin may affect the structural configuration of the binding site, which

Table I. Comparison of manufacturer's cyclodiene ELISA performance data with performance data from ground water monitoring study. IC_{50} is 50% inhibition concentration, LDD is least-detectable dose.

Compound Name	Manufacturer's Concentrations		Study Concentrations	
	IC_{50}	LDD	IC_{50}	LDD
Aldrin	84 µg/L	17 µg/L	250 µg/L	48 µg/L
Chlordane	30 µg/L	5 µg/L	34 µg/L	2.0 µg/L
Dieldrin	27 µg/L	2 µg/L	18 µg/L	1.8 µg/L
Endosulfan	6 µg/L	0.6 µg/L	12 µg/L	1.7 µg/L
Endrin	3 µg/L	0.15 µg/L	17.5 µg/L	1.7 µg/L
Heptachlor	33 µg/L	4 µg/L	140 µg/L	2.1 µg/L

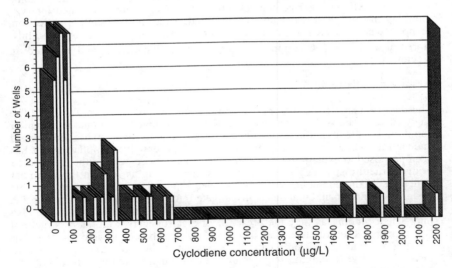

Figure 2. Frequency of cyclodiene concentrations in ground-water samples from the Rocky Mountain Arsenal.

results in the decreased sensitivity of the ELISA kit to this compound.

The overall least detectable dose (LDD) of the ELISA kit was rated by the manufacturer at a cyclodiene concentration of 5 µg/L (as chlordane). In actual practice, however, the method quantitation limit (calculated as 10x the standard deviation of the blank) was set at 15 µg/L (as chlordane) by the authors, to assure that an appropriate level of confidence in the values reported was maintained. Given this linear range of approximately 15 to 100 µg/L the ELISA kit was suited for application at the RMA, although the less contaminated samples would fall in the lower end of the quantitation range and may require some preconcentration if remediation limits are established below 15 µg/L. As stated previously, historical GC data available for wells in this study showed cyclodiene concentrations which ranged from non-detectable levels to about 100 µg/L (composite values from the four cyclodienes, aldrin, chlordane, dieldrin., and endrin, were evaluated separately) with the majority of the wells showing levels below 10 µg/L. The linear range of the ELISA kit therefore approximated the concentration range expected in the RMA samples. Given the manufacturing and disposal (manufacturing wastes and by-products) history of the RMA, it is expected that the application of this ELISA kit to other sites of cyclodiene contamination would be successful. The toxicity of the individual cyclodiene insecticides range from about 5 to 500 mg/kg of body weight (oral dosage, rat) so the sensitivity of the ELISA kit is more or less adequate based on health considerations. (Because of the toxicity of these compounds, all personal precautions were taken when handling either sample or reagent solutions.) As low concentration samples may benefit from preconcentration methods, a solid-phase extraction (SPE) procedure is being developed for the RMA samples. Due to the highly contaminated matrix, standard SPE procedures (*4,5*) are being modified to increase selectivity in retained compounds. The linear range of each of the six cyclodiene compounds listed was investigated. Dieldrin, endosulfan, and heptachlor showed a linear range of roughly 3 to 100 µg/L while the ranges for aldrin, endrin and chlordane (the calibrator) were shifted to slightly higher values.

The samples analyzed show a wide concentration range (see Figure 2), with more than 85% of the wells sampled exhibiting positive contamination values. The majority of contaminant concentrations identified were less than 120 µg/L, and less than 20% of the wells sampled having concentrations greater than 500 µg/L. Identification of potential interferents is currently underway for these high concentration samples. The relative percent difference (RPD) for all duplicate field samples was within 20%, and similarly, the RPDs for the analytical duplicates were all within 12% for the 56 samples analyzed. Quality control samples analyzed included both trip and sample blanks as well as the duplicate samples previously mentioned. All analysis blanks analyzed agreed to within 10% (coefficient of variance) of the negative control. GC/MS methods are currently under development to provide verification of the ELISA kit results. All analysis results were available within a short time, and at substantially reduced cost in terms of laboratory hours and equipment cost as compared to traditional GC and GC/MS analyses.

When the sample data obtained were plotted with the appropriate geographical orientation, contaminant plume boundaries could be identified through concentration gradients defined by the data points and showed strong relative agreement with those wells for which historical GC cyclodiene data were available. These unpublished GC data are consistent with, though lower than the ELISA results.

Conclusions

A wide range of contaminant concentrations at the RMA was identified using the ELISA cyclodiene kit. The data obtained exhibited a high degree of correlation among wells in close geographical proximity and with the available historical GC data for the wells sampled. It resulted in the identification of the current contaminant plume boundaries in a rapid and efficient manner. The only sample-preparation step required was dilution, which represents a considerable simplification of the many extraction, concentration, and analytical steps commonly required for the instrumental analysis of these cyclodiene insecticides.

From the data generated through the sensitivity studies performed, the method quantitation limit was calculated as 15 µg/L (as chlordane). The cross reactivity of the ELISA kit as evaluated using the procedure previously described parallels that given by the manufacturer. Compounds to which this kit showed the greatest sensitivity were endosulfan and endrin. While direct GC/MS conformation was not available for the samples analyzed using the ELISA kit, the data obtained compares favorably with historical GC data available for some of these wells. The unpublished GC data available, are consistent with, though lower than the ELISA results. However, the relative concentration trends are similar for both sample sets, and contaminant plume boundaries outlined are essentially the same for both methods of analysis. The ELISA results did display a positive bias when correlated with the available GC data for the same wells. This may indicate an interfering compound, or may be inherent in the correlation due to the fact that GC data is available for aldrin, chlordane, dieldrin, and endrin only, while the ELISA kit is sensitive to several other compounds in this same family. However, in comparison, the ELISA data clearly revealed the same relative concentration trends and plume boundaries as the GC data.

Acknowledgment: The authors acknowledge the U.S. Department of the Army for funding of this work.

Literature Cited:

1. Garlock, E.T. Ed.; *Eagle Watch* ; August **1992**, Vol. 4, No. 8, p. 1-19.

2. Vanderlaan, M.; Watkins, B.E.; Stanker, L.; Environmental Monitoring by Immunoassay; *Environmental Science & Technology*; **1988**, Vol. 22; No. 8; p. 247-254.

3. Van Emon, J.M.; Lopez-Avila, V.; Immunochemical Methods for Environmental Analysis; *Analytical Chemistry*; **1992**, Vol. 64, No. 2; p. 79A-88A.

4. Aga, D.S.; Thurman, E.M.; Coupling Solid-Phase Extraction and Enzyme-Linked Immunosorbent Assay for Ultratrace Determination of Herbicides in Pristine Water; *Analytical Chemistry*; **1993**, Vol. 64; No. 20; p. 2894-2898.

5. Thurman, E.M.; Goolsby D.A.; Meyer, M.T.; Mills, M.S.; Pomes, M.L.; Kolpin, D.W.; A Reconnaissance Study of Herbicides and Their Metabolites in Surface Water of the Midwestern United States Using Immunoassay and Gas Chromatography/Mass Spectrometry; *Environmental Science & Technology*; **1992**, Vol. 26; No. 12; p. 2440-2447.

Chapter 14

Maumee Area of Concern Sediment Screening Survey, Toledo, Ohio

Thomas J. Balduf[1], Jeff Wander[2], Philip A. Williams[2], Brent Kuenzli[1,3], and Patrick J. Heider[2]

[1]Division of Surface Water and [2]Division of Emergency and Remedial Response, Ohio Environmental Protection Agency, Northwest District Office, 347 North Dunbridge Road, Bowling Green, OH 43402

The Maumee Bay, located in the western basin of Lake Erie, was once known as the most prolific fish spawning ground in Lake Erie. The heavy metal and organic chemical contamination caused by agriculture and heavy industry such as oil refining, petrochemical, metal fabricating, auto parts and manufacturing resulted in the Maumee Bay being listed as an Area of Concern (AOC) in 1985 by the International Joint Commission.

In September, 1994, during the initial reconnaissance of sediment quality in the Ottawa River, Ohio, both screening and standard methods were used to analyze 29 sediment core samples. Total Polynuclear Aromatic Hydrocarbons (PAHs) and polychlorinated biphenyls (PCBs) were quantified to parts per million (ppm) level by enzyme immunoassay using a variation of EPA 4020 and 4035 draft methods. Contract Laboratory Program (CLP) analysis of the same samples for the parameters followed standard Statement of Work (SOW). This study indicates that immunoassays are valuable screening devices for PAHs and PCBs, especially when used with periodic laboratory confirmation sampling

The Maumee Bay is located in the western basin of Lake Erie (northwest section of Ohio and the southeast section of Michigan). It was once known as the most prolific fish spawning ground in Lake Erie. This area included what was known as the Great Black Swamp, which contained a faunal association requiring water free of clayey silts and containing aquatic vegetation. Habitat and water quality degradation began as far back as 1850, due to the effects of dams, channelization, over-fishing and pollution.

[3]Corresponding author

The heavy metal and organic chemical contamination caused by agriculture and heavy industry such as oil refining, petrochemical, metal fabricating, auto parts and manufacturing resulted in the Maumee Bay being listed as an AOC in 1985 by the International Joint Commission. It is one of 43 areas with pollution problems so severe that the 14 identified beneficial uses in the Great Lakes Water Quality Agreement are impaired. Because the Maumee River is the largest tributary to the Great Lakes, this pollution is readily carried into Lake Erie, contaminating water and sediment.

In the spring of 1991, the Ohio Environmental Protection Agency (EPA) produced the "Fish Tissue, Bottom Sediment, Surface Water, Organic and Metal Chemical Evaluation and Biological Community Evaluation" for the Ottawa River and Tenmile Creek, a tributary in the Maumee AOC. These data documented the pollution problems and identified areas needing further analysis, including grossly contaminated surface sediments. This report led to the Ohio Department of Health issuing a fish consumption/contact advisory for the Ottawa River from River Mile 8.8 to Lake Erie in April 1991. This advisory was based on the detection of extremely elevated levels of (PCBs) in sediments and fish tissues. A few highlights in this report include:

> **A)** Extensive PCB contamination in the lower 10 miles of the Ottawa River. PCBs were detected in all media sampled within the Ottawa River, with Ohio Water Quality Standard violations noted in both surface water and fish tissue samples. The highest PCB concentration **(1,200 ppm)** was documented in sediment sampled from a portion of the former river channel which now serves as a drainage swale for a storm water discharge into the Ottawa River. The highest total PCB levels in fish tissue occurred at River Mile 5.2, where common carp fillets and whole body PCB concentrations were 65 ppm and 84 ppm, respectively. Theses fish tissue samples had to be disposed of as Toxic Substance Control Act (TSCA) waste because they exceeded the regulator levels.

> **B)** A wide range of pesticides were detected in the fish tissue and sediment samples. Most pesticides appeared to be in low concentrations. However, heptachlor epoxide and dieldrin were considered extremely elevated.

> **C)** Five heavy metal contaminants (barium, cadmium, chromium, lead, and selenium) were measured in fish tissue fillet and whole body composite samples from the 1990 sampling sites. Eleven heavy metal elements were detected in the sediments from the Ottawa River between 1986 and 1990.

> **D)** Biological community results show non-attainment of the Warmwater Habitat aquatic life use designation for nearly the entire sampling area of the Ottawa River. Based on 1986 and 1990 Ohio EPA sampling results, the Ottawa River is in violation of Ohio Water Quality Standards.

The severe water quality noted in this document along with the health advisory prompted the first Ohio EPA sediment screening survey to be conducted in the Ottawa River.

The Division of Surface Water (DSW) and the Division of Emergency and Remedial Response (DERR) worked together during the various stages of this screening survey. The DSW focused their efforts on establishing a baseline of current water quality conditions in the AOC. Their efforts included the collection and chemical/physical analysis of the surface water samples and the chemical analysis of sediments at several stations throughout the AOC. DERR began their study efforts by initiating site inspections at 21 identified uncontrolled/unregulated hazardous waste disposal sites known within the Ottawa River watershed.

In September 1994 the Ohio EPA Maumee Area of Concern (AOC) Project team consisting of personnel from Ohio EPA, DSW and DERR began the initial reconnaissance of sediment quality in the Ottawa River, Ohio. In this reconnaissance both screening and standard methods were used to analyze 29 subsamples from 18 sediment cores. For screening, 12 metals were quantified at part-per-million (ppm) levels with energy dispersive X-ray fluorescence (EDXRF) and organics (total PCBs and PAHs) were quantified at part-per-billion (ppb) to ppm levels by enzyme immunoassay, using a variation of EPA 4020 and 4035 (1) draft methods. CLP analysis of the same samples for the parameters followed standard CLP SOW EPA methods (2).

The goal of this survey was to conduct the initial portion (Step 1) of a sediment screening survey to collect field data supported by laboratory confirmational data to be used to evaluate the extent of PCB, PAH and heavy metals contamination within the Ottawa River health advisory zone. This was to be accomplished by using field screening methods (with laboratory confirmation) to provide a broad survey of 8.8 miles of the Ottawa River. The survey is to determine the broad distribution of contamination in the sediment within the health advisory zone and to determine, within the constraints of this initial sediment sampling, the extent of the contamination , not to determine responsible parties. However the information obtained through this survey may aid in the direction of future remedial investigatory actions taken within the health advisory zone of the Ottawa River.

Table I is a summary of the 29 replicate sub-samples collected from various depths of the 18 cores that were analyzed by both CLP laboratory methods and by immunoassay methods for PCB, the constituent of most concern with respect to the use of immunoassay as a field screening method. Table I and Figure 1 displays the correlation between the laboratory methods and the field screening methods.

The data shown in this table have undergone validation procedures by a third party contractor. The validation results for semivolatiles, pesticides and PCB were developed in accordance with the U.S EPA Contract Laboratory Program National Functional Guidelines for Organic Data Review, February 1994. The validation results for metals were developed in accordance with the U.S. EPA Contract Laboratory Program National Functional Guidelines for Inorganic Data Review, February 1994.

All samples were analyzed by the laboratory using Contract Laboratory Program protocols with full packages as a deliverable.

The results from this screening survey were submitted to the U.S.EPA, Region 5, Immunoassay Assessment Team for their review of the immunoassay process. Hopefully, these data will demonstrate, to the Team, the information necessary to gain their confidence in this field screening methodology.

Table I.   Data from laboratory and field analysis of sediment at the Maumee AOC,
            Toledo, Ohio

OEPA SAMPLE #	ROSS LAB SAMPLE #	PCB FIELD (ppm)	PCB LAB. (ppm)	TOC (ppm)
OR1/4.9/9-15	101	3.9	6	56800
OR1/4.9/15-34	102	2.1	0.11	34000
OR1/6.0/0-6	103	244	1300	130000
OR1/6.0/30-57	104	1045	2000	112000
OR1/4.2/6-23	105	9.6	5	59400
OR1/4.2/23-40	106	8.4	0.79	54700
OR1/3.4/8-19	107	1.2	0.79	0100
OR1/3.4/19-41	108	5.8	0.17	120000
OR1/5.8/0-6	109	11	37	81200
OR1/6.4/0-6	110	2.7	3.7	37700
OR1/7.9/0-6	111	0.67	0.53	41200
OR1/7.4/0-6	112	0.35	0.65	28000
OR1/6.9/8-21	113	2.7	8.6	42700
OR1/6.9/21-31	114	1.3	0.7	40900
OR1/9.0/6-23	115	6.1	2.7	66400
OR1/10.0/0-6	116	4.7	10	43200
OR1/1.6/12-27	117	3.6	0.3	74000
OR1/1.6/0-12	118	1.1	0.62	60400
OR1/6.1/0-16	119	1.7	3.8	33800
OR1/6.1/16-50	120	<0.12	0	21500
OR1/6.0A/0-24	121	848	2500	135000
OR1/5.5/0-8	122	4.3	16	25600
OR1/8.8/4-17	123	<0.12	0.3	31000
OR1/12.0/0-6	124	1.4	3.3	51800
OR1/6.0/6-30	125	72	190	96300
OR1/11.0/8-17	126	<0.12	0.22	14400
OR1/11.0/0-8	127	<0.12	0.099	47300
OR1/0.0/0-11	128	1.8	0.66	25700
OR1/0.0/11-25	129	<0.12	0.29	5420

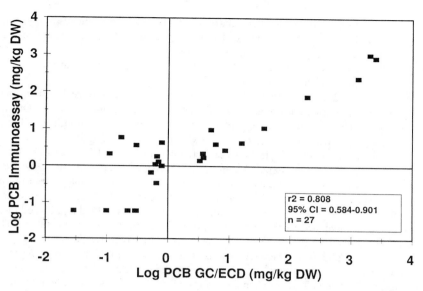

Figure 1. Log-log plot of data shown in Table I. DW indicates dry weight.

This initial reconnaissance of sediment quality in the Ottawa River is only the first step. Conceptually, the project team believes that a three stage approach would be the most effective means to characterize the Ottawa River sediments, with each stage implemented using the information (results) from the previous stage. During this initial reconnaissance of the sediment screening survey, Step 1 of Stage One (3) was the only step implemented. The three stages proposed by DERR will provide the most effective and efficient means to characterize the contaminants present in the sediments, the extent of these contaminants, and to aid in the delineation of the sources of the contaminants. This approach will also allow, given budgetary constraints, for the most thorough and extensive investigation efforts to be focused on the areas of greatest concern to the river ecosystem (areas of greatest contamination). The information gathered from Step 1 of the sediment survey, together with the information gathered from other surveys and assessments being performed as part of the Remedial Action Plan process will aid in future activities to be implemented in order to achieve the overall goal of swimmable and fishable water within the Ottawa River Watershed.

In May of 1995, as an extension to the Ottawa River Sediment Screening Survey, the Ohio EPA, together with the U.S. EPA, Great Lakes National Program Office, began conducting additional sediment screening survey investigative work in the Maumee AOC. The Maumee AOC Contaminated Sediment Screening Survey study area was separated into several distinct sampling zones based on historical data and characteristics of the locations of interest. This type of study area design allows for the prioritizing of sampling sites and aids in the tailoring of sampling plan to best serve the project objectives, which are;

- To determine the general distribution of PCBs, PAHs and metals in the study area of the Maumee Bay, in the depositional areas of the tributaries of the bay and the lower mainstem of the Maumee River.
- To gather sufficient sediment screening data to determine the presence of concentrated "hot spots", particularly PCBs.

The information gained from this sediment screening survey will be used to determine the distribution and extent of sediment contamination in the mouths of the Ottawa River, Swan Creek, Otter Creek and Duck Creek and in the mouth and lower mainstream of the Maumee River and in the Maumee Bay.

The sediment core samples were collected using the U.S. EPA research vessel, Mudpuppy, core sampling apparatus and/or a vibracore device using the Ohio EPA sampling vessel. The sediment cores were/will be subsampled and analyzed primarily by immunoassay techniques for PAHs and PCBs and XRF methods for metals. Select subsamples were/will be collected for confirmational analyses. To date 108 sediment core samples have been collected, from which 215 subsamples were taken. The analytical results from this screening survey are unavailable at this time.

The results from these two screening surveys should not be considered conclusive; however, the information obtained will give the Ohio EPA a better idea of the impact of pollutant loadings from the Maumee River and the other tributaries on the Bay. In addition, these results will aid in allocating sampling resources for the most efficient use of the more costly laboratory analysis.

In conclusion, screening level analysis can play a valuable role in the environmental field. These inexpensive test kits will likely not replace conventional laboratory analysis, but may serve to complement the traditional methods. Prior to selecting a screening method it is essential to establish the project objectives or data quality objectives in order to assure that results which are obtained are of desired quality and of use to the overall project.

Literature Cited

1.   Methods 4020 and 4035, in Test Methods for Evaluating Solid Waste (SW-846), U.S. Environmental Protection Agency, Office of Solid Waste and Emergency Response, Washington, DC, 1995.

2.   U.S. Environmental Protection Agency Contract Laboratory Program Statement of Work for Organics Analysis OLMO1.0, U.S. Environmental Protection Agency, Analytical Operations Branch, Washington, DC, 1990.

3.   Ottawa River, Lucas County, Ohio Health Advisory Zone Sediment Screening Survey Workplan, Step One of Stage One ONLY, Ohio Environmental Protection Agency, Northwest District Office, Bowling Green, Ohio, 1994.

Chapter 15

Validation of an Immunoassay for Screening Chlorpyrifos-methyl Residues on Grain

Brian A. Skoczenski[1,4], Titan S. Fan[1,4], Jonathan J. Matt[1], J. Terry Pitts[2], and J. Larry Zettler[3]

[1]Millipore Corporation, 80 Ashby Road, Bedford, MA 01730
[2]Gustafson, Inc., 1400 Preston Road, Suite 400, Plano, TX 75093
[3]Agricultural Research Service, U.S. Department of Agriculture, 2021 South Peach Avenue, Fresno, CA 93727

A competitive, enzyme labeled, heterogeneous immunoassay (EnviroGard Chlorpyrifos-methyl Screening Kit) was validated for the rapid detection of residues of the post-harvest insecticide chlorpyrifos-methyl (CPM). Grain samples were treated with CPM in a manner simulating commercial field applications and were analyzed by the immunoassay and simultaneously by a PAM II instrumental method. Analysis was repeated at t = 30 days after storage of the grain under field conditions. The initial phase was performed in the developer's lab (ImmunoSystems, Inc.). This was duplicated at the United States Department of Agriculture Stored-Product Insects Research and Development Laboratory, Savannah, GA. Field trials were performed by individuals selected to represent the typical user. Recently, the U.S. EPA, Office of Prevention, Pesticides and Toxic Substances, Analytical Chemistry Branch independently validated the method.

Chlorpyrifos-methyl (O,O-dimethyl O-3,5,6-trichloro-2-pyridyl phosphorothionate) is a broad range insecticide. Post-harvest application prior to storage is indicated for pest protection. Because of the need to limit grain application to a single treatment and the numerous times that grain may change hands between harvest and ultimate use, there exists a need to analyze grain samples to determine if treatment is appropriate.

We have developed an immunoassay which can be used at remote locations without dedicated instruments for screening grain samples for the presence of CPM residues. Results can be interpreted visually and the assay can reliably detect residues at 0.25 mg/kg (ppm).

[4]Current address: Beacon Analytical Systems, 4 Washington Avenue, Scarborough, ME 04074

Grain samples were treated at 0, 0.1, 0.25, 1.0 and 6.0 ppm active ingredient.

Treatment levels were verified by analysis of three to five replicates of each sample using the PAM II instrumental method (acetone extraction and gas chromatography/electron capture detector). Five replicates of each treatment level were extracted and analyzed using the immunoassay method. Samples were then stored under ambient conditions for a period of thirty days, at which time both the GC and immunoassay analyses were repeated.

The study was designed and performed in accordance with Pesticide Assessment Guidelines promulgated by the US EPA in compliance with Good Laboratory Practices Standard (1).

Immunoassay Development

Antisera were raised in rabbits against a derivitized chlorpyrifos compound attached to a protein carrier. These antibodies were purified and immobilized on 12x75 mm polystyrene test tubes. A similar derivitized chlorpyrifos compound was covalently attached to the enzyme horseradish peroxidase.

The design criteria for the assay was a rapid (< 30 minute) protocol which could be utilized at remote locations with a minimum of dedicated equipment or specialized training. The assay needed to be able to identify grain samples which had been previously treated with CPM; defined as containing ≥ 0.25 ppm CPM residues.

A method of sample extraction was developed which involves volumetric measurement of the grain sample, addition of rubbing alcohol (70% isopropyl alcohol) and shaking for two minutes. Studies performed by gas chromatographic analysis of extracts demonstrated that this extraction technique was approximately 50% efficient in the removal of CPM residues.

The actual CPM content of the calibrator was adjusted to allow for an easily visible distinction between the calibrator and samples which contain ≥ 0.25 ppm CPM. Because of this necessity, the assay will also yield a high percentage of positive results for grain samples containing ≥ 0.1 ppm CPM.

Kits are packaged to contain all materials required for the test, except the rubbing alcohol required for the grain extraction.

The immunoassay protocol can be summarized as follows:

1. Add grain sample to 15 mL mark on extraction vial (≈ 10 grams).

2. Add rubbing alcohol to the 45 mL mark, cap and shake for 2 minutes.

3. Add 5 drops of assay diluent to the antibody-coated tubes.

4.   Add 3 drops of calibrator or sample extract to the appropriate tubes.

5.   Add 5 drops of enzyme conjugate to the tubes and allow to incubate for 10 minutes.

6.   Wash the tubes, add 10 drops of substrate and allow to incubate for 10 minutes.

7.   Interpret results by comparing color of sample tubes to color of calibrator tube: If a sample tube contains more color than the calibrator tube, then the sample is negative for CPM. If a sample tube contains less color than the calibrator tube, the sample is positive for CPM.

Immunoassay Performance

One consideration in the design and use of immunoassays is the limited dynamic range compared to many instrumental methods. The dynamic range is the range of analyte concentrations over which the assay will respond with a corresponding change in signal generated.   This range is determined by a number of factors including antibody affinity, antibody concentration, enzyme conjugate concentration and sample size.   Since all of these factors are static for the normal immunoassay user, the situation would be somewhat comparable to an instrumental method where the detector gain can not be adjusted. Figure 1 shows the response of the assay for chlorpyrifos-methyl.   Generally, the dynamic range is considered to be limited to 20 to 80% of the signal generated at 0 concentration. For the current assay, this would correspond to approximately 0.5 to 10 ppb.

While the standard grain treatment level is 6 ppm, the design criteria required the ability to easily distinguish 0.25 ppm concentrations on grain as positive while also easily distinguishing untreated grain as negative.   Visual interpretation requires greater differences in color between calibrators and samples for unambiguous results as compared to that required for interpretation with a photometer.   Based on the design criteria, the dynamic range of the assay and the sample extraction; the calibrator was set at 5 ppb CPM.   Figure 2 shows the location of the calibrator and the theoretical content of extracts from 0.1 ppm and 0.25 ppm grain samples.   It is obvious from the response curve that the assay is not be able to distinguish, for instance 6 ppm from 1 ppm, but for this application, the important factor is that all of these can visually be distinguished as "positive" for chlorpyrifos-methyl.   A result of this approach is that the assay will potentially yield positive results for grain samples containing less than 0.25 ppm.   Theoretically, the assay would be able to identify grain samples with concentrations as low as 0.03 ppm as positive for CPM.

Grain Treatment.

The test commodity was corn grain, variety Funk's Blend, which had been harvested and stored without the application of post-harvest insecticides. The test substance was Reldan 4E and was GLP certified to contain 43.2% active ingredient.

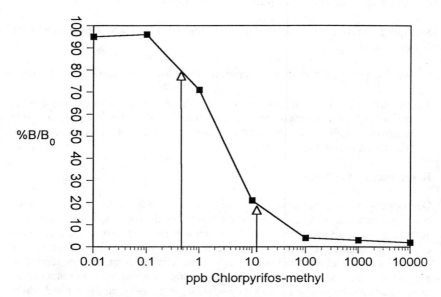

Figure 1. Dynamic Range for Chlorpyrifos-methyl

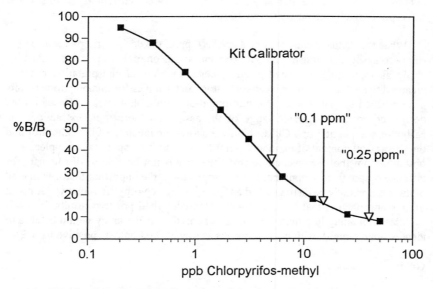

Figure 2. Theoretical Contents of Extracts for Grain Samples

The analytical standard was GLP certified to be 99.8% pure.

Ten pound samples of the test commodity were treated in a manner to simulate commercial field application. In addition to an untreated blank, samples were treated at 0.10, 0.25, 1.0 and 6.0 ppm (6.0 ppm is the standard application rate). The samples were split into two 5 pound aliquots and one was immediately frozen for shipment to the USDA facility. These frozen aliquots were shipped on dry ice to the USDA facility and were reportedly received frozen and in good condition. In addition to the treated grain samples, a vial of the analytical standard was shipped in a separate container.

Gas Chromatographic Analysis

Grain samples were analyzed for chlorpyrifos-methyl according to a modification of the method of Kuper, 1979 entitled "Determination of Residues of Chlorpyrifos-methyl in Grains" (DowElanco; unpublished).

Briefly, 10 g samples of grain were weighed and ground with 40 mL of acetone using a high-speed blender. The blended grain/acetone was then homogenized for 1 minute and mixed overnight on an orbital mixer at 200 RPM. The extracts were then centrifuged prior to injection.

Gas Chromatograph Conditions:

Column:	15m x 0.53mm methyl phenyl cyano silica
Detector:	ECD
Carrier Gas:	Helium @ 7 mL/min.
Make Up Gas:	Argon/Methane (95:5) @ 50 mL/min.
Oven Temperature:	140 - 260°C @ 8°C/min.
Injector Temperature:	250°C
Detector Temperature:	350°C
Injection Volume:	1 μL
Scale:	1 Volt

Fortified samples were analyzed prior to analysis of treated grain samples to demonstrate the functionality of the analytical method. In the initial phase, fortifications were at five levels and four replicates were analyzed at day 0 and again at day 30. Recoveries ranged from 92.5% to 103% and averaged 97.8%. In the validation phase, fortifications were at three levels and three replicates were analyzed at day 0 and day 30. Recoveries ranged from 95.6% to 109% and averaged 104%. This data are summarized in Table I.

In the initial phase, five replicates of each of the treatment levels were analyzed. In the validation phase, three replicates of duplicate samples were analyzed. Analysis was performed at Day 0 and repeated at Day 30. For the initial phase, recoveries averaged 88% at day 0 and 76% at day 30.

Table I - Recovery of Chlorpyrifos-methyl from Fortified Samples

I) Initial Phase (Average of four replicates)

	Recovery at Day 0		Recovery at Day 30	
ppm Added	ppm	%	ppm	%
0.008	0.0074	92.5	0.0082	102
0.04	0.041	102	0.0038	95.0
0.2	0.206	103	0.196	98.0
1.0	0.979	97.9	0.975	97.5
2.0	1.874	93.7	1.924	96.2

II) Validation Phase (Average of three replicates of duplicate samples)

	Recovery at Day 0		Recovery at Day 30	
ppm Added	ppm	%	ppm	%
0.137	0.131	95.6	0.150	109
1.37	1.45	106	1.44	105
6.56	6.73	103	6.78	103

Table II - Parts per Million of Chlorpyrifos-methyl on Treated Grain Samples

I) Initial Phase

Sample #(ppm)	Target Rate (ppm)	Day 0 (ppm)[1]	Day 30(ppm)
921025-14	0	<0.008	<0.008
921025-15	0.1	0.09	0.08
921025-16	0.25	0.21	0.17
921025-17	1.0	0.88	0.72
921025-18	6.0	5.63	5.24

II) Validation Phase

Sample #(ppm)	Target Rate (ppm)	Day 0 (ppm)[2]	Day 30(ppm)
921025-19	0	<0.03	<0.03
921025-20	0.10	0.12	0.06
921025-21	0.25	0.23	0.16
921025-22	1.0	0.98	0.78
921025-23	6.0	7.24	4.19

1 - Average of five replicates
2 - Average of three replicates of duplicate samples

For the validation phase, recoveries averaged 107% at day 0 and 69% at day 30. Recoveries at day 30 were lower than day 0 since, over time, CPM metabolizes to trichloropyridonal which is not of toxicological concern and is not measured in the analytical method. The results of these analyses are summarized in Table II.

Immunochemical Analysis

For both the initial and validation phases five replicates of each of the five treatment levels were analyzed at Day 0 and Day 30. Results were interpreted both visually and with the use of an optional differential photometer. Interpretation is based on the comparison of the amount of color generated in a sample tube compared to the amount of color generated in the calibrator tube. Sample tubes which contain more color than the calibrator tube are considered negative and sample tubes containing less color than the calibrator tube are considered positive. The differential photometer is a battery-powered unit which displays the difference in absorbance between a calibrator tube and a sample tube. A positive value on the photometer indicates a negative result and a negative value indicates a positive result.

In the initial phase, all 20 determinations of the four non-zero treatment levels were identified as positive and the five determinations of the zero treatment level were identified as negative by both visual and instrumental interpretations. The results were identical for the validation phase with the single exception that for one of the five replicates of the 0.1 ppm treatment level, the color of the tube was indistinguishable from the calibrator though the instrumental interpretation yielded the expected positive result. These results are summarized in Table III and IV.

Field Demonstration.

Nine operators were selected to be representative of potential users of the immunoassay. Each operator received four treated grain samples (0, 0.25, 1.0 and 6.0 ppm treatment levels) without knowledge as to the chlorpyrifos-methyl content and EnviroGard Chlorpyrifos-methyl Screening Kits. None of the users had any previous knowledge of or experience with immunochemical analysis.

Millipore Corporation prepared a 17 minute training video which described and illustrated the steps involved in the grain extraction and immunoassay procedures. A copy of this video and the standard kit package insert were the only training that the operators received. The optional instrumental interpretation was not used and only visual interpretations were recorded. Of the nine operators, all operators were able to identify all of the controls and treated samples properly. The data are summarized in Table V.

EPA Validation.

The Analytical Chemistry Branch of the Office of Prevention, Pesticides and Toxic Substances undertook an independent validation of the immunoassay.

Table III - EnviroGard™ Chlorpyrifos-methyl Screening Kit Results for the Initial Phase

Day 0 Determinations

Samples #	Treatment Level (ppm)	Visual Interpertation	Photometer Interpertation
921025-14	0.0	Negative	Negative
921025-15	0.10	Positive	Positive
921025-16	0.25	Positive	Positive
921025-17	1.0	Positive	Positive
921025-18	6.0	Positive	Positive

Day 30 Determinations

Sample #	Treatment Level (ppm)	Visual Interpertation	Photometer Interpertation
921025-14	0.0	Negative	Negative
921025-15	0.10	Positive	Positive
921025-16	0.25	Positive	Positive
921025-17	1.0	Positive	Positive
921025-18	6.0	Positive	Positive

Table IV - EnviroGard™ Chlorpyrifos-methyl Screening Kit Results for the Validation Phase

Day 0 Determinations

Sample#	Treatment Level (ppm)	Visual Interpertation	Photometer Interpertation
921025-19	0.0	Negative	Negative
921025-20	0.10	Positive	Positive
921025-21	0.25	Positive	Positive
921025-22	1.0	Positive	Positive
921025-23	6.0	Positive	Positive

Day 30 Determinations

Sample #	Treatment Level (ppm)	Visual Interpertation	Photometer Interpertation
921025-19	0.0	Negative	Negative
921025-20	0.10	Positive[1]	Positive
921025-21	0.25	Positive	Positive
921025-22	1.0	Positive	Positive
921025-23	6.0	Positive	Positive

1 - One of five replicates was visually indistinguishable from the calibrator.

Table V - Field Trial Results

Grain Treatment Level (ppm)	Aggregate Results
0	9 Negative/0 Positive
0.25	0 Negative/9 Positive
1.0	0 Negative/9 Positive
6.0	0 Negative/9 Positive

They were supplied corn grain samples which had been treated with chlorpyrifos-methyl at 0.0, 0.1, 0.25, 1.0 and 6.0 ppm; EnviroGard Chlorpyrifos-methyl Screening Kits; and a copy of the training video. After gas chromatographic analysis of the samples to confirm the treatment levels, three different chemists analyzed the samples using the immunoassay method. All three operators produced negative results for the controls and positive results for all of the treated samples.

Based on this independent validation and data submitted from the initial and validation phase, the EPA reviewer stated:

"..... immunoassay test for residues of chlorpyrifos methyl in corn grain has undergone successful EPA method validation. Residues in excess of 0.1 ppm can be rapidly detected by this method. They conclude that using the screening kit, one person can extract and visually analyze 6 whole grain corn samples for the presence of CPM (0.10 ppm) in less than 45 minutes."

Conclusion

An immunoassay for the detection of chlorpyrifos-methyl residues on grain (EnviroGard Chlorpyrifos-methyl Screening Kit) has been successfully validated. Initial validation was at the developers site and was supported by independent validation at two other sites. In addition, successful field trials demonstrated that with a minimum of training, persons with no prior experience in immunoassays could utilize the assays in remote locations. The method provides a rapid and inexpensive means to identify grain which has had prior treatment with chlorpyrifos-methyl.

Literature Cited

1. U.S. Code of Federal Register, Federal Regulations, Vol. 54, No. 158, 1989, U.S. CFR, Washington, D.C.

Chapter 16

An Evaluation of a Microtiter-Plate Enzyme-Linked Immunosorbent Assay Method for the Analysis of Triazine and Chloroacetanilide Herbicides in Storm Runoff Samples

Michael L. Pomes[1,3], E. M. Thurman[1], and Donald A. Goolsby[2]

[1]U.S. Geological Survey, 4821 Quail Crest Place, Lawrence, KS 66049
[2]U.S Geological Survey, Denver Federal Center, Building 25, Lakewood, CO 80225

Nine river sites in the midwestern United States were monitored with automatic samplers to assess temporal trends of herbicide concentrations in 1990. Microtiter-plate ELISA was chosen to detect triazines and chloroacetanilides in 1,725 storm runoff samples and to select 363 samples for confirmatory analysis by GC/MS. Evaluations of cross reactivity found that the 5-isopropyl secondary amine group determined the reactivity of the triazine ELISA, and the (methoxymethyl)acetamide group determined the reactivity of the chloroacetanilide ELISA. With a slope of 1.0 determined by least squares analysis with GC/MS results, and sensitivity and yield approaching 100 percent (found by Bayes's rule), the triazine ELISA accurately predicted atrazine concentrations in storm runoff samples. The chloroacetanilide ELISA was more problematic because of the finding of slopes greater than 1.0 and a specificity approaching 0.0 percent. Both indicated false-positive detections due to cross reactivity with a similarly-structured metabolite.

Storm-runoff at nine river sites in five Midwestern States was sampled to intensively monitor the temporal variation in herbicide concentrations during April to August 1990 (1, 2). Stage-activated samplers collected water samples for herbicide analysis during base flow and storm-runoff. An average of 200 samples were collected at each site (Michael Meyer, U.S. Geological Survey, oral commun., 1990) which emphasized the impracticability of analyzing a projected total of 1,800 samples by gas chromatography/mass spectrometry (GC/MS). A screening method was needed to select samples for GC/MS analysis.

The need to analyze large numbers of samples for triazine and chloroacetanilide herbicides inexpensively led to the selection of microtiter-plate, enzyme-linked immunosorbent assay (ELISA) for the analysis of base-flow and storm-runoff samples. Microtiter-plate ELISA analyses cost approximately one-fifth of GC/MS analyses.

[3]Current address: U.S. Geological Survey, 301 West Lexington Street, Independence, MO 64050

The purpose of this paper is to examine the utility of microtiter-plate ELISA as a semiquantitative screen for the GC/MS analysis of triazine and chloroacetanilide herbicides. The scope of this paper provides a framework for the evaluation of immunochemical techniques. This framework includes a discussion of cross reactivity of selected triazine and chloroacetanilide compounds with the antibodies used for ELISA methods. The discussion of cross reactivity should include consideration of which functional groups on the target analytes the antibodies bind with. This paper will also present the comparison of ELISA results with GC/MS results and an application of Bayes's rule for the evaluation of microtiter-plate ELISA as a screening method. Bayes's rule is a common technique used to evaluate the effectiveness of screening methods in the clinical sciences (3).

Experimental Methods

Sample Collection. Sample collection is detailed in (2). Briefly, water samples for the storm-runoff study were collected with automated samplers installed at nine U.S. Geological Survey (USGS) streamflow-gaging stations. The automated samplers were programmed to collect base flow samples every 2 days between storms. When a storm occurred and streamflow increased, the samplers were programmed to collect streamwater at specific time intervals or when streamflow changed. Samplers were serviced as frequently as twice a week at which time bottles were unloaded from the samplers in sequence and given sequence numbers that corresponded to the time interval between samplings.

All herbicide samples were filtered to remove particulates and colloids with 0.7-μm glass fiber filters. The filtrate was collected in precleaned 125-mL, amber-glass bottles. Sample bottles were chilled to 4° C, placed in insulated coolers, and sent to the U.S. Geological Survey (USGS) laboratory in Lawrence, Kansas.

Preparation of Standards. The use of firm, brand, or trade names is for identification purposes only and does not constitute endorsement by the U.S. Geological Survey. Atrazine (obtained from Supelco Bellefonte, PA) and alachlor (obtained from Environmental Protection Agency (EPA) Pesticide Chemical Repository, Research Triangle Park, NC) standards were prepared by spiking appropriate volumes of 1-mg/mL stock solutions into 100 mL volumes of organic-free water to yield 0.1-, 0.5- and 5.0-μg/L concentrations. The organic-free water is deionized and filtered with activated carbon prior to distillation. The organic-free water also was used as the blank. The atrazine, propazine, prometon, and simazine stocks used in the evaluation of cross reactivity were obtained from Supelco; cyanazine stock was obtained from the EPA Pesticide Chemical Repository, and triazine metabolite stocks, deethylatrazine, and deisopropylatrazine were obtained from Ciba-Geigy (Greensboro, NC). The chloroacetanilide herbicide and metabolite stocks were obtained from Monsanto Agricultural Company (St. Louis, MO).

Microtiter-Plate Method. Microtiter-plate ELISA procedures are detailed in (4). Briefly, all samples were analyzed for atrazine and alachlor using Res-I-Quant immunoassay kits (Immunosystems, Scarborough, ME). The immunoassay kits used for the ELISA analyses consisted of polystyrene microtiter plates (96 200 μl-wells to a plate) that are coated with polyclonal antibodies. An 80-μL aliquot of sample or standard was transferred to each well of the microtiter plate according to the plate template printout developed with the Softmax operating software of the VMAX microtiter-plate reader (Molecular Devices, Palo Alto, CA). Standards were placed on the plate in triplicate and samples in duplicate. After the addition of 80-μL aliquots of enzyme conjugate to all wells on the microtiter plate, the plate was covered with a paraffin film and allowed to incubate for 1 hour at room temperature 25° C while being shaken at 180 rpm on an orbital shaker. During the incubation period, atrazine or alachlor molecules in the sample and the enzyme conjugate competed for antibody binding sites. After 1 hour, the plate was emptied, flushed five times with organic-free deionized water, and dried. Next, 160 µL of a substrate and chromogen mixture was transferred to each well using a 12-channel pipette. The plate was covered with paraffin film and incubated at 25° C for 30 minutes while being shaken at 180 rpm. Finally, 40 µL of 2.5 N (normal) sulfuric acid was added to stop the reaction.

Results were quantified with three solutions of known atrazine or alachlor concentrations that ranged from 0.1 to 5.0 µg/L and blanks (no herbicides). Using the calibration curves, optical densities associated with calibration standards were measured. Wells with optical densities producing calculated values greater than 5 percent different than actual standard values were deleted from the plate template. The optical-density of a well was not determined if it was deleted from the template. Following this operation, the calibration curve was recalculated by rereading the plate. Samples were analyzed in duplicate, and the results averaged. The quantitation limits for ELISA were 0.1 µg/L for atrazine and 0.2 µg/L for alachlor. When concentrations found by ELISA exceeded the maximum standard, samples were reanalyzed as dilutions, ranging from 1:2 to 1:10.

Evaluation of Cross Reactivity. Organic-free water blanks and herbicide and metabolite standards in increasing concentrations were analyzed in triplicate on a single row of a microtiter plate, with each row devoted to individual standard curves for the analytes of interest. Optical-density values were measured and processed as individual standard curves. From the optical-density data, values of B_o (blank) and B (absorbance due to standard) were determined. The IC_{50} (inhibition concentration obtained at 50-percent of maximum absorbance) obtained at $B/B_o = 0.5$, measured the sensitivity of the immunoassay test. Lower IC_{50} concentrations generally result because the ELISA is sensitive to the particular analyte. The least-detectable dose (LDD) measured the lowest quantified concentration and was obtained at $B/B_o = 0.9$.

Gas Chromatography/Mass Spectrometry Procedure. For all sampling sites, water samples delineating highs and lows in triazine and chloroacetanilide concentrations determined by ELISA were chosen for GC/MS analysis. These samples were analyzed by GC/MS according to the method described by (5,6). Briefly, the method involved extraction of herbicides from 125-mL water samples with C_{18} (carbon-18) cartridges, elution of the cartridges, and spiking of extracts with phenanthrene-d_{10} using an automated workstation. The volume of the extract containing the herbicides was reduced to 100 µL with a stream of nitrogen. Concentrated extracts were analyzed using a Hewlett Packard (Palo Alto, CA) Model 5890 gas chromatograph and a Model 5970 mass-selective detector. Quantitation of herbicides was based on the analyte response ratioed to the response generated by the ion fragment of the internal standard, phenanthrene-d_{10}, that gives a mass-to-charge ratio of 188 mass units. Confirmation of analytes relied on the detection of the molecular ion, two confirming ions, and matching retention times relative to the internal standard. The quantitation limit for GC/MS analysis was 0.05 µg/L for all analytes of interest.

Method for Application of Bayes's Rule to ELISA Results. Bayes's rule provides a means to evaluate the operating performance of medical screening or diagnostic procedures (3). Bayes's rule determines the yield or predictive value of a screening test using the following equation adapted from (3):

$$\text{yield} = \frac{(\text{prevalence})(\text{sensitivity})}{(\text{prevalence})(\text{sensitivity}) + (1\text{-prevalence})(1\text{-specificity})} \quad (1)$$

where

prevalence = probability that the condition in population exists given that the screening procedure yielded positive result,

sensitivity = probability that the condition <u>would be indicated</u> by the screening procedure and subsequently confirmed positive,

specificity = probability that condition <u>would not be indicated</u> by the screening procedure and subsequently confirmed negative, and

(1-specificity) = probability that the condition <u>would not be indicated</u> by the screening procedure but <u>would be indicated</u> by confirmatory method resulting in a false-negative.

Results of a tabulation of confirmed positives, confirmed negatives, false-positives, and false-negatives are placed in a matrix (Table I). Formulas for the calculation of yield, sensitivity, specificity, false-positive, and false-negative rates are provided below the table.

Table I. Bayes's rule matrix for evaluation of ELISA screens for herbicide analytes modified from (3).

	Herbicide detected by GC/MS	Herbicide not detected by GC/MS	Total
ELISA positive	a	b	a + b
ELISA negative	c	d	c + d
Total	a + c	b + d	N

Explanation: a = confirmed positive; b = false-positive; c = false-negative; d = confirmed negative; and N = number of samples.
Reprinted with permission from reference 3. Copyright 1985. Prentice-Hall.

Formulas to calculate prevalence rate, sensitivity, specificity, false-positive rate, false-negative rate, and yield: prevalence rate = $(a + c)/N$; specificity = $d/(b+d)$; false-negative rate = $c/(a + c)$; sensitivity = $a/(a + c)$; false-positive rate = $b/(b + d)$; and yield = $a/(a + b)$.

Selected samples delineating highs and lows in triazine and chloroacetanilide concentrations found by microtiter-plate ELISA were confirmed by GC/MS. Tabulations were performed to find numbers of confirmed positives (detections by ELISA and GC/MS), confirmed negatives (nondetections by ELISA and GC/MS), false-positives (detection by ELISA, but nondetection by GC/MS), and false-negatives (nondetection by ELISA, but detection by GC/MS).

Results and Discussion

Cross Reactivity. The antibody used in the triazine ELISA procedure cross reacts with other related herbicides because of the affinity antibodies have for particular molecular structures; thus the test is nonspecific (7). Other workers reporting on triazine cross reactivity include (8,9). The Immunosystems antibody has the most affinity for atrazine, as signified by the lowest IC_{50} concentration (0.41 µg/L) and LDD (0.02 µg/L in Figure 1) Probable points of attachment for the antibody include the 5-isopropyl secondary amine and 3-ethyl secondary amine groups as well as a 1-chloro group. The difference in cross reactivity between deethylatrazine and deisopropylatrazine shows that the 5-isopropyl secondary amine group exerts significant control on the binding of antibodies. Decreased cross reactivity was observed for propazine, prometon, simazine, and deethylatrazine, which have molecular structures slightly different than atrazine. Alachlor evoked no response from the triazine ELISA.

The antibodies used for the chloroacetanilide ELISA procedure also demonstrates cross reactivity with metolachlor and some degradation products such as the ones used by (10). The antibodies used by Immunosystems had the highest affinity for alachlor as signified by the lowest IC_{50} concentration and LDD (Figure 2). Cross reactivity for chloroacetanilides appears to be based on the presence of (methoxymethyl)acetamide groups. The antibodies used by Immunosystems have the next highest affinity for the ethanesulfonic-acid metabolite of alachlor. Alachlor ethanesulfonic-acid (Figure 2) has been identified as the analyte responsible for producing the false-positive detections with the chloroacetanilide ELISA (11-13). Atrazine evoked no response from the chloroacetanilide ELISA. Alachlor ethanesulfonic-acid cannot be detected by the GC/MS because the metabolite occurs as an anion in solution. High-performance liquid-chromatography methods have been developed for the analysis of the metabolite (11, 14).

Comparison of ELISA to GC/MS. Results of ELISA analyses for triazine and chloroacetanilide herbicides in 1,725 storm runoff samples are listed in (2). Tables II and III lists the results of least squares analyses performed on a total of 360 triazine ELISA analyses confirmed by GC/MS and a total of 363 chloroacetanilide ELISA analyses confirmed by GC/MS, roughly 20 percent of the entire data set. Separate least squares analyses were completed to compare concentrations of atrazine determined by ELISA to concentrations of atrazine and total triazines [sum of triazine herbicides determined using the methods of (5,6)] determined by GC/MS in storm runoff samples collected from the 9 study sites (Table II). Similarly, separate least squares analyses were completed to compare concentrations of alachlor determined by ELISA to concentrations to concentrations of alachlor and total chloroacetanilides (alachlor plus metolachlor) determined by GC/MS (Table III).

Comparison of triazine ELISA results to GC/MS results yielded a median slope of 1.01 and a median correlation coefficient (r^2) of 0.52 for atrazine and a median slope of 0.56 and a median r^2 of 0.59 for total triazines (Table II). A similar comparison for chloroacetanilides yielded a median slope of 1.59 and a median r^2 of 0.54 for alachlor, and a median slope of 0.75 and a median r^2 of 0.58 for total chloroacetanilides (Table III). Generally, least squares analysis yielded greater slopes for the comparison of ELISA results to GC/MS determinations of atrazine and alachlor than respective totals of triazines and chloroacetanilides.

Cross reactivity can explain this decrease in slope. In the case of the triazine herbicides, decreased cross reactivity originated from the loss of the 5-isopropyl secondary amine group. Significantly greater concentrations of triazine herbicides and metabolites lacking this group were required to produce equivalent absorbances at

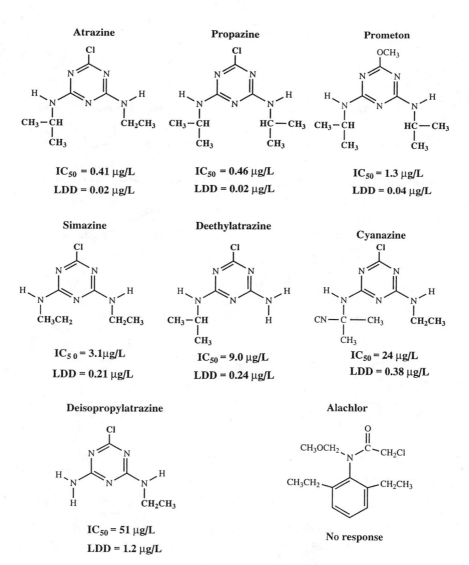

Figure 1. Selected triazine herbicides and metabolites in order of decreasing cross reactivity. Alachlor is included for comparison purposes. Explanation: IC_{50} = 50-percent inhibition concentration; LDD = least detectable dose; μg/L = micrograms per liter.

Alachlor

$IC_{50} = 0.87\ \mu g/L$

$LDD = 0.03\ \mu g/L$

Alachlor ethanesulfonic-acid

$IC_{50} = 1.7\ \mu g/L$

$LDD = 0.05\ \mu g/L$

Metolachlor

$IC_{50} = 26\ \mu g/L$

$LDD = 0.56\ \mu g/L$

Atrazine

No response

Figure 2. Selected chloroacetanilide herbicides and metabolites in order of decreasing cross reactivity. Atrazine is included for comparison purposes. Explanation: IC_{50} = 50-percent inhibition concentration; LDD = least detectable dose; $\mu g/L$ = micrograms per liter.

the 50-percent inhibition concentrations. This decreased cross reactivity became apparent with the smaller slopes in Table II which indicate that greater concentrations of triazine herbicides produced less response by the triazine ELISA method (4). Similarly, the addition of an isopropyl group to the (methoxymethyl)acetamide group of metolachlor leads to decreased cross reactivity (Figure 2) and the decreased slopes shown in Table III.

Table II. Results of least squares analysis performed to compare ELISA response to GC/MS response for analysis of water samples containing atrazine, selected triazine herbicides, deethylatrazine, and deisopropylatrazine. r^2 is the correlation coefficient.

River and State	Number	Atrazine Slope and Significance	Atrazine Correlation Coefficient	Total Triazines Slope and Significance	Total Triazines Correlation Coefficient
Iroquois (Illinois)	47	2.01***	0.780	1.05***	0.614
Sangamon (Illinois)	51	0.728***	0.436	0.466***	0.437
Silver Creek (Iowa)	29	0.770***	0.472	0.367***	0.349
Cedar River (Iowa)	43	1.24***	0.343	0.559***	0.220
Old Mans (Iowa)	51	1.03***	0.488	0.536***	0.586
Roberts (Iowa)	23	0.384***	0.884	0.258***	0.861
Delaware (Kansas)	30	0.753***	0.521	0.625***	0.508
Big Blue (Nebraska)	29	1.01***	0.648	0.826***	0.663
Huron (Ohio)	56	1.58***	0.731	0.820***	0.780

Level of Significance: *** -- $P < 0.001$.

Cross reactivity also explains the finding of slopes greater than 1.0 in Tables II and III. Least square analyses for the Iroquois, Silver Creek, Cedar, and Huron sites, and yielded slopes greater than 1.0 from instances that ELISA triazine analyses yielded concentrations two times or greater than GC/MS determinations of atrazine and total triazines in storm runoff samples (Table II). The ELISA method evidently cross reacted with some other triazine herbicide not included in the list of compounds listed in (2). Least squares analysis for the Iroquois, Silver Creek, Cedar River, Delaware, Big Blue, and Huron sites yielded slopes greater than 1.0 for comparisons of the results of chloroacetanilide ELISA analyses to GC/MS determinations of alachlor (Table III). Various sources (11-13) have suggested cross reactivity of the chloroacetanilide ELISA with alachlor ethanesulfonic-acid as the explanation for the disagreement between ELISA and GC/MS analyses.

   Another mechanism accounts for the finding of consistently low slopes for the least squares analysis of ELISA versus GC/MS results for storm runoff samples collected from Roberts Creek (Tables II-III). Slopes ranging from 0.268 to 0.384 were found because concentrations of atrazine and alachlor determined by GC/MS were much greater than concentrations of atrazine and alachlor determined by ELISA: 92 µg/L by GC/MS as opposed to 35 µg/L for atrazine, and 135 µg/L by GC/MS as opposed to 41 µg/L by ELISA for alachlor (2). Additional dilutions needed to be performed on these samples to find herbicide concentrations determined by ELISA that corresponded to those determined by GC/MS. This disparity between ELISA and GC/MS results suggests that the ELISA method has a threshold of saturation. Saturation occurs because each well contains a finite number of polyclonal antibodies bound to the inside of each 200 µL-well; samples containing larger concentrations overwhelm the antibodies because only a finite amount of atrazine or alachlor can bind with the antibodies (4). Careful and continued dilutions must be performed to assure that this level of saturation is not exceeded with the aim of finding concentrations of herbicides determined by ELISA that accurately represent those determined by GC/MS.

Table III. Results of least squares analysis performed to compare ELISA response to GC/MS response for analysis of water samples containing alachlor and chloroacetanilide herbicides. r^2 is the correlation coefficient.

River and State	Number	Alachlor Slope and Significance	Correlation Coefficient	Total Chloroacetanilides Slope and Significance	Correlation Coefficient
Iroquois (Illinois)	48	1.46***	0.510	0.572***	0.398
Sangamon (Illinois)	51	1.04***	0.540	0.399***	0.583
Silver Creek (Iowa)	32	2.46***	0.133	0.778***	0.913
Cedar River (Iowa)	43	1.96***	0.631	0.853***	0.605
Old Mans (Iowa)	51	0.922***	0.659	0.750***	0.674
Roberts (Iowa)	23	0.268***	0.467	0.281***	0.513
Delaware (Kansas)	30	2.00***	0.841	0.285***	0.585
Big Blue (Nebraska)	29	1.59***	0.452	1.33***	0.342
Huron (Ohio)	56	2.05***	0.605	0.798***	0.516

Level of Significance: *** -- $P < 0.001$.

Overall, the triazine ELISA effectively screened for atrazine in storm runoff samples; least squares analysis showed relatively good agreement between ELISA and GC/MS results with a median slope of 1.01 and a median correlation coefficient of 0.521 for atrazine concentrations less than 50 µg/L determined by GC/MS in storm runoff. However, the comparison of GC/MS to ELISA results for storm runoff samples show that allowances must be made for cross reactivity in finding instances of greatly enhanced ELISA response (slopes greater than 1.0) because of cross reactivity with analytes not found by GC/MS (alachlor ethanesulfonic-acid and the chloroacetanilide ELISA). Additionally, decreased cross reactivity caused by changes in functional groups resulted in the decreased slopes when least squares analyses were performed on total triazines and chloroacetanilides determined by GC/MS.

Application of Bayes's Rule. ELISA triazine and chloroacetanilide samples confirmed by GC/MS were tabulated to list the occurrence of confirmed positives, confirmed negatives, false-positives, and false-negatives. Out of a total of 360 samples, the triazine ELISA had the largest number of GC/MS confirmations of positives (357), no false-positives, and three false-negatives (Table IV). The three false-negatives represented respective detections of atrazine by GC/MS in the 0.13 to 0.16 µg/L range (2), close to the reporting limit of 0.10 µg/L for the triazine ELISA method.

The chloroacetanilide ELISA presented a different distribution of confirmed detections of positives and negatives (Table V). Of 363 samples, 23 false-positives were found in which alachlor as determined by GC/MS were not present in these samples. Concentrations of metolachlor as determined by GC/MS in these samples ranged from 0.08 to 1.7 µg/L. The chloroacetanilide ELISA is not sensitive to metolachlor as shown by an IC_{50} concentration of 26 µg/L and a least detectable dose of 0.56 µg/L (Figure 2) as opposed to an IC_{50} concentration of 0.87 µg/L and least detectable dose of 0.03 µg/L for alachlor. Thus, metolachlor concentrations between 0.08 and 1.7 µg/L found by GC/MS (Scribner and others, 1994) cannot account for the false-positive detections of alachlor determined by ELISA. The false-positives show that a compound, since found to be alachlor ethanesulfonic-acid (11-13), has cross reactivity with the chloroacetanilide ELISA, and is not detectable by GC/MS.

Table IV. Bayes's rule matrix for evaluation of ELISA screens for triazine herbicides.

	Herbicide detected by GC/MS	Herbicide not detected by GC/MS	Total
ELISA positive	357	0	357
ELISA negative	3	0	3
Total	360	0	360

Table V. Bayes's rule matrix for evaluation of ELISA screens for chloroacetanilide herbicides.

	Herbicide detected by GC/MS	Herbicide not detected by GC/MS	Total
ELISA positive	339	23	362
ELISA negative	0	1	1
Total	339	24	363

Table VI lists the prevalence rates, specificities, false-positive rates, sensitivities, false-negative rates and yields of the two ELISA methods under consideration. Of 360 samples confirmed for triazine herbicides by GC/MS, no confirmed negatives or false-positives were found (Table IV). With the exception of three false-negatives, the ELISA triazine method detected atrazine in nearly all the 1,715 samples of storm runoff listed in (2). No other negative detections of atrazine were identified. Accordingly, the specificity and false-positive rate could not be calculated (Table VI). Both the prevalence rate and yield of the triazine ELISA were 100 percent for the set of 360 samples. Because three false-negatives were found, the sensitivity equaled 99.2 percent, and the false-negative rate was 0.8 percent (Table VI). With GC/MS detections of atrazine in the 0.13- to 0.16-µg/L range representing thesefalse-negatives, the false-negative rate of 0.8 percent does not represent a drawback to the triazine ELISA method.

Table VI. Prevalence rate, sensitivity, specificity, false-positive rate, false-negative rate and yield for triazine and chloroacetanilide herbicides. [undef., Undefined due to division by zero.]

Variable	Triazine herbicides (percent)	Chloroacetanilide herbicides (percent)
Prevalence rate	100	93.3
Specificity	undef.	4.2
False-positive rate	undef.	95.8
Sensitivity	99.2	100
False-negative rate	0.8	0
Yield	100	93.6

In contrast, the chloroacetanilide ELISA yielded 23 false-positives and one confirmed negative (Table V) out of 363 samples confirmed by GC/MS. With one confirmed negative, specificity equaled 4.17 percent, and the false-positive rate equaled 95.8 percent. The false-positive rate resulted because 23 out of 24 analyses found to be negative by GC/MS were false-positive detections by the chloroacetanilide ELISA. The chloroacetanilide ELISA also yielded no false-negatives so a sensitivity of 100 percent resulted. Overall, the chloroacetanilide ELISA made positive detections in 1,724 out of 1,725 storm runoff samples (2).

Using Bayes's rule, the yield gives the predictive value of a screening procedure in which number of confirmed positives is divided by the sum of confirmed positives and false-positives. The triazine ELISA did not yield any false-positives out of the 357 positive detections under consideration; therefore a 100-percent yield was obtained. Of 339 positive detections found by the chloroacetanilide ELISA, 23 were false-positives, so a yield of 93.6 percent resulted. The prevalence rate denotes probability that the condition in population exists given that the screening procedure yielded positive result and is calculated by dividing the sum of positive and false-negative detections by the total number of samples. (Negative detections are included in the total number of samples.) With the absence of false-positives and negative detections, the triazine ELISA yielded a prevalence rate of 100 percent. In contrast, the chloroacetanilide ELISA yielded a prevalence rate of 93.3 percent because false-positives were found. Thus, the finding of false-positives decreases the prevalence rate and yield of a screening procedure, and increases the false-positive rate.

Ideally, confirmation of a screening method should result in no false-positives and no false-negatives. Under Bayes's rule, the absence of false-positives means specificity is 100 percent and the false-positive rate is 0 percent, meaning all positive detections were confirmed. The absence of false-negatives means sensitivity is 100 percent and the false-negative rate is 0 percent, meaning all negative detections were confirmed. Additionally, the yield of an ideal screening method must equal 100 percent. Thus, the effectiveness of a screening method can be evaluated by how closely specificity, sensitivity, and yield approach 100 percent. Although specificity could not be calculated because of the absence of negative detections, the triazine ELISA was an effective screening method because both sensitivity and yield approached 100 percent. When sensitivity approaches 0 percent (false-positive rate of 100 percent), a problem with cross reactivity is indicated. The chloroacetanilide

ELISA is less effective because of the finding of positive detections not confirmed by GC/MS.

Conclusions

The purpose of this paper was to examine the utility of microtiter-plate ELISA as a semiquantitative screen for the GC/MS analysis of triazine and chloroacetanilide herbicides and to provide a framework for the evaluation of immunochemical techniques. This framework included a discussion of cross reactivity of selected triazine and chloroacetanilide compounds with the antibodies used for ELISA methods, the comparison of ELISA results with GC/MS results, and an application of Bayes's rule for the evaluation of microtiter-plate ELISA as a screening method.

Evaluations of cross reactivity for the two ELISA methods presented in this study determined that the 5-isopropyl secondary amine group determined the reactivity of the triazine microtiter-plate ELISA, and (methoxymethyl)acetamide groups determined the reactivity of the chloroacetanilide ELISA. Loss of the 5-isopropyl secondary amine group on triazine herbicides decreased cross reactivity. The presence of the (methoxymethyl)acetamide group on the ethanesulfonic-acid metabolite of alachlor prompted cross reactivity with the chloroacetanilide ELISA. Decreased cross reactivity was observed for metolachlor because of the addition of an isopropyl group to the (methoxymethyl)acetamide group. The finding of decreased cross reactivity due to the loss or gain of certain functional groups and false-positive detections show that detailed structural analysis should be considered in any discussion of cross reactivity, and the cross reactivity of any ELISA method should be investigated in detail.

The triazine ELISA effectively screens for atrazine in storm runoff samples based on a comparison of ELISA and GC/MS results. Least squares analysis showed relatively good agreement between ELISA and GC/MS results with a median slope of 1.01 and a median correlation coefficient of 0.521 for concentrations of atrazine determined by GC/MS to be less than 50 μg/L. Results presented in this study show that cross reactivity influenced the departure of slopes from 1.0. Slopes less than 1.0 were found when least squares analysis were performed on total triazines and chloroacetanilides determined by GC/MS. Decreased cross reactivity caused by changes in functional groups produced the decreased slopes. Cross reactivity also accounted for finding slopes greater than 1.0 when the ELISA method reacted with analytes not detected by GC/MS.

Use of Bayes's rule in this study showed that the effectiveness of a screening method can be evaluated by how closely specificity, sensitivity, and yield approach 100 percent. Although specificity could not be calculated because of the absence of negative detections, the triazine ELISA is then an effective screening method because sensitivity and yield approach or equal 100 percent. However, the chloroacetanilide microtiter-plate ELISA is less effective because specificity approaches 0 percent because of cross reactivity with a similarly structured metabolite. With a slope of 1.0 determined by least squares analysis; and sensitivity and yield approaching 100 percent, the triazine microtiter-plate ELISA accurately predicts the occurrence of atrazine in storm runoff samples.

Literature Cited

1. Brown, D.E.; Meyer, M.T.; Pomes, M.L.; Thurman, E.M.; Goolsby, D.A. *EOS.* **1990,** 71, 1331.
2. Scribner, E.A.; Goolsby, D.A.; Thurman, E.M.; Meyer, M.T.; Pomes, M.L. *Concentrations of Selected Herbicides, Two Triazine Metabolites, and Nutrients in*

Storm Runoff from Nine Stream Basins in the Midwestern United States; U.S. Geol. Surv. Open-File Rep., 1994; 94-396, 144 p.
3. Remington, R.D; Schork, M.A. *Statistics with Applications to the Biological and Health Sciences*; Prentice-Hall: Englewood Cliffs, NJ, 1985; 415 p.
4. Pomes, M.L.; Thurman, E.M.; Goolsby, D.A. *Comparison of Microtiter-Plate, Enzyme-Linked Immunosorbent Assay (ELISA) and Gas Chromatography/Mass Spectrometry (GC/MS) for the Analysis of Herbicides in Storm Runoff Sample;* Aronson, D.A.; Mallard, G.E., Eds., U.S. Geological Survey Toxic Substances Hydrology Program--Proceedings of the technical meeting, Monterey, California, March 11-15, 1991: U.S. Geol. Surv. Wat.-Res. Invest. Rep. 1991; 91-4034, p. 572-575.
5. Thurman, E.M.; Goolsby, D.A.; Meyer, M.T.; Mills, M.M.; Pomes, M.L.; Koplin, D.W. *Envir. Sci. & Technol.* **1992,** 26, 2440.
6. Meyer, M.T.; Mills, M.S.; and Thurman, E.M. *J. Chromat.* **1993,** 629, 55.
7. Bushway, R.J.; Perkins, B.; Savage, S.A.; Lekousi, S.J.; Ferguson, B.S. *Bul. Envir. Contam. & Toxic.* **1988,** 40, 647.
8. Dunbar, B.; Riggle, B.; Niswender, G. *J. Agric. & Food Chem.* **1990,** 38, 433.
9. Thurman E.M.; Meyer, Michael; Pomes, Michael; Perry, C.A.; Schwab, A.P. *Anal.Chem.* **1990,** 62, 2043-2048.
10. Feng, P.C.C.; Wratten, S.J.; Horton, S.R.; Sharp, C.R.; Logusch, E.W. *J. Agric. & Food Chem.* **1990,** 38, 159-163.
11. Macomber, C. et al. *J. Agric. & Food Chem.* **1992,** 40, 1450-1452.
12. Baker, D.B.; Bushway, R.J.; Adams, S.A.; Macomber, C. *Envir. Sci. & Tech.* **1993,** 27, 562-564.
13. Aga, D.S.; Thurman, E.M.; Pomes, M.L. *Anal. Chem.* **1994,** 66, 1495-1499.
14. Pomes, M.L.; Holub, D.F.; Aga, D.S.; Thurman, E.M., [in press], Isocratic *Separation of Alachlor Ethanesulfonic-Acid, Alachlor Oxoacetic Acid, and Hydroxyatrazine by Reversed-Phase Liquid Chromatography.* U.S. Geological Survey Toxic Substances Hydrology Program--Proceedings of the Technical Meeting, Colorado Springs, Colorado, September 20-24, 1993: U.S. Geological Survey Open-File Report.

Chapter 17

Evaluation of an Automated Immunoassay System for Quantitative Analysis of Atrazine and Alachlor in Water Samples

Barbara Staller Young[1], Andrew Parsons[1], Christine Vampola[2], and Hong Wang[3]

[1]Millipore Corporation, 80 Ashby Road, Bedford, MA 01730
[2]EnSys Inc., Royal Center, 4222 Emperor Boulevard, Morrisville, NC 27560
[3]Environmental Health Laboratories, 110 Hill Street, South Bend, IN 46617

Two commercially available immunoassay kits were evaluated in conjunction with an automated microplate system for the detection of atrazine and alachlor in water. The results showed that excellent precision and accuracy were achieved. Eighty six water samples could be analyzed in approximately 2.5 hours, including sample and instrument set-up time. This study supports the concept that immunoassay technology would be useful for rapid, accurate screening of water samples for the presence of various pesticides. The automated system would be particularly effective when large numbers of samples must be analyzed.

Immunoassay technology has proven to be a useful tool for performing rapid, accurate, and cost-effective screening of field samples on site or in the laboratory (1-4). The goal of this study was to establish the feasibility of screening large numbers of drinking water samples for the presence of pesticides using commercially available ELISA (enzyme linked immunosorbent assay) plate kits (EnviroGard™, Millipore Corporation) in conjunction with an automated microplate system (ELs 1000, Bio-Tek Instruments, Inc.). Two EnviroGard™ kits (for triazines, and alachlor) were evaluated and found to give excellent precision and accuracy for the detection of atrazine and alachlor in fortified Milli-Q and drinking water samples.

General Description of Immunoassay Kits

Both the kits for triazines and alachlor are based on a standard microtiter plate competitive ELISA format. Calibrators or samples are added to the wells, followed by an enzyme conjugate. After a one hour incubation during which time the analyte present in the sample competes with enzyme conjugate for antibody binding sites, unbound reagents are washed away. Following the wash step, substrate is added and color is allowed to develop for 30 minutes. Color development is halted with the addition of stop solution, and the results are read at a wavelength of 450 nm. The ELs 1000 automated system is designed to perform all of these steps including data analysis. Millipore and Bio-Tek scientists worked together to design software programs tailored to the EnviroGard[TM] assays. Using the ELISA kits with the automated system, two full microtiter plates can be run at a time, allowing for the analysis of up to 86 water samples in approximately 2.5 hours, including set-up time. Clean-up time between runs takes less than 30 minutes.

Performance of the Alachlor Assay

Table I shows the calibration curve for the alachlor plate kit. The assay is calibrated during each run by including five standards that are supplied in the kit. The dynamic range for this assay is 0.1-2.5 ppb alachlor. Typically, standards and samples are run in duplicate, and the absorbance values are converted to $\%B_0$ by dividing the non-zero calibrator absorbance values by the value for the 0.0 ppb calibrator. There are specified $\%B_0$ ranges delineated in the product inserts, serving as a quality control check for assay performance. For instance, the acceptable ranges for the alachlor product are 64-86% for the 0.1 ppb calibrator, 33-55% for the 0.5 ppb calibrator, and 11-21% for the 2.5 ppb calibrator. The precision for duplicate calibrators shown in Table I is quite good, with % coefficient of variation (%CV is defined as (standard deviation/mean) x 100) values ranging from 0.26 to 1.56.

The alachlor plate assay was found to give very reproducible results over time. In Table II, precision was determined by running the calibrators in duplicate on five separate days. The $\%B_0$ values over the 19 day period were very consistent, yielding %CVs from 1.15 to 2.53.

The accuracy and precision of the assay were determined by measuring recovery of alachlor spiked into Milli-Q water at 1.0 ppb and 0.1 ppb. Table III summarizes data where triplicate samples were run on 5 separate days. The calculated %CVs for alachlor concentrations were excellent, averaging just over 7% for both spiked levels. The recovery ranges were 93-123% for 0.1 ppb alachlor, and 96-120% for 1.0 ppb. These data demonstrate excellent accuracy and precision for alachlor determinations in spiked reagent grade water samples.

Table I

Alachlor Plate Kit

Calibration Curve on Automated System

Standard Value (ppb)	Absorbance (450)	$\%B_0$	Mean $\%B_0$	$\%CV$ of Absorbance
0.0	1.949			
0.0	1.939		100.00	0.26
0.10	1.478	76.03		
0.10	1.511	77.73	76.88	1.56
0.25	1.217	62.60		
0.25	1.212	62.35	62.47	0.29
1.00	0.722	37.14		
1.00	0.731	37.60	37.37	0.88
2.50	0.424	21.81		
2.50	0.417	21.45	21.63	0.58

Table II

Alachlor Plate Kit

Calibrator Precision for 5 Runs

Conc. (ppb)	6/20/94 $\%B_0$	6/21/94 $\%B_0$	6/28/94 $\%B_0$	7/6/94 $\%B_0$	7/8/94 $\%B_0$	Av. $\%B_0$	$\%CV$ of B_0
0.10	77.88	80.49	78.96	78.36	78.37	79.13	1.91
0.25	62.47	66.13	63.75	65.28	64.50	62.85	5.46
1.00	37.37	39.07	38.47	38.48	36.88	40.08	9.32
2.50	21.63	22.69	22.38	21.15	22.27	21.95	2.55

$\%B_0$ = (Absorbance of calibrator or sample/Absorbance of Negative control) X 100

$\%CV$ = (Standard Deviation/Mean) X 100

Table III

Alachlor Plate Kit

Summary of Accuracy and Precision Data

Samples (ppb)	Average Conc. (ppb)	Conc, %CV	% Recovery	Average % Recovery
0.1	0.111	7.72	93-123	111
1.0	1.068	7.38	96-120	107

$\%B_0$ = (Absorbance of calibrator or sample/Absorbance of Negative control) X 100

$\%CV$ = (Standard Deviation/Mean) X 100

n=5 triplicate samples per level

Finally, in order to measure the precision and accuracy of the assay in drinking water samples, 11 water samples known to be negative for alachlor by GC analysis were spiked with 1.0 ppb alachlor and tested in the automated plate assay. Determinations were made in triplicate, and as summarized in Table IV, the percent recoveries range from 102.4 to 118.9. When the calculated concentrations for each of the triplicates were compared, the agreement was excellent, with CVs falling within the range of 1.5 to 6.9. This demonstrates excellent precision and accuracy for the determination of alachlor in drinking water samples.

Performance of the Assay for triazines

In order to meet regulatory requirements in Europe, a highly sensitive triazine assay was developed to achieve detection levels below 0.1 ppb. European regulations set a limit of 0.1 ppb for any one compound, therefore the highly sensitive triazines assay was designed to detect atrazine levels ranging from 0.01 to 0.5 ppb. Table V shows the calibration curve using this triazines assay run on the automated system. The calculated $\%B_0$

values were all within the recommended ranges of 76-90% for the 0.01 ppb calibrator, 51-66% for the 0.05 ppb calibrator, 38-54% for the 0.1 ppb calibrator, and 20-27% for the 0.5 ppb calibrator. This information is included in the kit insert, and is used as a quality control check for assay performance. The calibrators were run in duplicate and gave excellent precision, with the %CVs for absorbance values ranging from 0.1 to 3.

The data in Table VI demonstrates the stability of performance specifications of the triazines high sensitivity plate kit when run on four separate occasions, covering a period of 2 weeks. The $%B_0$ values remain quite constant, yielding %CVs from 1.16 to 5.52.

Table IV

Precision and Accuracy of Alachlor Determinations in Fortified

Environmental Water Samples

Drinking Water Sample	Rep-1 (ppb)	Rep-2 (ppb)	Rep-3 (ppb)	Mean (ppb)	%CV	Mean % Recovery
1	0.957	1.055	1.134	1.049	6.9	104.87
2	1.167	1.207	1.193	1.189	1.39	118.90
3	1.172	1.131	1.048	1.117	4.62	111.70
4	1.105	1.016	0.993	1.038	4.65	103.80
5	1.111	1.092	1.073	1.092	1.42	109.20
6	1.091	1.000	1.056	1.049	3.57	104.90
7	1.070	0.968	1.059	1.032	4.43	103.23
8	1.065	0.997	1.016	1.026	2.79	102.60
9	1.191	1.020	1.170	1.127	6.76	112.70
10	0.957	1.028	1.087	1.024	5.19	102.40
11	1.008	1.042	1.040	1.030	1.51	103.00

$%B_0$ = (Absorbance of calibrator or sample/Absorbance of Negative control) X 100

%CV = (Standard Deviation/Mean) X 100

Table V

Triazines High Sensitivity Plate Kit
Calibration Curve on Automated System

Calibrator Value (ppb)	Absorbance (450 nm)	%B_0	Mean %B_0	Absorbance %CV
0.00	1.683			
0.00	1.611		100.0	3.09
0.01	1.366	81.70		
0.01	1.364	81.58	81.64	0.10
0.05	1.000	59.81		
0.05	1.013	60.59	60.20	0.91
0.10	0.854	51.08		
0.10	0.860	51.44	51.26	0.50
0.50	0.364	21.77		
0.50	0.367	21.95	21.86	0.58

Table VI

Triazines High Sensitivity Plate Kit
Calibrator Precision for 4 Runs

Cal. Conc. (ppb)	4/26/94	4/26/94	5/10/94	5/11/94	Average %B_0	%CV of Absorb.
0.01	82.66	83.54	81.86	80.97	82.26	1.16
0.05	61.44	61.47	60.09	56.60	59.90	3.31
0.10	51.07	49.93	49.37	46.99	49.34	3.02
0.50	24.38	24.98	22.43	21.89	23.42	5.52

In order to further challenge the automated assay, and mimic a real-life situation where actual water samples would be placed randomly throughout the plate, an experiment was designed to run 11 replicates of samples placed randomly over two entire microtiter plates. Assay standards were always placed in the first two columns of the plate, tap water or Milli-Q water samples spiked with 0, 10, 20, 100 or 500 ppt atrazine were randomly distributed throughout the remaining wells in the two full plates. The two plates were run simultaneously in the automated system. Accuracy and precision results are shown in Table VII. The %CV values for the concentrations are somewhat higher than seen in previous experiments, especially at the 10 ppt levels. Since %CV is calculated by dividing the standard deviation by the concentration, very low concentrations tend to make the number larger. If a %CV value of 28.2 is considered unacceptably high, then the detection limit of this assay could be set at 20 ppt rather than 10 ppt. This value is still 5 times lower than the European regulated level of 100 ppt.

Table VII

Triazines High Sensitivity Plate Kit

Accuracy and Precision of Fortified Samples Placed Randomly in Two

Microtiter Plates

Spiked Tap Water Samples

Fortification (ppt)	%CV of Absorbance	%CV of Concentration	% Recovery Range	Average % Recovery
10	4.4	28.2	76-210	127
20	3.9	16.9	75-145	110
100	4.7	11.2	96-137	115
500	2.8	3.4	99-111	105

Spiked Milli-Q Water Samples

10	2.9	23.5	57-110	84
20	2.9	16.7	60-105	79
100	3.5	8.4	90-121	108
500	3.7	4.7	97-113	105

Conclusion

In summary, this study demonstrates that commercially available immunoassay kits used in conjunction with an automated plate reader can accurately determine atrazine and alachlor levels in spiked Milli-Q and tap water samples. The accuracy and precision are comparable to that generated by GC analysis (based on work generated for atrazine submission to the USEPA). The automated plate system used in this study accommodates two full microtiter plates, allowing for approximately 86 water samples to be screened at a time if calibrators and samples are run in duplicate. There is no sample preparation required for water analysis. Approximately 30 minutes of hands-on time is necessary for sample and reagent handling and loading into the instrument, and the run time for two plates is about 2.5 hours. This work demonstrates the use of ELISA technology and automated plate systems for the rapid, accurate, and cost effective screening large numbers of water samples for the presence of pesticides.

Literature Cited

1. Bushway, R.J., Perkins, L.B., Fukal, L., Harrison, R.O., and Ferguson, B.S. *Arch. Environ. Contam. Toxicol.* **1991**, *21*, 365-370.
2. Bushway, R.J., Perkins, L.B., Savage, S.A., Lekousi, S.J., and Ferguson, B.S. *Bull. Environ. Contam. Toxicol.* **1988**, *40*, 647-654.
3. Bushway, R.J., Young, B.E.S., Paradis, L.R., and Perkins, L.B. *Journal of AOAC Int.* **1994**, *77*, 1243-1248.
4. Moody, J.A., and Goolsby, D.A. *Environ. Sci. Technol.* **1993**, *27*, 2120-2126.
5. Goolsby, D.A. Thurman, E.M., Clark, M.L., and Pomes, M.L. In *Immunoassays for Trace Chemical Analysis;* Editors, M. Vanderlaan, L.H., Stanker, B.E. Watkins, and D.W. Roberts; ACS Symposium Series 451; **1991**, pp 86-99.
6. Thurman, E.M, Goolsby, D.A., Meyer, M.T., and Kolpin, D.W. *Environ. Sci. Technol.* **1991**, *25*, 1794-1796.
7. Thurman, E.M., Goolsby, D.A., Meyer, M.T., Mills, M.S., Pomes, R.L., and Kolpin, D.W. *Environ. Sci. Technol.* **1992**, *26*, 2440-2447.
8. Thurman, E.M., Meyer, M., Pomes, M., Perry, C., and Schwab, A.,*Anal.Chem.* **1990**, *62*, 2043-2048.

Chapter 18

An Evaluation of a Pentachlorophenol Immunoassay Soil Test Kit

Alan Humphrey

Environmental Response Team, U.S. Environmental Protection Agency, 2890 Woodbridge Road, Edison, NJ 08837

Pentachlorophenol (PCP) has been used extensively as a preservative in the wood treating industry. Nearly 1,400 wood preserving sites exist in the United States, 56 appear on the National Priority List (NPL) and hundreds more may have been abandoned. This study was conducted to determine effective utilization of the EnSys, Inc. semi-quantitative immunoassay test kits for on-site screening where PCP soil contamination is a concern. Analytical results from the kit and a Gas Chromatograph/Flame Ionization Detector (GC/FID) instrument were compared to ascertain if the kits were performing to manufacturer's claims. The GC/FID data were compared to GC/Mass Spectrometer (MS) data to assure its quality. Statistical analyses were performed both on the kit and GC/FID data to compare extraction efficiencies, develop possible explanatory models, determine dilution errors, and identify sources of variation inherent in the kit itself. In order to identify possible sources of error in the kit, the time component and the quantitation ranges developed by the manufacturer that are associated with its operating procedures were examined. Errors associated with kit operation and temperature were also considered. Statistical analyses indicated a statistically significant (p = 0.03) higher extraction efficiency by the laboratory method vs. the kit method; no statistically appropriate model could fit the GC/FID with the kit data; no statistically significant difference (p = 0.28) was observed between the GC/FID and the GC/MS data; and a good linear relationship was evident between the GC/FID and GC/MS data (r^2 = 0.89). The EnSys, Inc. PCP semiquantitative immunoassay test kits appear to be an effective screening tool for PCP soil contamination determination when utilized properly. Recommendations are made to ensure more reliable data and improve the kit's performance and similar immunoassay field screening kits.

Pentachlorophenol (PCP) has been used for over 50 years in the preservation of wooden utility poles and pilings and contamination in soil is common on wood treating sites. A recent listing of the wood-treating industry indicated that nearly 1,400 wood-preserving sites exist in the United States, of which more than 700 are inactive. Fifty-six wood-preserving sites appear on the U.S. EPA Superfund National Priority List (NPL); hundreds more may also have been abandoned(1). EnSys, Inc. developed a semi-quantitative immunoassay-based analytical screening method designed to detect PCP in soil. The PENTA RISc™ soil test system was developed as an efficient way to locate and map the extent of PCP contamination, screen samples in the field prior to laboratory testing, measure the effectiveness of remediation technologies, and ensure that cleanup levels meet state and federal regulations(2). The immunoassay technique utilized by this test kit is known as ELISA (enzyme-linked immunosorbent assay).

The following study was conducted by the U.S. Environmental Protection Agency (U.S. EPA) Environmental Response Team (ERT) and its prime contractor, Roy F. Weston, Inc. under the Response Engineering and Analytical Contract (REAC). U.S. EPA/ERT/REAC have performed almost 800 tests with the EnSys PENTA RISc test kit at four sites over a 14-month period. The EnSys PENTA RISc test kits have been used during extent of contamination studies, removal activities, and treatability studies. The data set utilized in this study was collected during Comprehensive Environmental Response Compensation and Liability Act (CERCLA) removal activities at a former wood-treating site. The site was chosen due to the relatively uniform sandy matrix which comprise the unconsolidated deposits, the considerable size of the data set, and its wide range of contaminant concentrations.

The purpose of this study was:

● To determine the usefulness of the kit under actual field conditions.

● To determine and make recommendations on the proper use of the kit to meet different site objectives.

● To statistically determine the accuracy of the kit.

● To determine if there is error inherent in the kit.

Theory

EnSys test for PCP is based on an analytical method in which an antibody recognizes and binds to a specific chemical or antigen. An antibody is a protein which can be designed to attach to a small organic molecule such as PCP at very low concentrations (ppb range) with a high degree of specificity. An antibody is analogous to a lock in the sense that there is a unique antigen shape, or key, that fits into it. However, this shape can be repeated on different analytes causing cross-reactivity. The test is a competitive assay in which the specially designed antibodies will bind both with PCP molecules in the unknown sample and with molecules of a PCP-enzyme conjugate.

The conjugate reagent is an enzyme to which molecules of PCP have been chemically attached. PCP molecules in the sample compete with the PCP end of the conjugate reagent for a limited number of antibody binding sites. The greater the number of sample-derived PCP molecules relative to enzyme-attached PCP molecules, the larger the proportion of antibody binding sites that are occupied by PCP molecules originating from the sample. After an incubation period during which the competitive binding occurs, unbound PCP and PCP- enzyme conjugate molecules are washed away and color-change reagents are added. The enzyme part of the bound conjugate molecules catalyzes the oxidation of a colorless substance to a colored (blue) one. The reaction is stopped by addition of dilute sulfuric acid (blue solution turns yellow), and the results are interpreted in the EnSys photometer. Figure 1 illustrates this analytical process. The degree of color development at the end of the test is proportional to the number of PCP-enzyme conjugate molecules bound to the antibody sites. Since the developed color intensity is inversely proportional to the number of PCP molecules in the sample, the concentration of a chemical in an environmental sample can be easily determined*(3)*.

METHODOLOGY

Analytical

PENTA RISc TEST KIT

The EnSys, Inc. PENTA RISc test kit that was studied contained the following: 4 sample extraction jars each containing 20 mL of methanol capable of extracting 10 grams of soil; 20 antibody coated tubes which are inserted into the photometer to measure the absorbance/transmittance of the solution; 12 buffer tubes containing a buffer solution to which a 100-μL aliquot of diluted extract is added; 8 standard tubes which contain known concentrations of PCP to be used as a reference absorbance when a differential photometer is used; 1 500 mL wash bottle containing a buffered wash solution; and 3 dropper bottles containing reagents identified as substrate "A", substrate "B", and "STOP," which is used to halt the reaction of the enzyme with the substrates. Depending on the site specific level of interest, standard tubes are provided at several concentrations by the manufacturer, usually from 5 to 500 parts per million. After dilution of the sample extract and the color reaction, a comparison is made with several different standard tubes, providing an approximate range of the sample PCP concentration. For example, the sample tube is compared with two standard tubes, 10 and 100 ppm, on the differential photometer. A negative reading is obtained by comparison with the 10 ppm standard and a positive reading for the 100 ppm standard, indicating the sample concentration is between 10 and 100ppm.

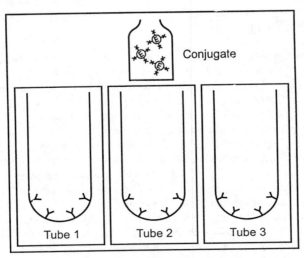

1. Components of ELISA Chemistry

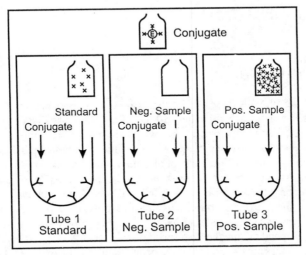

2. Enzyme Addition

Figure 1. EnSys PENTA RISc Analytical Process. (Reproduced with permission from ref. 4 Copyright 1991 Advanstar Communications, Inc.)

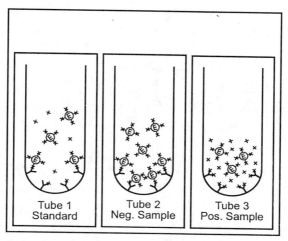

3. Incubation and Competitive Binding Reaction

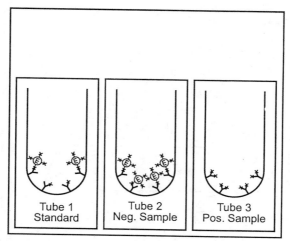

4. Wash

Figure 1. *Continued*

Continued on next page

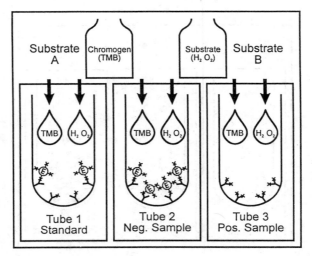

5. Color Development

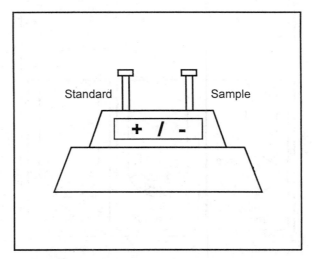

6. Read Sample

Figure 1. *Continued*

Photometers

Differential Photometer

The differential photometer used in conjunction with the EnSys PENTA RIS<u>c</u> Kit is a specific purpose photometer set at a 450 nanometer wavelength which gives an immediate direct comparison of the optical absorbance of two samples.

During analyses with the EnSys PENTA RIS<u>c</u> Kits, standards are supplied. Once the color development test is complete, both "standard" tubes are placed in the photometer, and the one with the greater amount of absorbance is used as the comparison standard. Then, samples are placed into the photometer with the single standard and the differential absorbance is read.

Spectrophotometer

Following analysis in the differential photometer, the samples were read in a Hitachi V-2000 Double-Beam Ultraviolet/Visible (UV/Vis) spectrophotometer. This spectrophotometer is used to analyze the samples against a calibration curve of standards run through the kits and transferred to a cuvet measured for absorbance at 450 nm. According to the Beer-Lambert Law, the absorbance of a given sample is proportional to the concentration of the analyte for a given absorption pathlength at any given wavelength*(5)*. This analysis will yield concentration values for a given sample and a measurable error.

GC/FID

Method 8270/SW-846 *(6)* was modified for field use, substituting a Gas Chromotograph/Flame Ionization Detector (GC/FID) for the Gas Chromatograph-Mass Spectrometer (GC/MS). An HP 5980 Series 2 Gas Chromatograph, equipped with a Flame Ionization Detector (FID) detector and controlled by a 3396A integrator was used to analyze the samples.

The instrument conditions were:

Column	Restek RTx-5 (crossbonded SE-54), 30 meter x 0.53mm ID, 0.50 μm film thickness
Injection Temperature	300°C
Detector Temperature	300°C
Temperature Program	70°C for 2 min., 8°C/min. to 285°C, hold for 15 min.
Injection Volume	2 μL

The GC system was calibrated using a 19 component creosote mixture and PCP at concentrations which ranged from 5 to 100 µg/mL (except for 2,4,6-tribromophenol which ranged from 25 to 500 µg/mL).

The concentration of the detected compounds was calculated using the following equation 1:

$$C_u = \frac{DF \times A_u \times V_t}{RF_{avg} \times V_i \times W \times D} \tag{1}$$

where:

$$
\begin{aligned}
DF &= \text{Dilution Factor} \\
RF_{avg} &= \text{Response Factor (unitless)} \\
A_u &= \text{Area of Analyte} \\
C_u &= \text{Concentration of Analyte (mg/kg)} \\
V_t &= \text{Volume of Extract (mL)} \\
V_i &= \text{Volume of Extract Injected (µL)} \\
W &= \text{Weight of Sample (g)} \\
D &= \text{Decimal Percent Solids}
\end{aligned}
$$

Response Factor Calculation

The response factor (RF) for each specific analyte is quantitated based on the area response from the continuing calibration check as follows in equation 2:

$$RF = \frac{A_c}{I_c} \tag{2}$$

where:

$$
\begin{aligned}
RF &= \text{Response Factor for a Specific Analyte} \\
A_c &= \text{Area of the Analyte in the Calibration Mixture} \\
I_c &= \text{Mass of the Analyte in the Calibration Mixture (ng)}
\end{aligned}
$$

An average of five values, RF_{avg} (at the five concentrations), was used.

GC/MS

An HP 5995C GC/MS equipped with a 7673A autosampler and controlled by an HP-1000 RTE-6/VM computer was used to analyze the samples (Modified Method 8270/SW-846) *(6)*.

The instrument conditions were:

Column		Restek RTx-5 (crossbonded SE-54), 30 meter x 0.32mm ID, 0.50 μm film thickness
Injection Temperature		290°C
Transfer Temperature	290°C	
Source Temperature	240°C	
Analyzer Temperature		240°C
Temperature Program		30°C for 3 min., 15°C/min. to 70°C, hold for 0.2 min.;8°C/min. to 295°C, hold for 12 min.
Splitless Injection		Split time = 60 sec.
Injection Volume		1 μL

The GC/MS system was calibrated using five Base, Neutral and Acid Extractable (BNA) standard mixtures at 20, 50, 80, 120, and 160 μg/mL. The calibration range was validated by evaluating the System Performance Check Compounds (SPCC) and the Calibration Check Compounds (CCC) as outlined in the Contract Laboratory Program (CLP) protocol. Before analysis each day, the system was tuned to decafluorotriphenylphosphine (DFTPP) and passed a continuing calibration check when analyzing a 50 μg/mL standard mixture in which the responses of the SPCC and CCC compounds were evaluated by comparison to the average response of the calibration curve.

The concentration of the detected compounds were calculated using the following equation 3:

$$C_u = \frac{DF \times A_u \times I_{is} \times V_t}{A_{is} \times RF \times V_i \times W \times D} \qquad (3)$$

where:

DF	=	Dilution Factor
RF	=	Response Factor (unitless)
A_u	=	Area of Analyte
A_{is}	=	Area of Internal Standard
I_{is}	=	Mass of Internal Standard (ng)
C_u	=	Concentration of Analyte (μg/Kg)
V_t	=	Volume of Extract (μL)
V_i	=	Volume of Extract Injected (μL)
W	=	Weight of Sample (g)
D	=	Decimal Percent Solids

Response Factor Calculation

The RF for each specific analyte is quantitated based on the area response from the continuing calibration check as follows in equation 4:

$$RF = \frac{A_c \times I_{is}}{A_{is} \times I_c} \qquad (4)$$

where:

RF = Response Factor for a Specific Analyte
A_c = Area of the Analyte in the Continuing Calibration Check
A_{is} = Area of the Internal Standard in the Continuing Calibration Check
I_c = Mass of the Analyte in the Continuing Calibration Check
I_{is} = Mass of the Internal Standard in the Continuing Calibration Check

Statistics

Four statistical analyses were run to evaluate the immunoassay kit, the GC/FID, and the GC/MS data. These were logistic regression, linear regression, pairwise comparison t-test, and Wilcoxon rank sum test. For all statistical analyses, the significance level was set at 0.05. The significance level is the probability of incorrectly rejecting the null hypothesis in a given statistical test, it is set *a priori* to running a test to ensure correctness. To determine the statistical significance of the analyses performed, a p-value was generated with each statistic. The p-value is the lowest level at which the significance level can be rejected; in this case, any p-value less than 0.05 would show that the test is statistically significant. All statistical analyses were run on SAS software V6.06[7,8].

Logistic Regression

Logistic regression analysis fits a model between categorical response data and explanatory data. It differs from typical regression analysis in that instead of the model predicting a value for the response data based on the explanatory data, it gives an associated probability of falling into a given category based on the explanatory data[9]. In this case, a model is fitted between the immunoassay kit data and the GC/FID data. The statistical parameter of interest in this test is the Score Test for Proportional Odds Assumption[7], which tests if the logistic regression model is appropriate in explaining the data.

Linear Regression

Linear regression analysis fits a linear model between a response variable and an explanatory model. The statistical parameter of interest in the linear regression analysis is the F test, which establishes if the model is statistically significant. If the model is statistically significant according to this value, the r^2 value is examined. The r^2 value gives the proportional amount of variability that is explained by the model. It ranges from 0, which is no variability explained by the model, to 1, which is all of the variability explained by it(8). From these criteria it can be determined whether the linear regression model adequately fits the data.

Pairwise Comparison

This hypothesis test determines if the mean difference between two sets of data is significantly different from 0. One data set is subtracted from the other to get a data set that is made up of the differences. If the test does not indicate rejection of the null hypothesis, it does not mean that the data sets are equal, but rather that they are not significantly different from each other *(8)*.

Wilcoxon Rank Sum Test

The Wilcoxon Rank Sum Test is a nonparametric alternative to the t-test for comparing two groups of data. The only assumption required for the test is that each observation is independent. This test is run when either sample size is small or the assumption of normality cannot be made(8).

STUDY

Field Data Comparison

GC/FID vs. Kit

Several hundred potentially contaminated soil samples have been analyzed with the EnSys PENTA RISc test kit at four different sites and compared to GC/FID results for the same samples. All soil samples used in this study were thoroughly sieved and homogenized prior to analysis. Poor correspondence between the kit data and the GC/FID data at these sites posed a need for this study.

GC/FID vs. GC/MS

Twenty samples analyzed with GC/FID were also analyzed by GC/MS to assure the quality of the GC/FID data.

In order to determine the different sources of error, the following variables were examined:

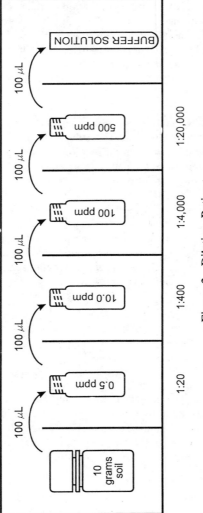

Figure 2. Dilution Ratios

Extraction
Sample Dilution
Color Reaction
Antibody Tube Variability
Operator Differences

To study each independently, the variables were studied either separately from the others or while the others were held constant.

Extraction Efficiency

To statistically compare the two different extraction methods used for the determination of PCP three site soil samples of different concentrations were each separated into eight equal aliquots. Four aliquots were extracted by Ensys and four were extracted by the GC (GC/FID and GC/MS) procedure. A total of 24 extracts were analyzed by GC/MS using Modified Method 8270 *(6)*. These samples were extracted using the following procedures.

- EnSys Extraction: Ten grams of soil were placed into the plastic extraction jar containing 20 mL of methanol/water solution and shaken vigorously for 1 minute so any clumps were adequately dispersed. The samples were allowed to settle for 15 minutes and then filtered with a supplied filtration device.

- GC/FID and GC/MS Extraction: Ten grams of soil were mixed with 10 grams anhydrous sodium sulfate and 40 mL of 1:4 acetone:methylene chloride in a glass extraction jar. The jar was placed on an orbital shaker for 30 minutes at 300 rpm. The extraction was repeated two more times with 30 mL portions of solvent. The extracts were combined and brought to a volume of 100 ml.

Sample Dilution

The kits that were used most often contained dilution vials labeled 0.5 ppm, 10 ppm, 100 ppm, and 500 ppm PCP in soil. These vials contained deionized water with a volume of 2 mL, 2 mL, 1 mL, 0.5 mL, respectively. When a 10 gram sample is extracted with 20 mL of methanol and 100 µl aliquots are transferred sequentially into each of these vials in series, the extracts are being diluted at ratios of 1:20, 1:20, 1:10, and 1:5 with an overall dilution of 1:20 (0.5 ppm), 1:400 (10 ppm), 1:4000 (100 ppm), and 1:20000 (500 ppm) as illustrated in Figure 2. These samples all dilute to a concentration of 12.5 ppb where the immunoassay test kit operates.

Color Reaction Error

A 50 ppb laboratory standard of PCP in water was run with the kit eight times. This procedure eliminated any extraction and dilution errors associated with the kit since no extractions or dilutions were performed. Other errors, such as operator, life expectancy of the kit, and vial differences, were eliminated by using the same operator, using the same kit, and by using optically matched spectrophotometer cuvets instead of the plastic tubes provided in the kits.

A 100 μL aliquot of a freshly prepared PCP standard was added directly to the buffer tubes. Then, the entire contents of the buffer tube was poured into a antibody coated tube and allowed to react for 10 minutes. After reaction the contents of the dilution tube were poured off and the tube was rinsed four times with buffered wash solution and patted dry. Five drops of "STOP" solution were added and the color development was read and the absorbance was determined. The contents of the tubes had to be diluted with 0.5 mL of deionized water to raise the level of the liquid so the light beam was able to pass through. This procedure was repeated and the reaction time was extended to 20 minutes. The resulting data from 20 tests was then statistically compared.

Antibody Tube Variability

The kit antibody tubes became scratched when the kit was used. These scratches can account for error in the absorbance readings of the samples. If the standard vial is scratched and the sample vial is not or vice versa, erroneous readings may be obtained. In order to determine the variability in the supplied vials, used vials were filled with deionized water and compared to a clear vial at 450 nm. Absorbance readings were taken for each vial.

RESULTS AND DISCUSSION

GC/FID vs. KIT

Logistic regression analysis determined that no appropriate model (p-value<0.05) could be fit between the kit data and the GC/FID data. This is a result of false negative and false positive values produced by the kit (Figures 3-6).

GC/FID vs. GC/MS

Results of pairwise comparisons between the twenty samples analyzed by GC/FID and GC/MS (same extraction procedure) showed no significant statistical difference (p value = 0.28). Regression analyses run on the same data indicated a high degree of linearity between the two data sets (p-value<0.05, r^2= 0.89). A plot of the data shows the high degree of linearity (Figure 7). The data is presented in Table I.

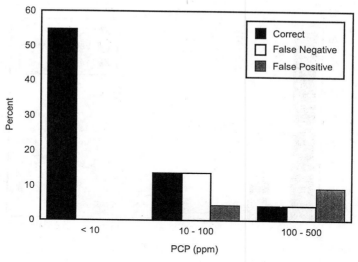

Sample Size = 22

Comparison Rate of Kit PCP Levels with GC/FID PCP Levels
Figure 3. Site 1

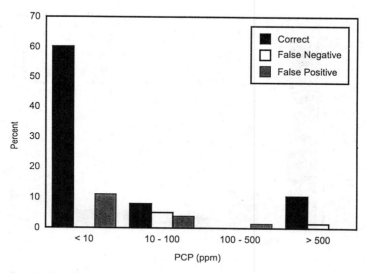

Sample Size = 92

Comparison Rate of Kit PCP Levels with GC/FID PCP Levels
Figure 4. Site 2

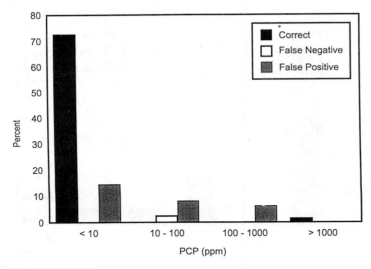

Sample Size = 89

Comparison Rate of Kit PCP Levels with GC/FID   PCP Levels
Figure 5. Site 3

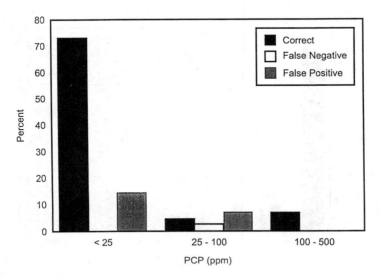

Sample Size = 116

Comparison Rate of Kit PCP Levels with GC/FID   PCP Levels
Figure 6. Site 4

Table I. GC/FID vs. GC/MS Data (mg/kg).

GC/FID	GC/MS		GC/FID	GC/MS
5	10		2500	2000
61	232		34	27
450	150		520	126
34	20		2400	2600
5	10		1900	2100
5	10		340	120
15	107		2400	620
380	380		120	29
460	140		1600	2600
3500	4300		18000	9500

Extraction Efficiency

As shown in Table II and Figure 8, the GC (GC/FID and GC/MS) extraction efficiencies were approximately four times higher than the kit extractions for this site and matrix. This is a significant error, and if these results were utilized during a removal action where the site cleanup goals were 10 ppm, the kits may have yielded two false negatives. Wilcoxon Rank Sum tests run on the GC and kit extract data indicated a significant statistical difference (p-value = 0.03) between the extraction methods. A bar chart of the means provides a good visual representation of this difference (Figure 8).

Dilution Errors

Errors associated with the balance used to weigh the soil sample (1 percent)*(10)*, with the accuracy of the mechanical pipet in transferring the sample (0.5 percent)*(11)*, with the amount of methanol in the extraction jar and water in the dilution vials (assumed to be 0.5 percent) accumulate over the range of the dilution and add to the overall error present in the kit. Figures 3-6 show the percentage of correct, false negatives

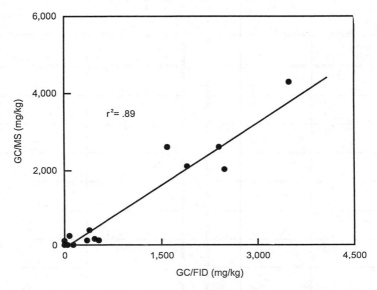

Figure 7. Linear Regression Plot, GC/FID vs. GC/MS for PCP

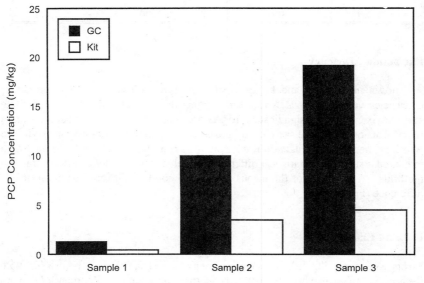

Figure 8. Mean Extraction Efficiency, GC vs. KIT

Table II. Extraction Efficiency - GC vs. Kit

Type	Extract (mg/kg)	Sample	Mean	Standard Deviation
LAB	0.80 2.20 1.40 1.13	1	1.38	0.60
KIT	0.30 0.24 0.29 0.25	1	0.27	0.03
LAB	10.41 09.18 10.23 10.63	2	10.1	0.64
KIT	3.66 3.62 3.52 3.72	2	3.63	0.08
LAB	19.55 21.25 16.40 19.20	3	19.1	2.01
KIT	3.66 3.47 5.14 6.38	3	4.66	1.37

and false positives of each kit level as compared to the actual GC/FID value from samples at four wood-treating sites. These figures show that the higher the concentration in the soil, the more difficulty the kits have in correctly identifying the PCP level.

Color Reaction Error

The 20-minute reaction time mean absorbance was marginally higher than the 10 minute reaction time mean. Results are presented in Table III. The Wilcoxon Rank Sum Test of this data set did not indicate a statistically significant difference (p = 0.22).

Table III. Reaction Time Absorbance, 10 vs. 20 Minutes.

REACTION TIME (Minutes)	ABSORBANCE	MEAN	STANDARD DEVIATION
10	0.3950 0.2360 0.2700 0.4030 0.3670 0.2570 0.3040 0.3130	0.318	0.06
20	0.3325 0.1540 0.3660 0.3325 0.5990 0.3450 0.5910 0.5040	0.403	0.15

Antibody Tube Variability

In order to check the antibody tube variability the percent absorbance was measured using deionized water in the tube. Ideally, mean percent absorbance should be zero. Eight vials were tested with deionized water and the average absorbance was 0.009 and the standard deviation was 0.006.

Operator Error

Errors associated with different operators were not determined but it is assumed that this may be a source of variability which may add to the errors associated with the kit due to a different technique. Although the kit instructions should be followed the same way, each person has their own way of performing the test. If the operator does not understand what may adversely affect the kit, a simple thing such as letting a drop hit the side of the tube may affect the results.

Kit Life

Each kit has an expiration date. This date is dependent upon the temperature under which the tests are stored and shipped. The closer the kits are to the expiration date the less accurate the data may be*(11)*.

KIT ADVANTAGES/CONCLUSIONS

After hundreds of tests on samples from wood-treating sites were run, several positive aspects of the EnSys PENTA RIS<u>c</u>™ kits are readily apparent. The ability to produce quick turnaround data in the field is desirable in extent of contamination studies and removal activities for several reasons. Typically, it can take 6 weeks to receive quality assurance/quality control (QA/QC) analytical data from a CLP laboratory; this time lag may necessitate several trips to a site for a full extent of contamination study. During removal activities the time lag would necessitate down time for heavy equipment and operators at a great expense. However, with quick turnaround data, the results produced in the field can help to steer extent of contamination studies and removal activities. Another good aspect of the kit is its cost relative to GC/MS analysis. A kit (enough for 4 tests) costs $225, and GC/MS analysis of one sample costs approximately $200-600 (not including shipping). A distinct advantage of the kits is that a technician can be trained in the proper operation of the kit in about one day. A disadvantage of field portable GCs and GC/MS instruments is that they require highly trained personnel to operate them.

KIT IMPROVEMENTS/RECOMMENDATIONS

Reaction Time

A test was conducted in order to determine if increasing the reaction time of the test will increase the consistency of the color reactor. Increasing the time showed a marginal increase in mean absorbance. Standard deviation was slightly higher for the 20 minute test. Increasing the reaction time did not increase the accuracy of the test.

Antibody Tube Changes

Transferring the final solution to optically matched cuvets after the color production has been stopped eliminates any errors present in the supplied antibody tubes.

Minimizing Dilutions

Because dilution errors compound with each dilution, it is suggested that the levels of concern with the kit be kept as low as possible to achieve the goals of the site. For example, a 10 ppm level requires a 1:40 dilution and a 200 ppm level requires a 1:8000 dilution.

Extraction Efficiency

Prior to any heavy usage of these kits for extent of contamination studies or for excavation activities the efficiency of the EnSys extraction with the site soils should be determined. This will allow the user to determine the dilution vials to be ordered that will best suit their needs. Eight tests should be performed and statistically compared as above(12).

Choice of Dilution Vials

Depending on the extraction efficiency and other errors inherent in the kit, dilution vials could be chosen based on the objectives of the screening.

For example: If the level of concern for PCP on the site is 50 ppm, then two dilutions above this level and two below it could be chosen. If the extraction efficiency and other errors added up to 25 percent, then dilution vials of possibly 10 (7.5 - 12.5), 30 (22.5 - 37.5), 75 (56.3 - 93.8), and 150 (112.5 -187.5) ppm PCP in soil could be chosen so that the dilutions do not overlap with the level of concern and more useable and conservative data is obtained. One conservative level can be chosen such as 30 ppm to obtain a greater than or less than value.

Calibration Curve

An objective of this study was to obtain more accurate concentrations of a sample, rather than a range, from a given absorbance knowing the extent of dilution. Several modifications were made to the Ensys, Inc. PENTA RISc test kit to meet this objective. Rather than using a differential photometer, a spectrophotometer calibrated to read 0 absorbance with deionized water was used, the use of the supplied dilution vials eliminated, and optically matched cuvets were used.

A variety of standards were used in an effort to determine the concentrations which produced a linear relationship when the absorbance was plotted against the log of the concentration. A linear range from 10 ppb to 250 ppb was obtained (Figure 9). Field samples could then be analyzed in the same manner and plotted to obtain a more quantifiable concentration.

SUMMARY

Initial examination of the EnSys, Inc. PENTA RIS<u>c</u> test kit indicates that it provides quick turnaround data, is field portable, operates easily, and can be utilized as part of a cost effective screening operation. Although these aspects of the kit are appealing, there are also identifiable sources of error involved with the kit and its operation. It has been determined in this study that sources of error include the extraction efficiency, sample dilution, and equipment limitations. Results from the statistical analyses indicate that when the kit data is compared with the quantifiable GC/FID data, the kit data has a substantial amount of error associated with it. Because of this error, results from the kit should be used with discretion. When selecting a level of concern for testing, the potential cumulative error should be considered. A conservative approach would utilize the soil action level for the site and subtract the potential cumulative error. A more quantifiable concentration could be determined using the EnSys, Inc. PENTA RIS<u>c</u> test kit by developing a calibration of sample absorbance ÷ standard absorbance -vs- PCP concentration and utilizing a portable

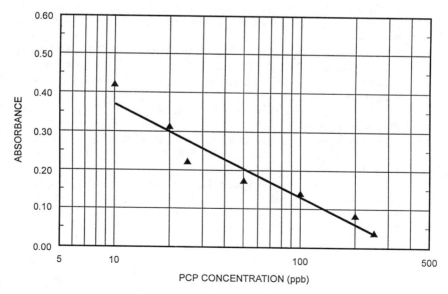

Figure 9. PCP Calibration Curve

spectrophotometer and different dilution procedures. Percent error in the kit can be reduced by utilizing a more precise scale and pipette, reducing the number of dilutions, and using optically matched cuvets. Possibly extending the extraction period and using a mechanical shaker may improve the Ensys extraction efficiency. Using the kits in a clean environment under controlled temperature conditions (60-80°F) could also eliminate other possible errors.

LITERATURE CITED

1.   U.S. Environmental Protection Agency. 1992. Contaminants and Remedial Options at Wood Preserving Sites, EPA/600/R-92/182.

2.   Rapid Immunoassay Screen PENTA RISc™ SOIL TEST SYSTEM, Instructions for Use, Ensys Inc., PO Box 14063, Research Triangle Park, NC, 27709.

3.   Validation Data for PentaRisc™ Soil Test Kit. 1991, Ensys, Inc.

4.   Roy, Kimberly A. December 1991. "Scientific Swapping". Haz Mat World. p. 28-31.

5.   User Manual, Model U-2000 Double-Beam UV/Vis spectrophotometer, Reorder No. ANO-0431-B, Hitachi Instruments, Inc.

6.   U.S. Environmental Protection Agency. Modified Method 8270. Manual SW-846; U.S. EPA, Office of Solid Waste and Emergency Response, Washington, D.C., 1990.

7.   SAS Institute Inc. SAS Technical Report P-200, SAS/STAT® Software: CALIS and LOGISTIC Procedures, Release 6.04, Cary, NC: SAS Institute Inc., 1990. 236pp.

8.   SAS Institute Inc. SAS/STAT™ User's Guide Release 6.03 Edition. Cary, NC: SAS Institute Inc., 1988. 1028pp.

9.   Hosmer, David W., Stanley Lemeshow. 1989. Applied Logistic Regression, John Wiley & Sons, New York.

10.   Cole-Parmer Instrument Company 1993-94 Catalog, p. 35. 7425 North Oak Park Avenue, Niles, IL 60714.

11.   Customer Service Representative. Lab Industries, Inc., February 9, 1993. 620 Hearst Ave., Berkley, CA 94710, (510) 843-0220. Personal Communication.

12.   Titan S. Fan, Ph.D., Immuno Systems Inc., December 2, 1992. 4 Washington Avenue, Scarborough, ME 04074, (207)883-9900. Personal Communication.

COMMUNICATION AND DATA INTERPRETATION

Chapter 19

An Immunochemistry Forum: A Proposal for an Immunochemistry Web Site

Donna W. Sutton[1] and Jeanette M. Van Emon[2]

[1]Lockheed Martin Environmental Systems, 980 Kelly Johnson Drive, Las Vegas, NV 89119
[2]Characterization Research Division, National Exposure Research Laboratory, U.S. Environmental Protection Agency, P.O. Box 93478, Las Vegas, NV 89193–3478

In response to requests for an easily accessible source of information about immunochemical methods, the U.S. Environmental Protection Agency laboratory in Las Vegas, Nevada, has been investigating possible electronic resources. This paper discusses these efforts and the current endeavor to develop this access under the umbrella of the EPA Office of Research and Development structure for technical information on Internet gateways. We illustrate a prototype Immunochemistry Forum World Wide Web page and related information links such as a list of commercially available immunoassays and vendors; A User's Guide to Environmental Immunochemical Analysis; an immunochemistry bibliography of the Characterization Research Division, Las Vegas, Immunochemistry Program; and a Fact Sheet about immunochemical techniques and method development. The discussion provides information about possible alternatives for the proposed Immunochemistry Forum link and the kind of information that may be added.

Immunochemistry has broad applications for a wide variety of environmental contaminants. The application of immunochemical methods to environmental studies is gaining acceptance among scientists in regulatory agencies, universities, and private laboratories. Researchers for all of these organizations are developing immunochemical methods for various applications. Developers, manufacturers, and regulatory agencies are testing and using these methods and techniques in real-world situations. For each of these applications, communications and technical networking are essential to the coordination of research, development, and applications (1).

At the 1993 Immunochemistry Summit II meeting, sponsored by the U.S. Environmental Protection Agency (EPA), in Las Vegas, Nevada, attendees identified a need for current information about developments in environmental immunochemistry. The representatives of chemical industries, immunoassay kit

0097–6156/96/0646–0216$15.00/0
© 1996 American Chemical Society

developers, instrument manufacturers, and regulatory agencies identified needs ranging from timely knowledge about technological advances to information about the pathways for regulatory acceptance of innovative technologies. At the EPA Characterization Research Division in Las Vegas, Nevada, the Immunochemistry Program has been investigating alternative means and the costs involved in providing a central source for this kind of information.

This report provides a brief review of the issues and user's needs identified in the initial survey, outlines the alternatives investigated, and describes subsequent efforts to develop a one-stop source for a variety of related immunochemical information.

What Are the Alternatives?

Attendees at the 1993 Summit meeting recognized that traditional routes of communication, such as fact sheets, journal articles, presentations, training courses, and manuals, are scattered resources that may not always be identified or available when a need arises. In addition, information about upgraded immunochemical methods, field studies, or state-of-the-art methods may not be readily available. Thus, interest centered on tools such as electronic bulletin board systems (BBSs) as a means of providing a forum for continuing the dialogues and networks begun at Summit meetings and a source of information about publications, regulations, data evaluation issues, applications, and validation protocols. These BBSs may be reached directly by a telephone number through a modem connection from a personal computer or through the Internet. The Internet has been described as the world's largest network--a collection or network of networks that allows computers at different locations to talk to each other. Thus, this capability can make a vast array of information available.

An initial survey (2) investigated users' needs and existing BBSs with related interests. As part of this survey, a set of evaluation criteria examined different aspects of a BBS to determine its effectiveness as a communication tool. These criteria include accessibility, services provided, and equipment required to access a BBS. Investigations into a number of BBSs examined them in terms of these criteria as well as each Board's stated objectives, audience, and estimated costs to support the service. Some points drawn from these investigations are that related BBSs:

- are usually accessible though Internet via Fedworld, an index of government information, although some are easier to find than others;
- have varying levels of ease of accessibility;
- have primary objectives that differ but often overlap;
- exhibit different levels of development (pathways or "doors" to other information such as data bases may not be available);
- provide varying levels of support for Special Interest Groups (SIGs), such as an Immunochemistry Forum; and
- have varying levels of costs generally based on the amount of support provided.

In addition, it should be noted that the usefulness of any resource like this depends to a large extent on the level of management attention to users' needs and the currency of the information available.

Two of the BBSs investigated demonstrate these differences and similarities--the EPA Office of Research and Development (ORD) BBS and the CLU-IN (Clean-up Information) BBS sponsored by the EPA Office of Solid Waste and Emergency Response. The ORD board is designed to facilitate the exchange of technical information and ORD products among EPA Headquarters, laboratories, and EPA Regional staff and contractors; the states; other federal agencies; and universities, industry, and the public--a broad range of information for a broad range of users.

The CLU-IN board, on the other hand, was designed for hazardous waste cleanup professionals such as those involved in Superfund and Resource Conservation and Recovery Act cleanup projects. This board provides a means to find current events information about innovative technologies, consult with others online, and access data bases--more narrowly defined information resources for a more narrowly defined audience. The potential audience includes government, contractor, and private industry personnel involved in cleanups as well as researchers and members of the affected public.

In addition to these two boards, mountains of information exist on the Internet not only through EPA but also through other government agencies, universities, and a multitude of other sources. Some help in locating information on the Internet is provided by newsletters, fact sheets with Internet addresses, and links to related resources. There are multiple ways to find different resources, each requiring that you remember what to say to which program to make it go where you want. But these sources may not be in hand or may have been overlooked.

From the perspective of these investigations, the question becomes how to make a choice that brings immunochemistry user needs and existing resources together to provide an easily accessible central site that leads to the various kinds of related news and information of interest to the immunochemistry world.

Internet Tools

In the meantime, as these investigations were progressing, organization of information on the Internet and improvements in means of access were also progressing. One of these access means is the World Wide Web, an online collection of documents that are interconnected by technologies known collectively as hypermedia. These interconnections or *links* form a virtual *web* that spans the Internet world. The Web is a major step in making the process of looking for information simpler and faster. Hypertext, or hypermedia, organizes data to help in information retrieval. It organizes information in relationship to other information by providing hyperlinks, a word or phrase usually highlighted or identified by an underline, that point to another document. Using a mouse button to click on a link that looks like it may take you to the information you need makes a nearly instantaneous connection with the document identified by the link. In turn, a new *page* may have more links that may attract your attention. The Web connects pieces of information from all around the world, on different machines, in different databases in a manner that provides ease of access to a multitude of resources.

To get around on the Web, software known as a *browser* interprets hypertext codes to display Web documents. Browsers such as Mosaic and Netscape have been undergoing almost continual improvements and may be available free on the Web or through a service provider. These enhanced features include the ability to jump to a new page before the current one has finished loading and to backtrack along your search path. Improvements in these tools decrease the costs of telephone bills by speeding up movement between links and the transmission of documents to your computer. Other tools such as Lycos and Yahoo provide indexes of Internet Web sites that help narrow searches for information.

EPA Resources

At this point in time, EPA has already established a number of Web pages. The EPA Home Page provides links to numerous internal Agency resources as well as to other government information servers. Internal Agency links include EPA Offices and Regions; consumer information; rules, regulations and legislation; and science, research and technology. The Office of Research and Development (ORD) page, for example, provides further links to its headquarters offices, national laboratories, and national centers.

In addition, ORD has developed a set of draft guidelines for information dissemination via the Internet. These guidelines provide a structure for electronic technical information on Internet gateways. This structure requires that all ORD information and data installed on EPA servers be routed through a central location to ensure that it conforms to technical information policy requirements and to provide appropriate organization and links to the ORD library of resources. This centralization and organization will improve user access and search and retrieval capabilities.

The EPA Internet resources are supported by several public access information servers:

- The **Gopher** server provides a hierarchy of menus that lead to files containing public access information.
- The **World Wide Web** (WWW) server provides access to almost all information that is made available via the Gopher server. This server can be accessed by browsers such as Mosaic and Netscape.
- The **Wide Area Information Server** (WAIS) is a system designed to retrieve information through searches of document contents by sets of words.
- The **File Transfer Protocol** (FTP) provides a means to copy files from one computer to another.

Immunochemistry Forum

The proposed ORD Home Page and its links are shown in Figure 1. Within the ORD organization, the Immunochemistry Program at the EPA Characterization Research

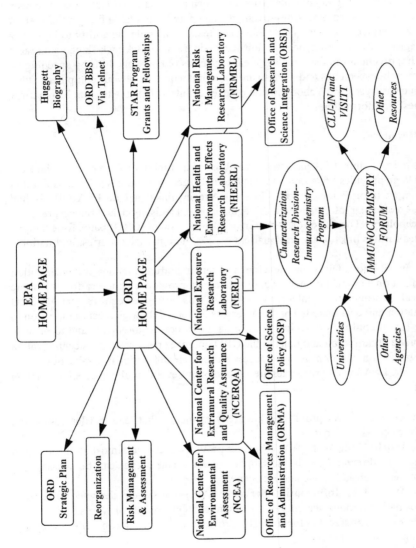

Figure 1. Proposed initial EPA Office of Research and Development (ORD) home page.

Division in Las Vegas has developed a prototype, one-stop, if you will, starting point to connect to the many kinds of immunochemistry information identified in the survey discussed earlier in this paper. This initial effort to develop an easily accessible Immunochemistry Forum focuses on immunoassays. Immunoassays allow field scientists to screen multiple samples in very little time and at relatively low cost. In the laboratory, immunoassays can provide a quick check for high concentration analytes. These analytical tools are under continuing development and evaluation. Current news about these ongoing efforts is valuable to those involved in research, development, and applications.

Figure 1 shows the proposed location for this Immunochemistry Forum within the ORD structure. The prototype Web page for the Immunochemistry Forum is shown in Figure 2 at the end of this paper. Underlined words represent links to other documents. For example, the approximately 120-page User's Guide to Environmental Immunochemical Analysis (3) is a tutorial designed to instruct the user in the use and application of immunochemical methods of analysis for environmental contaminants. A list of immunoassays and vendors and a bibliography developed by the Immunochemistry Program provide other sources of information. The Immunochemistry for Environmental Monitoring fact sheet describes efforts at the EPA Las Vegas laboratory to investigate the usefulness of several immunochemical techniques for monitoring the extent of contamination in various environmental and biological matrices.

Sharing data and research information is key to the successful absorption of new technologies by the environmental science community. The link called "Describe Your Experience with Immunoassays" is an initial effort to design a way to capture information about successes in the use of immunoassays. The Forum can also provide links within the EPA world, to other agency pages, and to the greater Web world. Other information included on the Forum page may include announcements about new information links, training opportunities, and meetings as well as a way for users to contribute comments and suggestions.

The effort to develop the Immunochemistry Forum and the prototype Web page were described at the Immunochemistry Summit IV in Las Vegas, Nevada, on August 2, 1995. During the poster session at this meeting, laptop computers gave participants the opportunity to visit the Forum prototype. Visitors could test the Forum links to other documents and make comments and suggestions related to future development efforts. Some ideas suggested by Summit attendees for additions to the Immunochemistry Forum include:

- a link to a chat group that would give individuals the opportunity to discuss problems and ask questions online to facilitate immunochemical research, development, and applications.
- a list of questions (related to the proposed form to collect information about the use of immunoassays) that should be answered when selecting an immunoassay to use in a project and questions to ask a vendor to ensure that a specific immunoassay will accomplish the required task. These questions might be developed from the experience of the Technology Support Center and the Immunochemistry Program of the EPA Characterization Research Division in Las Vegas.

Immunochemistry Forum
(Prototype)

FIELD ANALYTICAL TECHNOLOGIES AND METHODS

Welcome to the U.S. Environmental Protection
Agency Immunochemistry Forum. This prototype Home
Page is the initial effort to provide infor-mation
to interested parties about immunochemical field
techniques. Many of these field analytical
techniques are valuable tools in assessing contamination concentrations at a Superfund site.

Immunochemical methods have been developed for many compounds of
interest to the EPA. In response to interest expressed at
Immunochemistry Summit Meetings, the goal of this Forum is to
provide a source of information about immunochemical analytical
field techniques. This initial effort provides access to a
user's guide to immunochemical analysis and a list of
immunoassays that are currently available from several commercial
sources. A number of commercial companies are marketing
immunoassay test kits for the detection of various environmental
contaminants. This list of companies and their products does not
constitute endorsement or recommendation of their use. It is
provided only to assist potential users by providing information
about techniques that are available. The list will be updated
once a month.

You can see other information sources by clicking on the topics
listed below. These include a bibliography and a means to share
information about your experience with immunochemical field techniques. It may be possible to collect information in this manner
to add practical experience to this Forum.

Information Resources

- A User's Guide to Environmental Immunochemical Analysis
  (Abstract and Table of Contents)
- List of Immunoassays and Vendors
- Immunochemistry Program Bibliography
- Immunochemistry for Environmental Monitoring

Figure 2. Prototype World Wide Web page for the Immunochemistry Forum.

Describe Your Experience with Immunoassays

You can describe your experience using immunoassays by clicking this button Your Experience . Information on the form you complete will be collated with information on forms filled out by others and summarized here in this forum periodically.

ImmuNews

A **Quarterly Newsletter on Environmental Immunochemistry**, published by the Immunochemistry Program at the U.S. Environmental Protection Agency Characterization Research Division, Las Vegas, Nevada.

Meetings Pertaining to Immunochemistry

Oct. 15-18, 1995	Federation of Analytical Chemistry and Spectroscopy, Cincinnati, Ohio. Organized by Tuan Vo-Dinh. Oral Session: Electrochemical Immunosensors for Environmental Monitoring; Omowunmi Sadik and Jeanette M. Van Emon.
Nov. 12-17, 1995	Eastern Analytical Symposium, Somerset, N.J. Organized by Yan Xu. Session: Environmental Immunochemistry; Jeanette M. Van Emon, Jeffre C. Johnson, Omowunmi Sadik, Allan W. Reed, and Ben C. Hardwick.
Dec. 17-22, 1995	1995 International Chemical Congress of Pacific Basin Societies, Honolulu, Hawaii. Organized by V. D. Adams, M. J. M. Wells, J. M. Van Emon, J. N. Seiber, and S. Matsui. Oral Sessions: Practical Immunoassays, Jeanette M. Van Emon; Analysis of PAHs by On-line Immunoaffinity Extraction Coupled to Capillary Electrochromatography with Laser-Induced Fluorescence Detection (ACS), David Thomas, Viorica Lopez-Avila, and Jeanette M. Van Emon.
Mar. 3-8 1996	Pittcon'96, Chicago, IL. Organized by Jeanette M. Van Emon. Half-day Symposium: - Competitive Solid State Immunoassay on a Chip; Stephen L. Coulter, Devinder P. Saini, and Stanley M. Klainer

Figure 2 (continued).

Continued on next page

- Immunoaffinity Chromatography/Capillary Electrokinetic Chromatography with Induced Fluorescence Detection; David Thomas, Jeanette M. Van Emon, and Viorica Lopez-Avila
- Detecting Environmental Contaminants by Electroimmunochemical Methods; Omowunmi Sadik and Jeanette M. Van Emon
- Immunochemical Detection Using a Light-Addressable Potentiometric Sensor; Kilian Dill
- EPA Superfund Site Monitoring with a Mercury Immunoassay; Jeanette M. Van Emon and Jeffre C. Johnson

August 1996 Immunochemistry Summit V, Las Vegas, NV

Home Pages

(Possible Links)

- U.S. Environmental Protection Agency (EPA)
- EPA Office of Research and Development (ORD)
- National Environmental Research Laboratory (NERL)

Notice

This prototype Immunochemistry Forum is a preliminary draft. It has not been formally released by the U.S. Environmental Protection Agency and should not at this stage be construed to represent Agency policy.

U.S. EPA Characterization Research Division - Las Vegas
Immunochemistry program

This information resource has been developed by the Immunochemistry Program at the Characterization Research Division--Las Vegas (CRD-LV) of the National Exposure Research Laboratory (NERL), U.S. Environmental Protection Agency.

You can access the CRD-LV Immunochemistry Forum through:
http://www.epa.gov/ORD/nerl.htm

Figure 2 (continued).

DRAFT

IMMUNOASSAY USER EXPERIENCE

NAME: _____

AFFILIATION:_____

ADDRESS: _____

_____

_____

PHONE
NUMBER: _____

E-MAIL: _____

IS A PUBLICATION
AVAILABLE THAT
DESCRIBES THIS WORK: YES___NO___

IF YES,CITATION (authors, year,
title,publication/journal,city,
state,country,vol.,no. pp):

_____

_____

_____

_____

_____

_____

RETURN TO US AT:

(E-MAIL ADDRESS TO BE ANNOUNCED)

DATE: _____

BEGINNING/ENDING
DATES OF STUDY: _____

ANALYTES: _____

MATRIX (soil,
water, other): _____

TYPE/BRAND IA: _____

CONFIRMATORY
METHOD: _____

QUALITATIVE/
QUANTITATIVE
SPECIFY: _____

DECISION LEVEL,
IF APPROPRIATE: _____

OR,CONCENTRATION
RANGE: _____

NUMBER OF SAMPLES
ANALYZED: _____

FALSE POSITIVE
RATE (%): _____

FALSE NEGATIVE
RATE (%): _____

INTERFERENCES
FOUND: _____

_____

_____

SELECTIVITY/
SPECIFICITY
OF TEST: _____

Figure 2 (continued).

- a listserver which would allow individuals to e-mail questions to all members of the list and any member could then reply.
- regularly scheduled updates so that users know when to check in again to review new information.

Participants are looking forward to the availability of this resource and the timely immunochemistry information it can provide to answer questions and improve projects. Development of the Forum will facilitate networking and strengthen gateways to related information. Currently, Internet users can access the U.S. EPA National Exposure Research Laboratory home page at http://www.epa.gov/ORD/nerl.htm.

Implementation of the Immunochemistry Forum Web page will take place within the ORD Guidelines for technical information on Internet gateways. As development of this economical, accessible resource progresses, many publications will be readable online and most will be available for file transfer. In addition, the topics covered will expand from immunoassays to other immunochemistry areas of interest. Researchers, developers, manufacturers, and agency personnel will have access to timely information about technological advances. Links to other EPA sources of information and other agency sites will provide a well-rounded resource that disseminates much-needed information on innovative techniques and promotes the use of more cost-effective methods for onsite monitoring and measurement.

Acknowledgments

The U.S. Environmental Protection Agency (EPA), through its Office of Research and Development (ORD), funded and collaborated in the research described here. It has been subjected to the Agency's peer review and has been approved as an EPA publication. Neither the EPA nor ORD endorses or recommends any trade names or commercial products mentioned in this article; they are noted solely for the purpose of description and clarification.

Literature Cited

1. Van Emon, J. M., and C. L. Gerlach. *Immunochemical Exchanges. Environmental Lab Magazine*, April/May 1994.
2. Frost, K., Lockheed Environmental Systems & Technologies Co., 1994, "Immunochemistry: electronic bulletin board quick survey," unpublished internal report for U.S. Environmental Protection Agency, Environmental Monitoring Systems Laboratory, Las Vegas, Nevada.
3. Gee, S. J., B. D. Hammock, and J. M. Van Emon. *A User's Guide to Environmental Immunochemical Analysis*. EPA/540/R-94/509. U.S. Environmental Protection Agency, Environmental Monitoring Systems Laboratory: Las Vegas, Nevada, 1994.

Chapter 20

Screening Tests in a Changing Environment

Richard L. Ellis

Chemistry Division, Food Safety and Inspection Service,
U.S. Department of Agriculture, 300 12th Street Southwest,
Room 603, Cotton Annex, Washington, DC 20250

Practical screening tests for a variety of pesticides, environmental contaminants and veterinary drugs using immunochemistry and other technologies have been a mainstay of the Food Safety and Inspection Service (FSIS) methods development program for several years, primarily with the Agriculture Research Service (ARS) of the Deparment of Agriculture . Several tests have been developed by ARS over the past decade for a variety of applications for antibiotics and antimicrobials, internal cooking temperature for cooked beef products, species identification, trichina detection, beta-agonist detection and pathogen identification. As the regulatory climate is redefined, the role of screening tests is expected to change, placing more emphasis on ready-to-use, effective and efficient test systems. Major changes influencing food safety inspection programs include limited Congressional funding for FSIS methods development, new environmental requirements, a new FSIS proposal on Pathogen Reduction and Hazard Analysis Critical Control Point (HACCP) system as the new basis of its inspection program, and a top-to-bottom review on how the FSIS will carry out its food safety mandate. Although the HACCP proposal is focused on food microbiology issues, the concept will also apply to residue control. Producers and establishments will be taking more responsibility for proper use of agrichemicals and good agricultural practices to meet the food safety requirements. FSIS will focus on how to verify compliance with the individual HACCP plans and design changes in its National Residue Program. One option is new applications of cost effective, rapid test procedures that can be easily used for field, abattoir (slaughterplant), and laboratory testing.

The roles and responsibilities of federal regulatory agencies are changing. The National Performance Review (NPR) highlights new thinking on how, what and with how much, federal agencies are to deliver programs to the American public.

Some of the challenges are to reduce the size of the federal government and streamline regulations by, for example, reducing their redundancies and improving services. The goal is effective delivery of programs that are consistent with national priorities. This includes a strong environmental agenda addressing hazardous waste reduction at federal laboratories as stipulated by a recent presidential Executive Order. A major shift is also occurring in the regulatory agencies on how to carry out our food safety responsibilities. This is exemplified by recent proposals by the U.S. Food and Drug Administration for Hazard Analysis and Critical Control Point (HACCP) inspection of the foods they regulate and the National Marine Fisheries Service's HACCP program for seafood. These examples typify the challenges facing FSIS and its inspection program even while Congressional funding for food safety and other federal programs is reduced.

In February 1995, FSIS proposed a HACCP system for meat and poultry products (1). In developing its proposal, FSIS identified six key issues: HACCP plan approval, training/certification, phase-in, measures of effectiveness, compliance/enforcement, and the relationship and effect of HACCP with current inspection procedures. Of these, the most relevant issues for laboratory testing concern effectiveness, and compliance/verification. It will be necessary to determine that the individual HACCP plans continue to work effectively, and to identify the best ways to ensure compliance with the model plans. One option available to the industry and FSIS is some mode of analytical testing that is reliable and effective to confirm the safety of meat, poultry and egg products under FSIS inspection. With more than 6,700 establishments under inspection, HACCP verification may be a large task - strongly indicating that new inspection methods will be needed at all levels in the Agency. The recent acquisition of responsibility for egg products inspection will add to the FSIS umbrella for food safety inspection.

One basic change in the regulatory program is laboratory (or in-plant) testing. The HACCP system requires producers to take the primary responsibility for delivering products that are free of unacceptable amounts of microorganisms, chemical contaminants and residues. FSIS may have to develop or assess new procedures for federal inspectors to verify that producers are presenting animals and birds that are free of unacceptable residues of pesticides, veterinary drugs and environmental contaminants. This is consistent with the critical control point concept to assure that animals are safe for human consumption when entering a federally inspected abattoir or food processing establishment. An industry/producer focus for primary residue control may require the availability of simple, reliable and rugged screening tests for the above mentioned classes of analytes. In this situation, FSIS would have a verification (quality assurance) responsibility through an inspection control in the federal establishment or testing laboratory.   In the latter case, laboratory methods would serve as effective verification methods with a high degree of analytical confidence, to provide assurance that a producer's HACCP critical control point is operating in a state of control. This may require some shift from the historical dependence on quantitative methods in federal laboratories. This hierarchy of responsibility will make analysts scrutinize the effectiveness of quantification methods, although there will always be some confirmatory laboratory analysis methods employed.

Environmental issues continue to influence approaches to regulatory methods that are cost effective and reduce dependence on organic solvents. This may not seem significant for research purpose methods where 50 - 100 mL of solvent is used for each particular analysis. However, when procedures have to be replicated hundreds or thousands of times a year in a laboratory, environmental waste and analyst safety issues become significant matters of concern. Waste disposal can become more costly than reagent purchase or recovery. As noted below in more detail, a recent Executive Order on the federal government sector pollution prevention strategy reinforces the long range commitment to reduction in use of organic solvents in residue analysis. Use of new analytical procedures such as immunoassays are expected to contribute to the accomplishment of the Executive Order goals.

HACCP Inspection Issues

In developing FSIS's Pathogen Reduction and HACCP proposal, six key issues were identified and an agreement on them was an important factor in the Federal Register publication. Those six key issues were 1) HACCP plan approval, 2) training and certification, 3) phase-in, 4) measures of effectiveness, 5) compliance and enforcement, and 6) the relationship and effect of HACCP and the current inspection procedures. These key issues were also primary discussion points in the numerous public meetings following publication of the proposal. It is not the purpose of this paper to describe the proposal in detail, but it is relevant to comment briefly on the key issues and the consensus that was developed, particularly on the issues that might impact analytical testing.

With the first key issue, the HACCP plan approval, the dominant factor focused on finding the best way was to ensure that HACCP plans would effectively incorporate the key issues of the HACCP principles. There was agreement that the plans must be individually tailored to the specific establishment and that the establishments would have to adequately incorporate the seven principles, though some flexibility may be allowed among the establishments based on their size and complexity of operations. The HACCP documentation would only be expected to reference analytical testing in a general manner.

The training and certification issue was primarily concerned with what role FSIS ought to have with industry HACCP training. There was agreement that substantial and consistent training for both FSIS inspectors and industry personnel was necessary for a successful HACCP program. Nonetheless, it was clearly recognized that there would have to be substantial agreement in the materials provided to all key individuals responsible for managing and evaluating HACCP programs. The certification was for those individuals responsible for developing, managing and evaluating individual plans.

On the issue of phase-in, the main focus was whether or not to make the HACCP quidelines mandatory . There was consensus that, as noted above, the size and complexity of inspection establishments would be an important factor. In particular, it was recognized that very small establishments (as defined by the Small Business Administration, for example) would be given longer amounts of time, as they would be unlikely to have sufficient personnel or expertise to develop information on the critical control points for their HACCP plan and therefore would rely on outside technical experts. Another key factor for implementing HACCP plans focused on those meat and poultry products that are most likely to present a significant public health concern (e.g., fresh ground beef).

On measures of effectiveness, the issue was how to determine these measures initially and monitor them continuously to ensure HACCP plans are effective. There is an implicit question about type of testing or verification used by the industry since they would have the primary responsibility for determining compliance with their critical control point(s) for individual HACCP plans. This may suggest some microbiological or chemical analytical testing. The role of reliable and effective screening tests is clear for the industry because these tests may save the HACCP system and, for microbial contaminants, they may save lives. It is not as clear cut for FSIS verification or quality assurance systems. There are arguments for using screening tests (immunoassay tests are one option) for reliable quantitative analysis procedures in FSIS laboratories. It is not the intent of FSIS to comment on the pro or con of any specific option.

The compliance and enforcement issue is significant in that it addresses analytical testing. The challenge has been to define the best ways to adequately enforce and ensure compliance with HACCP models and the production of a safe product. This is one of the principle roles for which FSIS will have responsibility. This again, addresses the quality assurance component of industry HACCP plans. FSIS considerations include, appropriate sampling locations and procedures. Sampling at retail markets is a shift from previous testing programs that have focused almost exclusively on federal establishments for residue analysis of pesticides, environmental contaminants and veterinary drug residues. Sampling at retail will continue for compliance cases where issues of adulteration are concerned (e.g., excess water added to a product). What is clear, is that it is easier to *prevent* a food-borne pathogen or chemical residue contamination than to *control* it after it has occurred. This implies that the focus will be on the front end of the HACCP system. Nonetheless, immunoassay methods, among others, are expected to be employed more extensively in the future.

There is a great deal of concern pertaining to what the relationship and effect of HACCP systems and the current inspection procedures will be. FSIS is particularly interested in the extent that the HACCP system and its possible changes in the regulated industry will impact the current inspection system. The issue goes to the core of how to make an effective transition to HACCP systems while meeting existing regulatory statutes along with the variety of key interested and impacted groups. This matter does not focus on test systems.

Pollution Prevention Strategy

Although FSIS conducts inspection activities in approximately 6,700 meat and poultry abattoirs and processing plants throughout the U.S. and its territories, inspection procedures generate very small quantities of hazardous waste. Three current in-plant examples of analytical testing and sample processing procedures include antimicrobial residue testing using agar gel plates, the Sulfa-On-Site® test at approximately 65 swine slaughterplants, and use of 10% neutralized buffered formalin for the collection of pathological samples. FSIS performs tens of thousands of residue chemistry, food chemistry, food microbiology and antimicrobial tests among its testing services at its four national laboratories, and has some analytical testing performed by non-Federal laboratories, the environmental impact may be significant, even through a review of past and current operations, FSIS does not meet the requirements of laboratory facilities under Agency control that contain or release quantities of hazardous chemicals that would trigger reporting as required by the Emergency Planning and Community Right-To-Know Act (EPCRA).

As a federal agency, FSIS has assumed a leadership role in the field of pollution prevention through the management of its facilities. By implementating pollution prevention technologies designed to reduce environmental pollutants FSIS will continue its leadership role to meet the 1999 objectives. Where FSIS is a tenant at other laboratory locations such as the ARS facilities, information necessary to fulfill the reporting requirements of the lead agency are provided as requested. Screening methods with reduced organic solvent demands will continue to be part of this initiative.

The pollution prevention strategies are in response to federal legislation and Executive Order. Specifically, the legislation is known as the Pollution Prevention Act of 1990 (42 U.S.C. 13101-13105) and guidelines for accomplishing this legislative requirement are in Executive Order 12856 (August 3, 1993). The focus of this legislation is source reduction. FSIS assumed this as the first step in a hierarchy of options for reducing the generation of environmental contaminants. This is realized through reduction in the generation and off-site transfer of toxic chemical pollutants at its laboratories, analytical services contracted by FSIS and its procedures carried out at federally inspected meat and poultry plants. By the end of 1999, FSIS plans to achieve a 50% reduction in the release of toxic pollutants based on 1994 levels. The specific pollutants in the 1994 baseline include chemicals subject to the provisions of Sections 313 and 329 of EPCRA and hazardous waste as defined in the

1976 Resource Conservation and Recovery Act (RCRA). Partnerships are being encouraged between FSIS and other governmental, academic and private entities to develop and implement new environmental technologies.

The FSIS program to accomplish the federal mandates on pollution prevention targets 17 priority chemicals for achieving a 50% reduction at all government owned-contractor operated federal facilities by the end of 1999 and a 20% reduction in energy use in federal buildings by the year 2000.

The 17 compounds and elements are listed in Table I. Primary focus will be given to five of these based on the FSIS analytical methods and pathological examination use. Those five are chloroform, dichloromethane, mercury, toluene and xylene. Specifically, these reagents are used in pesticide, veterinary drug, environmental contaminant residue analysis, protein determination, and pathological tissue evaluations. Others that will have lesser impact are cyanide used in bone determination and silver for chloride (salt) determination.

Table I. 17 EPA Priority Chemicals Targeted For Reduction By 1999

Benzene	Dichloromethane	Tetrachloroethylene
Cadmium	Lead	Toluene
Carbon tetrachloride	Mercury	I,I,I-Trichloroethane
Chloroform	Methyl ethyl ketone	Trichloroethylene
Chromium	Methyl isobutyl ketone	Xylene
Cyanide	Nickel	

One predominant approach to reducing organic solvent waste is the application of immunochemical methods under development by the ARS and commercially developed test systems. Over the past decade, FSIS has supported this initiative for immunochemical methods for the synthetic pyrethroids, nitroimidazoles, β-agonists, salinomycin, halofuginone, hygromycin B, benzimidazoles (multiresidue method), pirlimycin, ceftiofur, carbadox, sulfonamides and monensin. Those either in use or expected to be in use within the year are synthetic pyrethrolds, nitroimidazoles, β-agonists, halofuginone and the benzimidazole multiresidue method (2). Others may have use in HACCP applications such as analysis of animal feeds, on farm or in-plant testing.

Research related to immunochemistry is supported through the ARS for applications other than chemistry end point use. Examples include (3):

- Determination of acute phase reactants in bovine haptoglobin.

- β-lactams using protein affinity columns.

- Attachment of immunochemical sensors for salmonella on vitreous carbon electrodes.

- Antibody binding to gold piezoelectric crystals for residues and pathogens.

- Immunoaffinity reagents attached to polystyrene beads for aminoglycosides (e.g., spectinomycin).

- Organic polymer molecular imprints as antibody mimics for residue isolation and clean-up procedures.

Evaluation Parameters For Analytical Tests

The primary parameters for evaluating analytical method performance are common to a variety of testing systems. They include: specificity, precision (variability), systematic error (bias), accuracy, sensititvity and limit of quantification, and are common to a variety of testing systems. They have been well described and are familiar to those who routinely work in a regulatory environment. There are also a number of method characteristics that are specific to Data Quality Objectives (DQO) used in regulatory analysis of test materials. These include reliability (robustness), cost effectiveness, simplicity, versatility, use of commercial equipment and reagents, defined QC and QA, defined performance limits, safety and critical control points. These method characteristics are important because of the manner in which they must be applied. A regulatory method must be capable of producing quality results by large numbers of analysts and laboratories on a variety of test materials in a broad array of environments (e.g., on farms and feed lots, in meat and poultry abattoirs and laboratories of varying kinds). These methods must be able to withstand peer review and other scrutiny when it becomes necessary to take adverse regulatory action on a product. Immunochemical methods are no different when it comes to meeting the rigor of regulatory programs applications. Brief descriptions of some key method characteristics provide the necessary perspective.

The ruggedness characteristic is a measure of the robustness of a method. It refers to the ability of a method to be relatively unaffected by small deviations from the written method protocol for use of materials, equipment, reagents and time and temperature factors for extractions or reactions. Obviously, extremes in temperature and humidity may affect the stability of reagents and solvent suitability. They must be sufficiently nonvolatile to minimize safety and disposal concerns yet maintain acceptable solvating properties as may be necessary. In addition, when biological reagents are used, maintaining their potency is critical to test performance. Inherent with ruggedness is the portability of the test system. The ability to transfer a test method from one location to another without losing performance may significantly influence a test methods acceptance and use.

The need for simplicity is obvious for methods intended for field, abattoir or laboratory use, where analytical skills may vary between operators. Simplifing preparation of documentation is necessary for conducting the method, training the analyst, following adequate QA and QC procedures, making measurements and interpreting and reporting test results. For novice users, the clarity of the procedure and the written documentation is critical to the correct performance of the test and correct interpretation of results. The obvious difficulty is that residue analysis at the trace levels requires rigorous procedures. Nonetheless, even a novice user can produce good results if methods and instruments are easy to use and clearly explained.

The need for cost effective methods for regulatory use is quite clear. With the emphasis on cutting costs, it is imperative that regulatory agencies operate smarter in the future. Cost effectiveness implies many of the attributes of ruggedness and simplicity noted above. It means other things as well, such as the use of common reagents and supplies, minimizing the use of reagents, solvents and nondisposable items, and where possible, employing commonly used instrumental techniques and instruments. It also indicates the need for handling large numbers of test materials simultaneously so that analysis can be performed in larger batches. The ability to analyze a larger batch of test samples reduces the average analytical time per sample. This is a significant factor when FSIS must analyze large numbers of samples for a particular analyte or analytes.

Safety issues for regulatory methods used in non-laboratory environments using chemical or microbiological reagents create different concerns.   For example, procedures that may be obvious to a laboratory trained analyst for disposal of reagents may not be routine to field personnel. Common waste disposal systems may not be available outside a laboratory environment. In addition, tests used in areas of food production need to be designed to avoid potential contamination of food products from the inadvertent contact with test materials and reagents. An important feature of screening tests is the provision of clear instructions to advise users of adequate safety and disposal procedures. Liability issues are a major concern for test manufacturers.

Although there may be similarities between methods intended for regulatory use and research- based methods, there are also differences. The method performance characteristics that are important in regulatory use are:

- The ability to run multiple samples.

- Flexibility to accommodate a variety of tissues or other matrices.

- Capacity for analyzing several related analytes.

Regulatory methods must demonstrate these performance characteristics in multilaboratory studies.

Another key fact is that regulatory methods unlike traditional residue analysis for risk, are not necessarily designed to test the limits of analytical detection. Most regulatory methods must demonstrate their optimum performance at a regulatory tolerance or action limit. When there is no regulatory limit, quantification is less important than identification. However, confirmation of identity is required to indicate misuse of an agricultural chemical, veterinary drug or other analyte.

Regulatory Screening Methods

Screening methods in regulatory programs have been an on-going objective of many of our regulatory testing schemes. Rather than being perceived as less important than the quantitative and confirmatory methods, they have been and will continue to be a most important first line approach by which residue testing for food safety determination will be made. Assuming that they have the desired method performance , can meet specific program objectives, and are properly employed, screening methods offer the potential to identify test samples that contain negligible residues at or below a regulatory limit. These key features also identify samples that require further regulatory analysis. Screening tests must perform at a high level of sensitivity and selectivity to avoid an improper disposition of product, minimize the inadvertent use of expensive laboratory quantitative and confirmatory methods, and avoid the potential of having a detrimental or discrediting influence on a regulatory program.

Screening tests must meet the method performance characteristics noted above. Perhaps most importantly, they must be capable of analyzing a relatively large number of samples in a given unit of time in a variety of environments. Screening tests should be based on the availability of reliable, commercially available and adequately characterized reagents. They should perform in a manner that can be adequately validated in a multilaboratory study or end user environment at indicated quality assurance parameters. By definition, well-designed immunoassays methods commonly demonstrate the desirable properties of using simple procedures for isolation, purification and analytical determination of target analytes. The ability to use a surrogate sample matrix, such as a biological fluid, may make sample preparation easier and simpler. Confirmatory tests must be available so that an acceptable level of confidence may be determined for regulatory use.

Screening methods provide reduced analysis time and reduced operating costs, for analyzing samples that contain no detectable residues, on sampled product. They reduce the number of laboratory tests required to establish that samples contain no detectable levels of residue. They are able to provide greater confidence in residue control programs by either estimating the prevalence of a residue in a homogeneous population of animals or providing a higher degree of confidence in the determination of prevalence. Table II indicates this feature (4). For example, if the objective is to determine a five percent prevalence with a 99 percent confidence limit, then 90 samples are required, assuming there is an adequate sample population. If the objective is to determine a 0. 1 percent prevalence with a 95 percent confidence interval, 2,995 samples would be required.

Typically, FSIS has used a one percent prevalence rate with a 95 percent confidence interval in a homogeneous population of animals (e.g., a slaughter class such as formula fed calves) for its National Residue Program. FSIS is studying ways to use this in a HACCP-based inspection system where producers are responsible for assuring that animals presented for inspection are free of adulterating residues.

Table II. Number of Samples Required to Detect Residue Violations With Predefined Probabilities in a Population Having a Known Violative Prevalence

Prevalence Rate (% in a population)	Minimum number of samples required to detect a violation with a confidence limit of		
	90%	95%	99%
10	22	29	44
5	45	59	90
1	230	299	459
0.5	460	598	919
0.1	2302	2995	4603

There have been some success stories for using screening tests for field inspections, most notably with antimicrobials using kidney extracellular fluid extracts and agar gel plate technology. There are some limitations, however, to the application of some (immunoassay) screening tests. An unpublished study conducted by FSIS scientists a few years ago highlight some deficiencies that make screening tests for residues of veterinary drugs and pesticides challenging. The following example does not imply deficiencies or validity of the test system. Rather, it demonstrates why development and implementation of immunoassays for the analytes of interest in meat and poultry tissues has not proceeded at a faster pace.

The test kit had been designed for determining chlorinated hydrocarbon pesticide residues in foods, particularly fruit and vegetable commodities for which an aqueous wash of the product was sufficient for the test kit analysis (5). To apply the test kit to the analytes in animal fat, the sample had to be heat rendered to collect an aliquot of fat, diluted with acetonitrile, vortexed to mix and extract the residue, eluted through a solid phase extraction column, evaporated to remove solvent, redissolved in methanol and, finally, diluted with 0.01% Tween 20™ before the extract was ready for the immunoassay procedure. Following a typical sequence of additions of enzyme conjugate, substrate, chromogen reagent and stop solution, the result was determined by measuring absorbance at 450 nm and comparing absorbance to daily standards employing blank and known fortification samples. With adequate quality controls, a set of standard operating procedurees were developed to provide a workable system for laboratory use. However, with the extensive preparation steps it would not be a practical test for use in a non-laboratory application. This fact negated many of the potential benefits of the immunoassay procedure.

The evolution of assay systems that are more tolerant to aqueous organic solvent systems is promising, and other sample extraction systems using an aqueous based medium may also be helpful. Another viable option is use of a biological fluid such as urine or plasma, but using these matrices would require tissue fluid correlation data in order to make an appropriate assessment of residues in animal tissue where a regulatory tolerance exists. To be clear, this does not diminish the potential application of immunoassay test kits. They do, and perhaps must, hold promise for residue testing in the new regulatory environment.

Analyte Testing Options

Given the implications of resources, environment, new inspection systems, and immunoassay characteristics, new approaches to applying immunoassay and other contemporary testing systems are imminent. Laboratories will have to critically assess processes for optimizing the efficiency of their operations, adapt or modify methodologies to reduce use of organic solvents, or introduce newer methods compatible with the current regulatory environment. Examples of such changes include methods for high volume analyses such as protein analysis (to reduce the waste streams containing mercury), chlorinated hydrocarbon pesticide residue analysis (to reduce or eliminate chlorinated and other organic solvent use) and lower sample volume when using any of the 17 EPA priority pollutants. Some potential exists for immunoassay methodology for the lipid soluble organohalide pesticides but other technologies reported elsewhere in this symposium are still being developed for analysis of some inorganic analytes.

More potential for immunoassay methods exists for detecting other more lipophobic analytes in the FSIS NRP. Traditionally the NRP has focused on statistically-based random sampling for a large number of analyte:species (and tissue) pairs using quantitative laboratory methods for most of the analyses, with some very satisfying public health results. Unfortunately, there are some limitations to continue with this sampling method because, most of the analyses (98-99 percent) are either non-detects or below a regulatory limit. The only category of analytes where residues are routinely observed with a prevalence of one or more percent is with antimicrobials such as penicillins, tetracyclines and aminoglycosides (6). These are often detected through carcass lesions noted by field inspectors and tentatively identified by agar gel plate antimicrobial inhibition assays. There are many categories of either analyte or analyte:species pairs where no residue violations have occurred in one or more years.

It is premature to predict the outcome of the new regulatory strategy. Nonetheless, there is extensive potential application for using screening tests (including immunoassay) in FSIS laboratories as part of a more effective integrated approach to residue control programs. New approaches are being considered, such as :

- Application of HACCP systems to residue control.

- Putting more emphasis on directed sampling for specific compound:species pairs based on historical records.

- Special studies for unusual applications.

With a focus on sampling programs for specific compound:species pairs, there would be an expected but undetermined increase in the frequency of residue positive and violative samples.With this directed sampling approach, laboratories, expecting a higher prevalence of positive or violative samples, could use rigorous and time-consuming methods more efficiently. The statistical random sampling portion of the NRP could be used less frequently, possibly to gather information on new analytes for inclusion in future residue control programs. The dynamic nature of the annual NRP would be maintained with this type of approach. Without putting rigorous quantitative limits and method performance restrictions on immunoassays, such methods would have much broader application for meeting public health needs.

Using the HACCP principle, and consistent with good agricultural practices, meat and poultry presented for inspection would be free of violative residues. This scenario could be accomplished by producers using good quality control production practices to prevent violative residues. Maintaining good veterinary records on the prophylactic or therapeutic use of veterinary products on animals is one control system. A regular on-farm testing program using reliable screening tests is another. Federal establishments could also use some in-plant testing for residue control as a prerequisite for presenting animals or birds for antemortem inspection. This would enable inspection and any subsequent sampling to be a verification system for determining whether the producer quality control program is operating in a state of control. Any samples collected for verification could be analyzed in a Federal laboratory using an approved method. These could be either reliable immunoassay tests or traditional quantitative methods. Immunoassays would be appealing , because of their relatively high sample throughput and the small volumes of reagents and solvents used.

Confirmatory analyses would be required on screening results indicating residue violations. Immunoassays are advantageous because of the relatively rapid turnaround for results to inspectors and abattoirs for appropriate disposition of animal or bird carcasses. The degree of application of this approach will depend primarily on the quality, availability, ease of use and cost of immunoassays for producers and the FSIS.

Conclusion

The shift in residue control responsibilities to producers (quality control systems) and regulators (quality assurance systems) will call for new thinking in the delivery of systems for ensuring food safety mandates. The selection of immunoassays and other cost effective methodologies for the new inspection systems will depend on factors noted above. It will depend also on the quality production, reproducibility and availability of the new test systems. It will be prudent to ensure that users have a residue control system that will withstand peer and regulatory review.

The opportunity to move immunoassay methods into the food safety regulatory agenda is brighter now than it ever has been. It will be the responsibility of producers and users to make this a successful venture.

Literature Cited

1. Federal Register. *Pathogen Reduction; Hazard Analysis Critical Control Point (HACCP) Systems;* 60 FR 6774, Docket 993-016P: Washington, D.C., February 3, 1995.
2. USDA, ARS. *1995 Progress Report of Food Safety Research Conducted by ARS;* Washington, D.C., December 1995; pp 84-103.
3. USDA, ARS. *1995 Progress Report of Food Safety Research; Conducted by ARS;* Washington, D.C., December 1995; pp 90-93.
4. *Tables of the Cumulative Binomial Probability Distribution;* Annal 35; Harvard University Press, Cambridge, MA, 1955.
5. ImmunoSystems, Inc.; Scarborough, ME., Product Information Brochure.
6. USDA, FSIS. *1993 Domestic Residue Data Book, National Residue Program;* Food Safety and Inspection Service; Washington, D.C., 1993.

Chapter 21

Considerations in Immunoassay Calibration

Thomas L. Fare, Robert G. Sandberg, and David P. Herzog

Ohmicron Corporation, 375 Pheasant Run, Newtown, PA 18940

The appeal of immunoassays has been their ability to provide precise and accurate quantitative results for a specific analyte at a low cost per test. This quantitative capability has been the motivation for analytical chemists to evaluate the technique for use with environmental samples. While some regulators have initially viewed immunoassays as a qualitative method, more recently they have expressed interest in extending the application of immunoassay beyond screening results to quantitative analyses. Although many of the principles underlying qualitative and quantitative techniques are similar, qualitative evaluation of data, particularly when applied to the detection of complex mixtures, is sufficiently complex to deserve separate treatment. This paper will cover calibration of small molecule immunoassays for quantitative analyses. A general approach to important considerations will be given along with a set of recommendations for the analyst using immunochemical methods.

The goal of immunoassay calibration is to estimate the concentration of an analyte as accurately as possible while understanding the practical limitations of this estimation. To calibrate an immunoassay requires obtaining the assay response as a function of known concentrations (or calibrators). Besides the contribution to calibration from analytical sources (e.g., pipette or calibrator accuracy), practical considerations that also affect the accurate estimate of concentration include 1) the confidence level required for the given application and 2) the economics of the analysis. Many problems associated with immunoassay calibration could be reduced by an increased number of more closely spaced calibrators, each analyzed with greater replication. Unfortunately, costs may preclude this approach and compromises are made at the expense of higher certainty.

0097–6156/96/0646–0240$15.00/0

Immunoassay calibration does not relieve the analyst of responsibility for reliability of his measurements; it provides statistically sound results and a means to assess reliability.   Several elements go into the development and support of a sound and reliable immunoassay: an understanding of the chemistry, careful validation of the method, and a comprehensive quality control program.   Faithful execution of the protocol developed from such a program should yield high quality results; however, the analyst is ultimately responsible for rational application of the calibration method. Internal, institutional, or government regulations (e.g., GLPs) should be reviewed for practical effects on the quality of these elements.

Characteristics of Immunoassay Calibration Curves

There are a few characteristics that are constant for all immunoassay calibration curves, regardless of the immunoassay technique being employed.   First, the measured assay response has a nonlinear relationship to analyte concentration.   A simple, straight-line analysis cannot be applied over the working range of the assay (typically 2 to 3 orders of magnitude).   If an immunoassay working range were limited to a "linear" portion of the calibration curve, it would result in the loss of large amounts of valuable analytical data.   Data can be transformed to yield a linear relationship (e.g., logarithmic, logistic); however, the analysis is essentially nonlinear. Since the calibration curve is nonlinear and a limited number of calibrators are used, there are many curves that could pass through a given set of calibrator points.   A choice of fit must be made, which introduces a risk of bias.

Under some measurement conditions, assay errors may be large relative to the analyte levels being measured.   When measuring the calibrators, these errors may contribute a significant uncertainty in determining the <u>relative</u> position of a calibration line, even when its general shape is defined.   Errors are not constant in every region of the assay's working range and, as a result, there is less confidence in the calibration curve in some parts of the concentration range than others.   Since calibration is not constant in every batch, a new curve may need to be determined for every run.   For these reasons, method developers should specify recommended procedures and operating conditions to minimize the potential for error, including the use of calibrator replicates and curve-fitting methods for the calibration curve.

Curve Fitting Methods

Numerous mathematical methods to adjust the calibration curve have been proposed and are well characterized (*1–3*).   Some examples are given in Table I.   These methods can be divided into three major groups; first, the empirical methods, so named because their use is based on practical success, not on some physicochemical model for the assay process.   In several of these methods (e.g., point-to-point, spline functions, polygonal interpolations), the calibration curve will closely fit the experimental data, regardless of how unlikely the data are on chemical grounds.   The position of each segment of the calibration curve is largely independent of the rest of the curve and it is possible that some segments will be accurate and others not. Consider the polynomial interpolation, for which an n-th order polynomial is made to fit a given set of n+1 data points. The fit can result in highly erratic oscillations between data points. Such oscillations are typical of higher order polynomial fits, and, consequently, a cubic fit is the most commonly used with immunoassay data. In general, the order selected should be much lower than the available number of points to be fitted.

Table I. Curve Fitting Methods

Empirical methods	Semi-empirical methods
Manual curve fitting	Log-Log
Point-to-point linear	Reciprocals - 1/B, T/B, F/B, Bo/B
Polygonal interpolation	Logistic - two, three, four, five, six
Spline function interpolation	parameter
Polynomials - straight line, parabolic, cubic,	
quartic, adjustable order	**Model-based methods**
Rectangular hyperbola	Scatchard
Exponential function of concentration	Two, three, four, five, N-parameter
Log concentration	
Log response	

The second group of methods can be referred to as semi-empirical because there are theoretical justifications, under very rigid, simplifying assumptions, that predict the calibration curve. The most common of these is the logistic function, first named by Berkson in the 1920's and used in population studies, tumor growth, and economic models. This method was first introduced for immunoassay calibration by Rodbard in the late 1960's (*4*). The simplest form of logistic function is the popular log-logit. The log-logit model produces two parameters: the slope and the intercept of the linear regression fit to the transformed data.

The final group of methods is based on equations derived from the Law of Mass Action applied to antibody-antigen binding systems at equilibrium. This approach is attractive because it is based on sound chemical theory and is therefore likely to be more reliable than any arbitrary model. In practice, however, these models are partly empirical because the actual mechanism of the reaction is more complex than the assumptions.

Log-linear Curve Fitting. A plot of a typical immunoassay calibration curve is shown on linear axes in Figure 1. Since data from immunoassays may form a straight line when plotted as the log concentration versus response, some investigators have referred to it as "linear." Clearly the relationship is not linear in the same sense that absorbance and concentration are linearly related by Beer's Law. One drawback to the log-linear transform is that unphysical response values are predicted at extreme concentrations. At very low concentrations, the transform will result in responses approaching infinity; at high concentrations, negative responses would be obtained. As a result, the analyst must be careful to use this type of transform across a limited range of concentrations. At higher concentrations, the responsiveness (the ability to quantify small changes in concentrations, see also section on **Precision Profiles**) of an immunoassay decreases, so the upper limit imposed by the transform itself will not be unnecessarily restrictive. Limiting the calibration curve at low concentrations, however, may result in the loss of useful information where the assay might still provide accurate *and* precise results. In general, kit developers should recommend appropriate lower limits for their protocols and discourage extrapolating concentrations beyond the standards.

Log-logit Curve Fitting. In practice, the shape of the immunoassay calibration curve is sigmoidal (Figure 2). Unlike a log-linear relationship, the actual calibration curve of an immunoassay has a maximum and minimum that are approached asymptotically at extremes of concentration. The maximum response is referred to as B_0 and the calibration curves are typically given in terms of B/B_0, where B is the assay response at a given concentration, c. The value of B/B_0 has a maximum of 1 and can approach 0 at its minimum.

In general, a sigmoidal formula for competitive immunoassays can be written as

$$\frac{B}{B_0} = \frac{1}{1 + (c/c_0)^b} \tag{1}$$

where c_0 is the concentration at which B/B_0 is 0.5 and b is a fitting parameter for the model (where typically $0 < b < 1$). This formula expresses the essentials of a sigmoidal relationship: as c/c_0 goes to zero, B/B_0 approaches one; as c/c_0 becomes large, B/B_0 approaches zero. This behavior is characteristic of adsorption phenomena in which adsorbates compete for a limited number of binding sites on a surface.

A practical way to take advantage of Equation 1 for the normalized assay response, B/B_0, is to use a logit transform, defined as

$$\text{logit}\left(\frac{B}{B_0}\right) = \log\left(\frac{B/B_0}{1 - B/B_0}\right) \tag{2}$$

In this case, the relationship between B/B_0 and concentration, c, can be expressed using Equations 1 and 2 as

$$\text{logit}\left(\frac{B}{B_0}\right) = -b\log(c) + b\log(c_0) \tag{3}$$

In this formulation, $-b$ is the slope and $b\log(c_0)$ is the y-intercept of a linear fit. The concentration c_0 corresponds to a B/B_0 value of 0.5; c_0 is also referred to as ED50. A better approximation of the calibration curve to general immunoassay

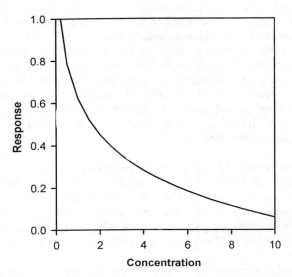

Figure 1. Typical immunoassay calibration curve: An immunoassay calibration curve for which a linear relationship exists between the response and the log concentration is shown plotted on linear axes.

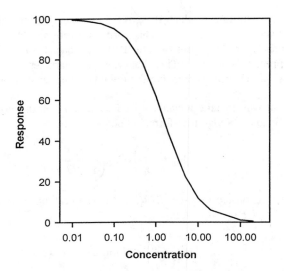

Figure 2. Immunoassay calibration curve for a log-logit transform fit: A calibration curve is shown as a percentage of B/Bo versus log concentration. This curve would be "linear" if plotted on logit response versus log concentration axes.

behavior is predicted by the log-logit transform. Immunoassay data that would fit a log-logit relationship is shown in Figure 2.

   This simple model fits many immunoassays well, hence the popularity of this curve fitting technique.   In its simplest implementation, B_0 and the infinite concentration response (due to non-specific binding, NSB) can have no influence on the shape of the curve. The normalized calibration curve is assumed to be symmetric about c_0, although there are cases (e.g., assays optimized for very low concentrations) that yield asymmetric curves.

   Healy (5) added two more parameters to develop a four parameter logistic function as a refinement of the log-logit approach. The equation that describes the behavior of the assay for this analysis is given by

$$B = \frac{[a-d]}{\left[1+(c/c_0)^b\right]} + d \qquad (4)$$

   Equation 4 can be written as a logarithmic relation that parallels Equation 3 and is given by

$$\log\left[\frac{(B-d)}{(a-B)}\right] = -b\log(c) + b\log(c_0) \qquad (5)$$

   The four parameter model includes **b**, the log-logit slope; **a**, a fitted value for the zero dose response, B_0; **d**, a fitted value for infinite dose response (NSB); and c_0, the concentration at which **(B-d)/(a-B)** is 0.5 (ED50). It should be noted that Equation 4 can be re-written as

$$\frac{(B-d)}{(a-d)} = \frac{1}{\left[1+(c/c_0)^b\right]} \qquad (6)$$

Equation 6 can be viewed as an NSB-corrected version of the log-logit fit of Equation 1.

   The values for the four parameters can be obtained by an iterative algorithm: initial estimates of **a** and **d** are used to produce estimates of **b** and c_0, which are then used to obtain new estimates of **a** and **d**, and so on. The iterations are halted when a pre-set fit criteria has been reached.   This formulation can adequately describe immunoassay data which are not linearized by the log-logit method using Equation 1 (2 and references therein).

Assay Design

Having discussed some mathematical considerations of immunoassay response, let's examine some implications of these on the development of an optimized assay. Figure 3 displays the calibration curves from two different assay systems which use the same response parameter and are designed to detect the same analyte. The question is: Which of these assays is better (3)? The correct answer depends on comparing the *precision* of the two methods for the intended application. Often an analyst answers that Assay A is better because the slope is higher and, therefore, the sensitivity is better. This is a common misconception because, without knowing the relative precision, the practical sensitivity of the two methods cannot be determined. One could easily multiply the response of either of these assays and the appearance of the calibration curves would change, but the error would also be multiplied by the same factor.

Clearly, we must understand the nature of assay error when we evaluate performance. Even if Assay A is more sensitive, it may not be better than Assay B for a particular application. In some cases, the level of error at a particular concentration range may be the appropriate characteristic and the less-sensitive assay may be more accurate and precise at this concentration. Other criteria which may affect the answer include cross-reactivity, speed, cost, and ruggedness.

Having recognized the need for optimum assay performance, the choice of curve fit now needs to be considered (3). Figure 4 illustrates a set of points to which three different curves have been fitted: a linear least-squares line, a nonlinear least-squares fit of a four-parameter logistic, and a manually drawn curve. Now the question is: Which is the best curve fit? Again the analyst needs to consider additional information. Very little can be concluded from this single set of data since it is unclear whether deviations observed are a result of random error in the determinations or bias inherent in the model. If the deviation seen in the first curve is the result of very precise and repeatable data, then this curve fit is clearly unacceptable (despite the almost mystical properties that some investigators attribute to a straight line).

We must keep focused on the goal of the highest quality information and reject principles that, although attractive, may be misleading when critically scrutinized. The manually drawn curve (C) is appealing since it goes perfectly through each point. If we evaluate a curve fit based on performance at calibrator concentrations, the manual method along with interpolation curve fits will appear to have no bias. If they are evaluated at points between the calibrators, however, the results are often more variable than when using an automated, analytical curve-fitting technique.

The usual way for the analyst to evaluate whether, say, the logistic curve fit is best is to collect a large number of sets of data between the calibrators. From these data sets, one can plot the difference between the expected values and the mean observed values (the residuals) versus concentration. If this plot reveals a random distribution of residuals above and below zero, then it can be concluded that the curve fit is unbiased.

Errors

Types. Classifying errors into three groups, systematic, random, and outliers, can be instructive when considering the challenge of determining the best curve fit.

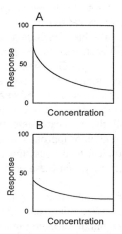

Figure 3. Comparison of the calibration curves from two different immunoassay systems. See text for a description of potential applications for which the precision of each curve would prove more appropriate.

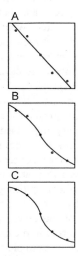

Figure 4. Comparison of three different curve fits to the same immunoassay data. These curves depict three potential fits for the data. See text for a discussion of the criteria for choosing the best curve fit.

Systematic errors or bias cannot be corrected by replication. No matter how many replicates are run of a sample or a set of calibrators, assay bias will result in the same systematic error. In addition to bias caused by an inaccurate curve fit, there are a number of factors that can create bias, including standards that do not behave identically with unknowns, cross-reactions, interferences, and assay drift.

Random errors generally follow a Gaussian distribution and create uncertainty in the determination of calibrator points. These errors are generally small but occur frequently and are due to the cumulative effect of small random errors at each stage of the assay and response measurement. The best estimate of the actual value for a calibrator is the average of all the replicate determinations. As the number of replicates increases, the average should approach the true value (in the absence of bias). In practice, it is uneconomical to process more than a few replicates, so it is essential to understand the extent to which these errors can influence the position of the calibration curve.

All analysts have observed infrequent error that causes deviation from an expected value to a (seemingly) improbable extent. These errors are said to result in outlier responses. Such gross errors differ from random errors in that they are much greater in size and much less frequent. The magnitude of the error can cause a value to be reported that may lead to undesirable or possibly harmful decisions.

The rejection of outliers is an old problem for statisticians that has never been adequately resolved. Of specific concern here is the effect of outliers on least-squares regressions applied to calibration curve-fitting. Techniques that reduce disproportionate effects of random errors and outliers on regression curve fits are referred to as weighting. By appropriately weighting response data, the resulting calibration is less affected by standards that may vary or deviate greatly from expected responses. Specific methods to weight responses for random error come from the study of the precision profile of immunoassay calibration curves (see also section on **Effects of errors and their treatment**).

Precision Profiles. A typical precision profile for an immunoassay is shown in Figure 5. It should be noted that the variability, expressed as a coefficient of variation (CV), is highest at the extremes of the assay. The increase in variability affects the responsiveness of the assay at the extremes; consequently, when examining calibration data, greater weight is generally given to points in the central part of the response range. This is difficult to do when manually plotting calibration curves.

One way to evaluate variability is to consider the effect of small changes of the measured response on the changes of the quantity to be determined. Because the calibration curve is nonlinear, small errors in measured values can create large errors in the calculated concentrations. As an example, consider the logit-log function of Equation 1 (or 3): if the expected response is B^* and the error in measurement is δ, then the calculated concentration, c_{cal}, can be expressed in terms of the expected concentration, c^*, as

$$c_{cal} = c^* \left(1 + \frac{\delta/B_0}{B^*/B_0} \right)^{(-1/b)} \left(1 - \frac{\delta/B_0}{1 - B^*/B_0} \right)^{(1/b)} \qquad (7)$$

For the case when $b = 0.75$, the values for the percent error, $100\%\left(\dfrac{c^*-c_{cal}}{c^*}\right)$, are given in Table II for $\mathbf{B^*/B_0} = 0.1$, 0.5, and 0.9. As can be seen from the table, small relative errors in the measured response (1%–5%) can lead to large errors when calculating concentrations. It should be noted that the error is asymmetric and worsens at the more extreme $\mathbf{B/B_0}$ values. This example illustrates the usefulness of replicate measurements to obtain a good approximation to an expected value.

Table II. Effect of relative measurement errors on calculated concentrations, c_{cal}, relative to expected concentrations, c^*, (in percent) for mid-range and extreme values of $\mathbf{B/B_0}$.

Percent Error in $\mathbf{c^*}$, $100\%\left(\dfrac{c^*-c_{cal}}{c^*}\right)$, $b=0.75$			
$\delta/\mathbf{B_0} = 0.01$	$\delta/\mathbf{B_0} = -0.01$	$\delta/\mathbf{B_0} = 0.05$	$\delta/\mathbf{B_0} = -0.05$
$\mathbf{B^*/B_0} = 0.1$ -13	+17	-46	+170
$\mathbf{B^*/B_0} = 0.5$ -5	+5	-23	+30
$\mathbf{B^*/B_0} = 0.9$ -14	+15	-63	+85

Confidence limits. In Figure 6, a calibration curve is shown as the solid line bracketed by calculated confidence intervals (dashed). The confidence intervals shown for the position of the calibration line are curved inward toward the center reflecting the greater confidence in curve position in the center of the calibration range (*3*). In contrast, at the extremities of the calibration line, the confidence intervals widen dramatically. As a result, the analyst should be careful interpreting results at the limits of the calibration curve. A more subtle feature of the curved confidence intervals is that the confidence limits for analyte concentration are asymmetric and the asymmetry becomes more pronounced at the extremes of the concentration range. As illustrated by the arrows in Figure 6, the minimum and maximum of the confidence interval are clearly asymmetric about the expected concentration value.

Effects of errors and their treatment. Figure 7 demonstrates the effect of an outlier on a linear regression analysis (*1*). In a least squares analysis, the sum of the squares of the differences between the ordinates of a calculated line and the measured data is minimized. Since data in a linear regression are equally weighted, outliers at extremes of the data set have a disproportionate influence on the slope compared to points in the middle. This is clearly illustrated in Figure 7 where a single outlying point causes a huge change in slope (from -5.0 to -2.35).

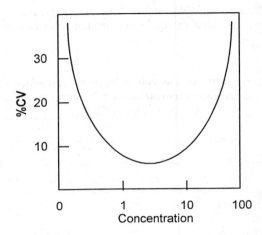

Figure 5. Typical precision profile for an immunoassay. Variability is expressed as the percent coefficent of variation (%CV) at a given concentration.

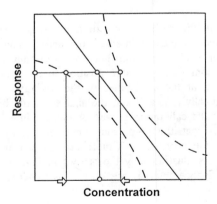

Figure 6. A typical immunoassay calibration curve (solid line) shown with confidence limits (dashed lines around calibration curve).

To improve the robustness of curve fitting methods, techniques have been developed to deal with errors in response data. When the sum of squares is calculated for the linear regression, one may either give each data point the same weight (simply adding all the square of the differences together) or each point may be weighted in some way (depending on its reliability). One way to weight calibration data is shown in Figure 8 (*1*). With this technique, a fixed function is used to determine a weighting factor whose value depends on the distance of a given data point from its expected value; data are then multiplied by this weighting factor during the regression analysis. More sophisticated weighting schemes have been developed that vary by response level and are customized for a given immunoassay's response-error relationship. One such method uses a weighting factor inversely related to the respective standards' variances (*6*).

Prior to applying these weighting factors, it is generally recommended that outliers be removed from the data set. The simplest technique is to delete those data points greater than some number of standard deviations away from the expected value, as determined from the regression fit. As shown in Figure 9, this can be viewed as applying a weighting factor of zero to these data points. Again, more sophisticated techniques have been developed that automatically remove outliers by taking into account the response-error relation.

Recommendations for Processing Immunoassay Data

In summary, there are several recommended steps for processing immunoassay data. First, the analyst should carefully consider an assortment of mathematical models when implementing an immunoassay. The data transform chosen must be experimentally shown to reflect the response of the method. As discussed above, no single model fits the entire range of an immunoassay; however, the logistic methods generally have wide applicability Validation of any curve fitting technique should include characterization of assay precision and bias at points between the calibrators. For commercial immunoassay kits, developers should recommend a minimum number of standards to be used and a transform for the calibration curve fit.

Given the availability of microprocessor-based systems, an appropriate automated technique for curve fitting is the method of choice. Not only does a microprocessor-based system allow the analyst to determine unbiased calibration curves efficiently but, most importantly, these systems have been found to produce better results repeatedly (*7*). In addition to fitting standards to a calibration curve automatically, an instrument can provide statistical information (e.g., standard deviation of calibrator replicates, correlation coefficient for the fitted curve) to accept or reject any or all of the data based on pre-set criteria. These criteria can be incorporated as recommendations for data review in actual applications.

Next, the analyst must develop an understanding of the source and nature of errors in the immunoassay under development. These characteristics can be described through the experimental determination of the method precision profile or the response-error relation, leading to the development of confidence intervals for the curve fit and concentration results. Finally, all of these considerations must be applied

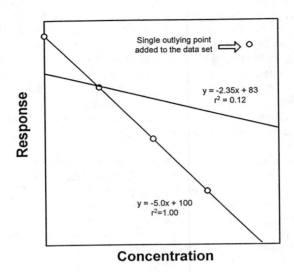

Figure 7. Effect of a single outlier on the slope of a linear regression analysis.

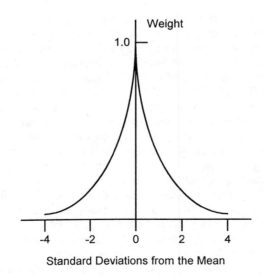

Figure 8. Example of a weighting function that weights in favor of values close to the mean.

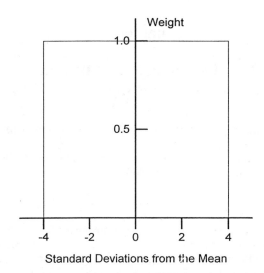

Figure 9. Example of a uniform weighting function to delete outliers.

to the development of a comprehensive quality control program that is able to provide information on the reliability of the result and identify whether the assay method is under control (*8*).

Literature Cited

1. Raggatt, P. Data Processing. In *Principles and Practice of Immunoassay*; Price, C.P.;Newman, D.J., Eds.; Stockton Press: New York, NY, 1991; pp. 190–218.
2. Rodbard, D. *Clin. Chem.* **1974**, *20*, 1255–1270.
3. Rodgers, R.P.C. Data Analysis and Quality Control of Assays: A Practical Primer. In *Practical Immunoassay*; Butt, W.R., Ed.; Clinical and Biochemical Analysis Series, Vol. 14; Marcel Dekker, Inc.: New York, NY, 1984; pp. 253–308.
4. Rodbard, D.; Lewald, J.E. *Acta Endocrinol.* (Copenhagen) **1970**, *64*, Suppl. 147, 79–83.
5. Healy, M.J.R. *Biochem. J.* **1972**, *130*, 207–210.
6. Rodbard, D., Munson, P.J., DeLean, A., Improved Curve-Fitting, Parallelism Testing, Characterization of Sensitivity and Specificity, Validation, and Optimization for Radioligand Assays. In *Radioimmunoassay and Related Procedures in Medicine 1977*, Vol. 1, IAEA: Vienna, 1978, pp. 469–514.
7. Dudley, R.A.; Edwards, P.; Ekins, R.P.; Finney, D.J.; McKenzie, I.G.M.; Raab, G.M.; Rodbard, D.; Rodgers, R.P.C. *Clin. Chem.* **1985**, *31*, 1264–1271.
8. Hayes, M.C., Dautlick, J.X., Herzog, D.P., Quality Control of Immunoassays for Pesticide Residues. In *New Frontiers in Agrochemical Immunoassay*, Eds. Kurtz, D.A., Skerritt, J.H., Stanker, L., AOAC International: Arlington VA, 1995, pp. 237–250.

Chapter 22

Quality Assurance Indicators for Immunoassay Test Kits

William A. Coakley[1], Christine M. Andreas[2], and Susan M. Jacobowitz[2]

[1]Environmental Response Center, U.S. Environmental Protection Agency, and [2]Roy F. Weston/REAC, 2890 Woodbridge Avenue, Edison, NJ 08837

Increasing costs associated with environmental site investigations have led to the emergence of various field screening techniques to streamline the process and help reduce analytical costs. In keeping with this trend, the U.S. Environmental Protection Agency (U.S. EPA), Environmental Response Team (ERT), is currently employing immunoassay test kits at a variety of sites. Critical to using these test kits are the Quality Assurance (QA) indicators used to establish data of known and acceptable quality. When considering QA indicators of confidence for the test kits, consider both generic and core indicators. Generic indicators are requirements which are common to all analytical data-generation methods. Core indicators are method-specific requirements established just for the immunoassay test kits. Criteria must be included as part of the QA evaluation when determining overall quality of the data. This paper discusses how to apply these QA indicators to generate data of known and acceptable quality for immunoassay test kits.

Increasing costs associated with conducting environmental site investigations have led to the emergence of various field screening techniques to streamline the process and reduce analytical costs. These field screening techniques are typically procedures capable of providing the project manager with near real-time data, at lower costs than those incurred with standard laboratory analytical methods. Lower analytical costs also allow the project manager to collect data from a greater number of locations, increasing the sample pool size for selection of more focused samples for traditional laboratory analysis, thus speeding up and improving the site characterization process. One of the field screening methods currently being employed is the immunoassay test kit.

0097–6156/96/0646–0254$15.00/0

While immunoassays have been employed by the medical diagnostics industry for years, their applications for the environmental field were not developed until the late 1980s. Numerous immunoassay test kit applications have recently been proposed as draft or have received final approval as part of SW-846 methodologies. These particular methods are considered semiquantitative screening methods. It should be noted however, that some manufacturers have developed quantitative assays which may also be employed.

The U.S. Environmental Protection Agency (U.S. EPA), Environmental Response Team (ERT), is currently employing immunoassay test kits at a variety of sites. Critical to using these test kits are the identification and application of QA indicators used to establish data of known and acceptable quality. Recent U.S. EPA Superfund Program guidance has established a baseline set of criteria which are applicable when generating data with immunoassay test kits. Major components of this process include development of data quality objectives (DQOs) and preparation of a site-specific quality assurance program plan (QAPP) to ensure the generation of data of known and acceptable quality.

Superfund activities involve the collection, evaluation, and interpretation of site-specific data. As part of Superfund requirements, the U.S. EPA developed and implemented a mandatory QA program with respect to the generation of environmental data. This program also includes a process for developing DQOs. The DQO process is a planning tool which helps site managers determine what type, quantity, and quality of data will be required for environmental decision-making. Guidance on the DQO process is described in "Data Quality Objectives Process for Superfund".(1) This guidance superceded an earlier guidance document which described the DQO process and five associated analytical levels for remedial response activities.(2)

Superfund data requirements include the development of DQOs as well as a site-specific QAPP. The overall goal is to generate data of known and acceptable quality. Benefits of developing DQOs and incorporating them into the data generation process include: 1. scientific and legally defensible data collection; 2. establishment of a framework for organizing existing QA planning procedures; 3. production of specific data quality for specific methods; 4. assistance in developing a statistical sampling design; 5. a basis for defining QA/Quality Control (QC) requirements; 6. reduction of overall project costs; 7. and elucidation of two data categories. Achieving these benefits is contingent upon clearly defining the qualitative and quantitative DQOs that will be applied to the process. Related to these specific DQOs are specific QA/QC requirements. Superfund has developed two descriptive QA/QC data categories: 1. screening data with definitive confirmation, and 2. definitive data. A wide range of analytical methods are available that meet the requirements of these two data categories. Immunoassay test kits fit into the first category.

The screening data with definitive confirmation category shown in the "Data Quality Objectives Process for Superfund", is described as data generated by rapid, less precise analytical methods. It provides analyte identification and quantification, even though the quantification may be imprecise. A minimum of 10% of the screening data samples must be confirmed by a rigorous analytical method, and QA/QC procedures, typically associated with definitive data. Screening data are not considered data of known quality unless associated with confirmation data. QA/QC elements associated with screening data are summarized in Table I(1).

TABLE I. Screening Data QA/QC Elements

- Sample documentation
- Chain-of-custody, when appropriate
- Sampling design approach
- Initial and continuing calibration
- Determination and documentation of detection limits
- Analyte(s) identification and quantification
- Analytical error determination
- Definitive confirmation

The definitive data category is described as data generated using rigorous analytical methods, typically EPA-approved reference methods. Definitive methods produce analyte-specific data with confirmation of analyte identity and quantification with tangible raw data output. Definitive data requires determination of analytical or total measurement error. QA/QC elements associated with definitive data include those identified in Table I for screening data, in addition to those elements identified and summarized in Table II.

Superfund guidelines require the use of quantitative immunoassays (Table I). Although immunoassay test kits meet the requirements of the screening with definitive confirmation data category, in order for data from test kits to truly fit into this category, some type of analyte quantitation procedure must be employed. Test kit results that simply indicate the presence or absence of an analyte relative to a standard or control sample, do not satisfy the criteria set forth by the Agency. A calibration procedure, preferably with a hard copy output from the instrument, must be performed, accompanied by the appropriate documentation. However, the authors acknowledge that numerous immunoassay test kit methods accepted by the

SW-846 methods manual allow for the semi-quantitative interpretation of results. When using immunoassay test kits for the Resource Conservation and Recovery Act (RCRA) Program, data generated may simply indicate presence and greater than or less than concentrations relative to some predetermined analyte standard(s).

However, assuming that the immunoassay test kits can be used as required in the "Data Quality Objectives Process for Superfund"(1), there are a number of confidence indicators that should be evaluated. Quality assurance indicators for immunoassay methods, must be considered indicators of confidence from an overall, method perspective. In general terms, any analytical method may be looked at in terms of a subset of core, method-specific indicators within a set of generic, overall indicators of confidence. In addition to the generic indicators required by the Agency(1), the authors consider the core indicators specified in Table III as necessary in determining overall data quality.

TABLE II. Definitive Data QA/QC Elements

- Sample documentation
- Chain-of-custody required
- Sampling design approach
- Initial and continuing calibration
- Determination and documentation of detection limits
- Analyte(s) identification and quantification
- Analytical error determination
- Definitive confirmation
- **QC blanks (trip, method, rinsate)**
- **Matrix spike recoveries**
- **Performance Evaluation (PE) sample, when specified**
- **Analytical error or total error determination**

The generic indicators of confidence include: blanks, documentation, matrix spikes, calibration standards, sample preparation, representativeness, comparability, confirmation analysis, and replicates. These are QA indicators of confidence that are associated with all analytical methods and must be evaluated in order to determine whether data generated meet QA/QC objectives outlined in the project-specific DQOs generated at the commencement of the project.

Conversely, the core indicators reflect method-specific indicators of confidence which vary based on the analytical method employed for the data generation activity. For immunoassay test kits which employ antibodies as the mode of detection and quantitation, at this point in time, the core indicators of confidence include: temperature, target analyte specificity, interference, moisture content, dilutions, stability, reaction time, and user friendliness. While the authors acknowledge that specificity and interference may be generic indicators, they were included with the core indicators to emphasize their importance relative to immunoassay test kits.

Generic Indicators of Confidence

A review of the generic QA performance indicators shows how these elements apply to any method, and should be evaluated in order to generate data of known and acceptable quality - a main focus of the Superfund Program.

TABLE III. Indicators of Confidence	
Generic Indicators	**Core Indicators**
Blanks	Temperature
Documentation	Analyte Specificity
Matrix Spikes	Non-analyte Interference
Calibration Standards	Moisture Content
Sample Preparation	Dilutions
Representativeness	Stability
Comparability	Reaction Time
Confirmation Analysis	User Friendliness
Replicates	

Blanks. Blanks of various types may be included with field sample-collection activities, but must be included with confirmation samples being sent for laboratory analyses. These may include trip, field, method, or rinsate blanks. Data generated from blanks may be used to assess contamination error associated with sample collection, sample preparation, and analytical procedures.

Documentation. Sufficient documentation must be maintained for all aspects of the sample collection and analysis process. Documentation verifies adherence to procedures specified in the site-specific QA plan or documents any deviations from the plan with an explanation for the occurrence.

Matrix Spikes. Matrix spike results are used primarily to determine matrix interference by calculating the percent recovery (%R) and comparing this value to an established acceptance range. For this reason, it is also an indicator of accuracy.

Calibration Standards. Method sensitivity, detection limit, and linearity are evaluated by analyzing calibration standards. Proper calibration procedures ensure accurate results.

Sample Preparation. Sample preparation should adhere to established procedures to ensure homogeneity of the sample. This is especially critical for splitting samples or taking replicate aliquots from the same sample.

Representativeness. In terms of representativeness, samples collected must adequately characterize the area under investigation.

Comparability. In order for data generated to be comparable, sample handling, preparation, and analytical procedures employed for one sample, must be maintained for all samples.

Confirmation Analysis. As dictated by EPA guidance, a minimum of 10% of the screened samples must be confirmed by a more rigorous analytical method in order to obtain data of known quality. Confirmation ensures verification of identification and quantitative accuracy by an approved method. Sample preparation may play a major role. Field screening immunoassays employ extraction procedures that may differ greatly from those suggested in the definitive data category.

Replicates. And lastly, replicates should be analyzed as an indicator of precision. Results generated are used to assess error associated with sample heterogeneity, sampling methodology, and analytical procedures.

All generic indicators must be considered and, depending on the field analysis procedure employed, must be incorporated into site activities.

Core Performance Indicators of Confidence

Whereas the generic indicators focus on the overall performance of sampling and analysis, the core indicators are more refined and focus on errors associated with the mode of analytical detection. The QA indicators of confidence/error are no longer applicable across the board, rather the indicators become more exacting and precise to the method being performed. Because they employ antibodies, the core indicators determined to be of major significance for immunoassay test kits at this time include: temperature, specificity, interference, moisture content, dilutions, stability, reaction time, and user friendliness. This list may not be all inclusive and is subject to change at any time, depending on manufacturer's modifications to existing products and any future findings by EPA. A clear understanding of these core indicators is essential to accurately interpret data generated by the immunoassay procedure. A subsequent discussion on each indicator follows.

Temperature. Both reagents and equipment should be used at ambient temperature. Manufacturer's recommendations include storing the kit and reagents at 4°C to 8°C. Immunoassay reactions are equilibrium reactions and are sensitive to temperature. Therefore, kits should be given enough time to equilibrate to ambient conditions before performing the analysis. Extreme cold decreases the concentration range of the assay, while excessive heat may affect maximum antibody binding ability. A simple thermometer can be used to monitor the temperature; these readings should be documented. A standard practice of allowing reagents and equipment to equilibrate to room temperature for one hour is recommended. For example, never use a standard at ambient temperature with samples that have been refrigerated, and not allowed to equilibrate to ambient temperature prior to analysis.(3)

Analyte Specificity. Depending on the particular test kit being utilized, specificity or cross-reactivity, may contribute significantly to the final result. Before determining whether a particular kit will provide useful data, the site manager should review the manufacturer's information on other, chemically-similar compounds, which the immunoassay kit cannot distinguish from the primary contaminant of concern. For example, if pentachlorophenol (PCP) is the primary contaminant of concern at the site, a PCP kit may be selected. However, if other closely related compounds, such as di- or tri-chlorinated phenols are also present, the kit may not differentiate between PCP and these related compounds. Information provided by manufacturers includes a list of cross-reacting compounds and the concentration required for a positive response.

Depending on the analyte and the immunoassay, in order for cross-reactivity to be of concern, these chemically similar cross-reactants may need to be present in concentrations that are orders of magnitude greater than the target analyte or may need to be present in just slightly greater concentrations. It is important to have some site background information, prior to determining whether immunoassay test kits meets your particular site data generation requirements. If one, or some of the cross-reactive compounds are present in significant quantities at the site and immunoassay test kits are used, the end data user should consider the potential impact that the presence of these substances may have on the data, and carefully weigh decisions made based solely on immunoassay data. Using a spiked sample can aid the end user in determining how to interpret the results obtained. Cross-reactivity can also be checked by confirming results with an approved U.S. EPA method. It must be remembered that reliance is being placed on an antibody to detect an analyte. There is no spectrum or chromatogram as proof of identification.

Non-analyte Interference. The effect of fuel oil in concentrations greater than 10% in the sample, has not yet been determined. Method 4010(4) indicates that no interference was observed in samples with up to 10% oil contamination. Whenever fuel oil is suspected of being present, regardless of concentration, it is wise to run a matrix spike to check for interference. If interference occurs, use clean-up procedures (e.g., fluorisil or gel permeation chromatography) to eliminate the fuel oil from the extract prior to performing the immunoassay.

Moisture Content. Moisture content of samples will vary with the type and location of sample collected. When possible, samples should appear to be dry to minimize any potential error introduced by the presence of water. If it does not affect the analyte of concern (e.g., volatile organics), samples should be air dried prior to preparation for analysis. However, if this is not possible, currently approved immunoassay methods (Method 4010, 4030, 4031, and 4035)(4) state that up to 30% water in soil had no detectable effect on the resultant analytical data. If the analyte(s) of interest is volatile, such that air drying is not an option, a determination of the percent moisture should be performed and the appropriate correction factor applied to the results. Percent moisture should always be factored into the final results, even if the percent moisture is less than 30%, to ensure data that are comparable to the definitive confirmatory method results, which are generally based on dry weight.

Laboratory prepared soil samples, spiked to adjust the pH to 2-4 and 10-12, indicated that samples with pH ranging from 3 to 11 had no detectable effect on the performance of the method. If there is reason to suspect very acidic or basic conditions on site, soil pH should be determined prior to immunoassay analysis.

Dilutions. For some of the test kits, it is important to accurately dilute the standard and sample extract to the level of interest. One will need to perform serial dilutions that can compound error especially when performed by an inexperienced technician. Clearly written procedures for serial dilutions can both document and avoid any dilution errors.

Some test kits may require the preparation of reagents before use. In these cases, the use of clean equipment and measuring devices is crucial. Also, during preparation, solutions must be thoroughly mixed.

Stability. Test kits should not be used beyond their expiration dates. Typical shelf life for test kits is 12 months with some kits extending to 18 months. Components from one test kit should not be interchanged with components from another kit.

Reaction Time. Timing of the immunochemical incubation between individual samples is critical as color intensity is being compared to a standard. Assay drift may occur from deviations in timing of the immunochemical incubation between samples. Immunoassay tests are run in batches that contain standards, controls, and samples. When the procedure uses sequential pipetting steps, there can be a significant difference in the timing between the first and last sample. However, the incubation for all the samples is terminated at the same time by performing the separation steps as a group in a tray or rack. The magnitude of this error depends on the reaction time difference between samples and the rate of the binding reaction. Therefore, it is important to remember to be thorough and consistent throughout the test procedures, and incorporate the use of an electronic timer.[3]

User Friendliness. Although user friendliness may not be a truly measurable indicator of confidence, ease of use related to test kits plays a role in generating quality data. Due to the use of different reagents and dilutions, immunoassay test kits involve many manipulations that can lead to errors. Less steps, reagents, pipettes, and glassware would help to minimize user errors.

In addition, test kits require training prior to field use, especially if the individual performing the immunoassay procedure is not an experienced chemist. Training should include a good understanding of the principles behind immunoassay and familiarity with steps involved in conducting the test, such as pipetting procedures. It is highly recommended that anyone performing the immunoassay procedure receive training from an experienced co-worker or directly from the manufacturer, if available. If the manufacturer does not provide training, a dry run through the test kit procedure should be completed prior to field activities. All potential users should participate in this training, and all training should be documented. To decrease the error due to operator variability, it is recommended that one operator complete all procedures associated with a particular batch.

Current Status of ERT Activities Relative to Test Kits

Currently, ERT is in the process of developing an Immunoassay Technical Information Bulletin. This bulletin will contain general information on immunoassay techniques, equipment/apparatus, sample preparation, documentation and reporting, QA/QC criteria, interferences and potential sources of error, and limitations. In addition, standard operating procedures (SOPs) have been developed for test kits routinely used by ERT.

Conclusions

Immunoassay methods have great potential for field analyses. Ideally antibodies should be analyte specific; methods should be capable of detecting analytes in low parts per billion concentrations; and kits should be usable with complex physical and chemical matrices. However, at this time, the present test kits do not have all these characteristics. Some methods are not analyte-specific, but instead may react with analytes of the same class or functional group; the kits can detect ppb levels if the solvent (usually dilute methanol) can extract the analyte efficiently. In addition, complex matrices such as soils impregnated with fuel oils, tars, and other organic material can interfere with the antibody binding activity.

When considering QA indicators of confidence for the test kits, the user should consider both generic and core indicators. Generic indicators of confidence are those requirements which are common to all analytical data-generation methods. These include: blanks, documentation, matrix spikes, calibration standards, sample preparation, representativeness, comparability, confirmation analysis, and replicates. Core indicators are those method-specific requirements established just for the immunoassay test kits. Criteria must be included as part of the QA evaluation when determining overall quality of the data. These include: temperature, analyte specificity, non-analyte interference, moisture content, dilutions, stability, reaction time, and user friendliness. Clearly, as test kits are refined, these core indicators may be modified. It is the combination of generic and core indicators that is necessary for data to be in compliance with Superfund Program requirements for generating data of known and acceptable quality.

An ideal immunoassay method would have the following attributes: 1. direct use in the field with no need for an on-site lab; 2. capability of analyte quantification; 3. utilization of calibration curve; 4. minimal steps (i.e., extract, react, and measure concentration); 5. no dependence on reaction timing; 6. greater method specificity; 7. a self-contained "black box" that incorporates a measurement detector; 8. minimal interference from matrix effects; and 9. greater efficiency in analyte extraction from solid matrices. Of course an immunoassay "dip stick" method, calibrated to specific concentration ranges, similar to sugar in urine test kits would be widely welcomed for field use.

In searching for this ideal method, the ERT is presently evaluating other means of utilizing antibodies. These methods include electrochemical and fiber optic techniques which incorporate most of the attributes mentioned above.

Literature Cited

1.   U.S. EPA,"*Data Quality Objectives for Superfund - Interim Final Guidance,*" EPA540-R-93-071, U.S. Environmental Protection Agency, Washington DC, 1993.

2.   U.S. EPA. "*Data Quality Objectives for Remedial Response Activities,*" EPA 540-G-87/003, U.S. Environmental Protection Agency, Washington DC, 1987.

3.   Hayes, Mary C., Joseph X. Dautlick, and David P. Herzog. (1993). "*Quality Control of Immunoassays for Pesticide Residues,*" Ohmicron Corporation.

4.   U.S. EPA. "*Test Methods for Evaluating Solid Waste (SW-846),*" Methods 4010, 4030, 4031, and 4035, U.S. Environmental Protection Agency, Washington DC, 1995.

Chapter 23

Maximizing Information from Field Immunoassay Evaluation Studies

Robert W. Gerlach[1] and Jeanette M. Van Emon[2]

[1]Lockheed Martin Environmental Systems, 980 Kelly Johnson Drive, Las Vegas, NV 89119
[2]Characterization Research Division, National Exposure Research Laboratory, U.S. Environmental Protection Agency, P.O. Box 93478, Las Vegas, NV 89193–3478

The U.S. Environmental Protection Agency (EPA) and other regulatory agencies are beginning to accept data from environmental immunoassay field evaluation studies. This paper focuses on practical problems and suggested solutions to a variety of statistical and data analysis issues related to analytical method evaluation problems encountered with environmental immunoassays. We propose that multiple estimates of performance parameters be obtained from independent parts of an evaluation whenever possible, and that confidence intervals or range estimates are better descriptors of expected performance than point estimates. Methods for minimizing false negative and false positive rates are discussed and guidance is provided when calculating confidence intervals of small rates and proportions. Examples using nonlinear calibration curves and limits of detection demonstrate the importance of understanding experimental design and variance sources when interpreting field evaluation results. Experimental factors and scientific assumptions must match statistical assumptions in order to produce results which correctly characterize an evaluation.

The EPA's Characterization Research Division – Las Vegas has been active in several field evaluation studies of immunoassay test kits at Superfund and other hazardous waste sites. These studies are an important phase in building confidence in the environmental analysis community for this new methodology. One feature of many of the field immunoassays is their relative ease of use. However, easy-to-use methods do not necessarily generate easy-to-understand data sets. Each evaluation also tends to have many phases which generate a number of sets of data. Our experience in developing and evaluating environmental immunoassays has resulted in improved understanding of many aspects of immunoassay method validation. Our review of data packages sub-

0097–6156/96/0646–0265$15.00/0
© 1996 American Chemical Society

mitted to EPA has also identified several areas where inappropriate statistical treatments are commonly made. The following discussions illustrate some of the data analysis problems we have encountered and recommendations to enhance the analyses and maximize the information derived from these studies.

Multiple Statistical Estimates

Statistical parameters can often be estimated from more than one subset of data in a single study. Table 1 shows the results from two independent measures of false positive rates from an evaluation of a pentachlorophenol immunoassay (*1*). The immunoassay used an 8-well microtiter strip in each analysis batch to analyze a blank control consisting of sample dilution buffer, four calibration curve standards, and three unknowns. The first row in Table 1 summarizes results from blank control samples, which gave a false positive rate estimate of 20%.

The same study also included a daily equipment wash sample, which was a distilled water rinse through the sampling equipment after cleaning between sampling events. The wash sample was intended to monitor for cross-contamination between sampling runs, and immunoassay analyses were performed on wash samples with no dilution and after a 10-fold dilution. The rate of positive results was similar for both the no-dilution and 10-fold dilution samples, and there was no statistically significant correlation between the two dilution levels for positive results. If some wash samples were contaminated, one might expect more positive results from the no-dilution samples and /or a correlation between positive results from the no-dilution and 10-fold dilution samples. Thus, a false positive rate estimated from the wash sample results appeared to be justified. At worst, the false positive rate might be slightly high if a few samples were contaminated. However, the false positive rate from the wash samples was 12%, which was less than the rate based on blank control samples.

Table 1. Independent false positive rate estimates from pentachlorophenol immunoassay kit evaluation

Sample type	*n*	False positive rate (%)	95% CI[A] (%)
Blank control	98	20	13 - 30
Equipment wash	102	12	6 - 20
Total	200	16	11 - 23

[A] 95 percent confidence interval.

The difference in the above rates is not statistically significant. Hence, they were pooled to give a false positive rate of 16% with a smaller 95% confidence interval

(CI) than either individual estimate would have produced (Table 1). The pooled rate is also a more robust estimate, as different factors contributing to the production of false positives are affecting each individual estimate.

Table 2 shows the results from three independent estimates for the false negative rate from the same immunoassay evaluation. In this case, false negative results were estimated from field, performance, and audit samples. The field sample rate was based on the number of samples classified as positive by a confirmatory method, so it could be biased if the confirmatory method generated biased results. However, all performance and audit samples were known to be positive. Performance samples had analyte concentrations known to the analyst while the audit samples were known to be positive by the analyst, but without an expected concentration level. The three false negative rate estimates of 6.0, 2.6, and 1.8% for performance, field, and audit samples, respectively, were pooled to give a 3% false negative rate with 95% CI from 1.2 to 6%. The respective (nonsymmetrical) 95% CI estimates for each sample type were 1.3 to 17, 0.3 to 9, and 0.2 to 6.3 percent. Note that the false negative rate from the audit samples (*1.8%, n = 112, 95% CI = 0.2 - 6.3%*) has almost the same 95% CI as the pooled false negative rate (*3%, n = 237, 95% CI = 1.2 - 6%*). This illustrates the fact that both the sample size and estimated rate are important in determining the size of the confidence intervals. Smaller rates have smaller CIs for the same sample size.

Table 2. Independent false negative rate estimates from pentachlorophenol immuno-assay kit evaluation

Sample type	n	False negative rate (%)	95% CI[A] (%)
Performance standards (20 µg/g, conc. known)	49	6.0	1.3 - 17
Field samples	76	2.6	0.3 - 9
Audit samples (20, 25 µg/g, semi-blind)	112	1.8	0.2 - 6.3
Total	237	3	1.2 - 6

[A] 95 percent confidence interval.

In cases where very different conditions are associated with a large range of results from individual data sets, one would not want to pool results by degrees-of-freedom. (We use the term degrees-of-freedom in the statistical sense (*2*). It does not refer to the number of variables or states one might associate with the term in physical chemistry or thermodynamic usage (*3*).) The situation would be analogous to an analysis-of-variance (ANOVA) design where many replicates are made within each of several groups and where the within-group variance is much smaller than the between group variance. As in an ANOVA, the average response across all tested conditions should be calculated from group means. We suggest combining rate estimates with equal weight if the between estimate differences are greater than the 95% CIs.

Our experience shows that multiple estimates for a summary statistic often provide a more realistic picture of performance than a single point estimate. An interval estimate, such as a confidence interval or range, more appropriately describes future method performance results which are in agreement with those found in the evaluation. Evaluation designs which allow multiple estimates should always be considered and additional estimates, such as our use of contamination check samples for a false positive rate, should be used whenever the opportunity arises.

Proportion Estimates: Sample Size and Statistical Confidence Intervals

We have just seen that an important aspect of developing statistical estimates is the effect of sample size on the width of the confidence intervals. False positive and false negative rates are typically presented in units of percentages or fractions, usually referred to as proportions in statistical terminology. If there were x false negatives in n measurements, the false negative rate (proportion) estimate is $\hat{p} = x/n$. Confidence intervals for a proportion are based on the binomial distribution (4). A simple formula (5) for the lower confidence limit is:

$$L = \frac{x}{x + (n - x + 1)F_{\alpha(2), v_1, v_2}} \tag{1}$$

where $v_1 = 2(n - x + 1)$ and $v_2 = 2x$, and the corresponding formula for the upper confidence limit is:

$$U = \frac{(x + 1)F_{\alpha(2), v_1', v_2'}}{n - x + (x + 1)F_{\alpha(2), v_1', v_2'}} \tag{2}$$

where $v_1' = 2(x + 1)$, $v_2' = 2(n - x)$, and $F_{\alpha(2), v_1, v_2}$ is a critical value from the F distribution for a two-sided test with a confidence level of $1 - \alpha$ and v_1 and v_2 degrees-of-freedom. Note that the confidence limits are not symmetrical about the proportion.

Table 3, column 2 illustrates the effect of sample size on the 95% CIs for a proportional estimate of 3 percent. The value of U is about 5 times the proportion when $n = 33$. One needs over 200 samples to get a 95% CI whose bounds are between 0 and 6 percent (i.e., less than $p \pm p$), and one needs about 1000 samples to get a 95% CI from 2 to 4 percent. Estimates such as these can be made for any targeted percentage of false positive or false negative rates, and provide useful guidance on designing evaluation studies and on the expected limitations of the results. When statistics such as false positive or false negative rates have low values, (the desired goal of method developers), the level of uncertainty is usually high unless a very large (and expensive) study is performed. Reporting the 95% CIs for these statistics helps to place the results in perspective. The effect of higher sample numbers on reducing CIs is also an incentive to combine multiple estimates.

Approximations. Many statistical texts only have approximate formulas or provide a graphical treatment for CIs for a proportion. Each of these options is questionable when very low proportions are being estimated. The recommended cutoff for most approximate formulas is when $n \cdot p = 5$. In our example with $p = 0.03$, this means there should be about 167 measurements before the confidence interval is calculated with an approximate formula. For the simplest approximation using the normal distribution, the confidence interval is given by: $\hat{p} \pm t_{[\alpha/2, n-1]} \hat{s}_p$, where t is a value from a t-distribution at the $\alpha/2$ percentile with n-1 degrees-of-freedom and $\hat{s}_p = \sqrt{\hat{p}(1-\hat{p})/(n-1)}$ is the normal approximation to the standard deviation of a proportion (see column 4 of Table 3) (6). One can see in Table 3 that these CIs are inaccurate below $n \cdot p = 5$. Thus, simple CI approximations require inordinately high sample numbers when proportions are small.

Other approximate estimates for the CIs of a proportion have also been proposed and Blythe has evaluated several of the most accurate (7), including:

$$\frac{n}{n + z_{1-\alpha/2}^2}\left[\hat{p} \mp \frac{1}{2n} + \frac{z_{1-\alpha/2}^2}{2n} \mp z_{1-\alpha/2}\sqrt{\frac{[\hat{p} \pm 1/(2n)][1 - \hat{p} \mp 1/(2n)]}{n} + \frac{z_{1-\alpha/2}^2}{4n^2}}\right] \quad (3)$$

where $z_{1-\alpha/2}$ is the ordinate position of a normal distribution at the $1 - \alpha/2$ percentile (6). (Equation 3 from Dixon and Massey (6) is equivalent to Blythe's Approximation A (7).) The approximation includes numerous terms which attempt to bridge the gap between the discrete nature of the true distribution and the continuous nature of the normal distribution. Equation 3 is less accurate than equations 1 and 2, and should only be used if one is limited to using normal probability tables. Applying Equation 3 to the example with $p = 0.03$ (Table 3, column 3) gives CIs much closer to the actual CIs (column 2) compared to the normal approximation CIs (column 4). At low values of n, the CIs from Equation 3 are just slightly larger than the true interval, especially if one changes all negative lower bounds to zero.

Recommendations. Our recommendation is to use the accurate formulas given in equations 1 and 2 for determining CIs for a proportion. An approximate formula may be used if the true CIs cannot be estimated because the necessary statistical tables are unavailable, but analysts should investigate whether the approximate formula is appropriate for the data set being studied (7). Approximate formulas tend to break down near the boundaries of the estimated statistic, such as when x is 0 or 1.

An important observation with respect to interpreting results is that confidence intervals for proportions are relatively large when the proportion is low and the degrees-of-freedom are below 100. The above example where a 3% rate gave a 95% confidence range from 0.6 to 8.5% reflects the high relative error of proportional estimates that are less than 10%. Thus, appropriate interpretation and use of low proportional estimates are best made while knowing the CIs for these statistics. Either the CIs or the degrees-of-freedom (or n, which can be used to construct the CIs,) should be reported whenever low proportional estimates are given. **Technical note**: Though our applications have always involved proportions which are small, it should be pointed out that these same arguments hold for proportional estimates that are high. Letting $p + q = 1$, the statistics associated with a low value of p are the same as those for $q = 1 - p$.

Table 3. 95% confidence limit estimates for a proportional estimate of 3 percent for different sample sizes

Sample size n	True[A] 95% CI (%)	Approximate[B] 95% CI (%)	Normal Approximation[C] to 95% CI (%)	$n \cdot p$
33	0.08 - 15.8	-1.7 - 15.6	-4.4 - 10.4	1
67	0.4 - 10.4	-0.2 - 10.6	-1.9 - 7.9	2
100	0.6 - 8.5	0.3 - 8.7	-0.9 - 6.9	3
200	1.1 - 6.4	1.1 - 6.5	0.4 - 5.6	6
500	1.7 - 5.0	1.7 - 5.0	1.4 - 4.6	15
1000	2.0 - 4.2	2.1 - 4.3	1.9 - 4.1	30

[A] True 95 percent confidence limits, Zar, p. 524 (5).
[B] Approximate 95 percent confidence limits ($n \cdot p > 5$), Dixon and Massey, p 246 (6).
[C] Normal approximation for 95 percent confidence limits ($n \cdot p > 5$), Zar, p. 523 (5).

Conflicting Results

In our analysis of numerous data sets associated with field and laboratory analytical method evaluations, we have seen several occasions where repeated analysis of the original data using different techniques have produced conflicting results. An example of this type of behavior was observed during a recent field evaluation of an immuno-assay for benzene, toluene, and xylene(s) (BTX) (8). Figure 1 displays a series of sample absorbance to reference absorbance values for a 250 ng/g BTX performance evaluation sample analyzed with each batch of samples during the study. The results are plotted by index number, which is related to batch analysis sequence. For this immunoassay, the developer claimed a decision level of 0.85 (the dashed line in Figure 1) would identify samples that were above or below 25 ng/g BTX. Figure 1 also shows the mean response at just above 0.7 and two heavy lines at the 95% confidence limits for the mean.

The results in Figure 1 might be interpreted in two ways. First, all 250 ng/g samples were measured below 0.85, suggesting the assay is performing correctly. On the other hand, the upper 95% confidence limit is above 0.85. This suggests there is a significant probability of generating a false negative result for samples with 250 ng/g toluene. Situations like this, where all the data are acceptable yet the statistics suggest less than optimum performance, indicate the method may perform poorly for a certain category of samples. This example shows how one might produce statistics from a single data set that are favorable to opposing positions of a technical issue. The appropriate conclusion requires an understanding of the technical question being asked, experimental design factors affecting the data, and the assumptions behind the statistical treatment.

Since the results from a single study are only a snapshot of what responses one might get, the correct conclusion is that performance is not as good as desired. The wide confidence levels imply that variability is high enough to produce false negative results at the rate of several percentage points when sample concentrations are at 10 times the decision level. This conclusion was also supported by results from performance evaluation samples at lower concentrations and from examination of confirmatory data for field samples (Figure 2). However, even out of context, the conflicting results suggest a performance problem. Evaluators of field or laboratory method performance studies should watch for conflicting results. When this happens, it usually means careful review of the scientific issue, the experimental design, and the statistical assumptions are needed in order to reach a scientifically defensible conclusion.

Minimizing False Negative and False Positive Rates

False negative and false positive rates are often reported as part of a field evaluation, always with the hope they will be low. False negative and false positive results can be evaluated with the assistance of plots such as Figure 2. Figure 2 shows field sample results from a BTX immunoassay evaluation where low concentration samples gave high responses and high concentration samples gave low responses (*8*). The principal goal of the immunoassay was to classify samples as above or below 25 ng/g BTX. No standard curves are generated with this assay, as the intention was to provide a positive or negative decision. The area associated with false positive results is where responses are below the 0.85 decision level on the vertical axis and to the left of 25 ng/g on the horizontal axis. Similarly, the area associated with false negative results is where responses are above the 0.85 decision level on the vertical axis and to the right of 25 ng/g on the horizontal axis.

The overlap between the above two areas and the area within the 95% confidence limits for response as a function of concentration (similar to the 95% CI about a calibration curve) is useful for assessing false positive and false negative concerns.

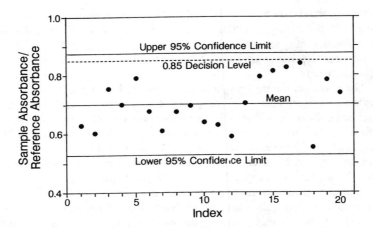

Figure 1. Benzene, toluene, xylene (BTX) immunoassay (sample absorbance)/ (reference absorbance) for 250 ng/g toluene performance standard vs time index for run batches.

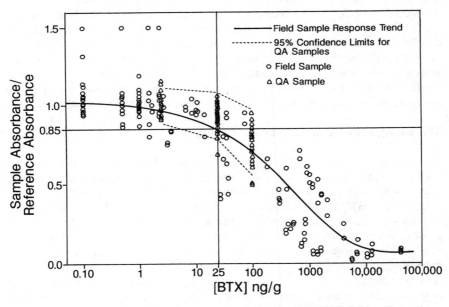

Figure 2. Response versus concentration behavior for BTX field and quality assurance samples. The decision level for the (sample absorbance)/(reference absorbance) ratio is 0.85, used to classify samples as having BTX levels above or below 25 ng/g.

In this example, the area susceptible to false positives is much smaller than the area susceptible to false negatives because sample responses are biased high at the concentration associated with the decision level. Fewer false positives than false negatives would be expected if sample concentrations were uniformly distributed across the graph near the 25 ng/g concentration level. It should be emphasized that the expected false negative and false positive rates for a particular study will depend both on the response characteristics of the immunoassay as outlined above and on the distribution of analyte levels along the concentration axis.

Theory. For any situation in which there is a reproducible response (i.e., calibration) where the response-concentration curve shape is like those in figures 2 and 3, the false positive and false negative rates can be formulated as:

$$FPR = \int_{-\infty}^{C_D} \int_{-\infty}^{R_D} f(r,c)\, dr\, dc \qquad (4)$$

$$FNR = \int_{C_D}^{\infty} \int_{R_D}^{\infty} f(r,c)\, dr\, dc \qquad (5)$$

where $f(r,c)$ is the joint distribution of results as a function of response, r, and concentration, c. The concentration corresponding to the decision level response (R_D) for the test is C_D. In the example shown in Figure 2, C_D is 25 ng/g and R_D is 0.85. If response curves for a particular immunoassay change shape or location between runs, the above equations still apply, but on a case-by-case basis. That is, samples associated with different response curves would have different joint distribution functions. (For analytical methods with positive slopes, the limits of integration in equations 4 and 5 have to be adjusted appropriately.)

Evaluation. The main difficulty in using equations 4 and 5 is identifying $f(r,c)$. For each unique source of samples or study conditions, $f(r,c)$ is different. The joint distribution function can be written as $f(r,c) = h(r|c) \cdot g(c)$, where $h(r|c)$ is the conditional response distribution for a given concentration and $g(c)$ is the distribution of samples along the concentration axis. The conditional response distribution is the statistical distribution of responses one would get for repeated analysis of samples with analyte concentration c. Changes in response characteristics due to alterations in temperature, timing, reagent lots, etc., change the form of h. If a method is under control and the response function is stable, then h is the same for each sample. However, as noted above, the sample distribution, g, is study or application dependent since it represents the distribution of sample concentrations for a particular study. This dependence on sample distribution limits the usefulness and predictive power of all reported false positive and false negative rates. Reporting additional information about the distribution of sample concentrations associated with false negative and false positive results would permit more appropriate decisions to be made about future applications of a method.

Rate Control. Since one cannot control the distribution of sample concentrations, the only adjustable factors affecting the false negative and false positive rates are related to $h(r|c)$. There are three principal factors of the response function which can affect the false negative and false positive rates. One factor is whether the response curve is biased with respect to the decision point (C_D, R_D). In this context bias is not used to mean the difference between the true value and the limiting mean. This bias denotes the difference between the chosen decision point and the true response at C_D. Biased responses will either increase the expected false negative rate and decrease the expected false positive rate or vice versa. For instance, the expected response at C_D might be $R_D + \Delta$, which would tend to reduce false positive rates and increase false negative rates compared to a similar curve passing through the decision point. Commercial screening tests are often designed with a bias because it is important to minimize either false negatives or false positives. For example, if one was testing at a hazardous waste site during a remediation project, it would be better to detect a few false positives and slightly over clean the site than to have false negatives and stop the cleanup with contamination remaining at levels above the goals for the site. This would avoid the potential of a costly second visit to the site if postremediation sampling revealed unacceptably high contamination levels.

Another factor affecting false negative and false postive rates is the shape of the response (calibration) curve. Increasing the slope of the response curve at the decision level changes the mean value of $h(r|c)$ as a function of concentration. If the true response curve passes close to (C_D, R_D), a steep slope will reduce the concentration range over which false positive and/or false negative results are most likely to occur. The third, and usually most important, factor for reducing false positive and false negative rates is to minimize the variability of $h(r|c)$, especially near the decision level concentration, C_D. The larger the variability, the greater the concentration range over which misclassified results are likely to occur. These last two factors become increasingly important when bias between the true response curve and the claimed decision level at (C_D, R_D) becomes smaller.

Based on the results in Figure 2, it appears that *reducing response variance will result in the most improvement* for the BTX immunoassay. Once the variance level is reduced, the bias appears to be the next most important factor affecting false positive and false negative rates.

Calibration and Detection

Many environmental immunoassays use a four-parameter curve or some variation on this function to develop calibration functions for quantitative estimates (*9-11*). An example is shown as the fitted curve in Figure 3. This curve relates the absorbance response, Y, to concentration, X, with the non-linear relationship:

$$Y = D + \frac{(A-D)}{\left(1 + \left(\frac{X}{C}\right)^B\right)}$$
(6)

where A, B, C, and D are fitted parameters. Linear calibration equations relate response to a function of concentration with linear parameters, e.g., $Y = M \cdot f(X) + B$, where the parameters M and B, the slope and intercept, are constant coefficients (*12*). Both $Y = M \cdot X + B$ and $Y = M \cdot \log(X) + B$ are linear equations. Linear equations can be represented as lines when response is plotted versus the appropriate transformation of concentration. The four-parameter equation cannot be written so that each of the parameters is a simple constant coefficient, which makes it nonlinear (*13*). Linear equations such as $Y = M_1 \cdot X + M_2 \cdot X^2 + B$ are also possible, but cannot be represented as a line. However, the additional complications that arise due to the presence of correlated variables (e.g., X is not independent of X^2) are beyond the scope of this discussion. Our discussion involving calibration or response functions will be limited to linear functions which can be plotted as a line versus nonlinear functions which can only be plotted as a curve. A non-linear response function complicates the development and interpretation of summary statistics (including the assumption of normality) for an analytical method. Much of this section will utilize the detection limit as an example, but similar arguements can be applied to other statistical parameters.

Detection Limit. The detection limit is a term notorious for having multiple definitions and a variety of formulas (*14*). Thus any reporting of detection limits should include a reference to, or description of, its calculation. The USEPA has defined a method detection limit as:

$$L_D(EPA) = t_{[0.01, n-1]} \cdot s_c$$
(7)

where t is a Student's t-value at the 99th percentile with $n-1$ degrees-of-freedom, and s_c is the standard deviation from n analytical spikes at concentration c (*15,16*). This definition means that the chance of a Type I error (a false positive) for samples with no analyte in them is 1% if the standard deviation of the blank equals the standard deviation for samples with concentration $L_D(EPA)$. However, it also means that there is a 50% chance of a Type II error (a false negative) when analyzing a sample having concentration $L_D(EPA)$. Due to analytical variability, half the samples with concentration $L_D(EPA)$ would be expected to have analysis results above and half below this detection limit value. Thus, one cannot reliably detect the analyte of interest when the sample concentration is $L_D(EPA)$.

Decision or Detection. Most practitioners in the field distinguish between a detection limit and a decision limit (*17*). The decision limit (or critical level), L_C, is defined as the lowest estimated concentration which is significantly different from the blank and the detection limit, L_D, is defined as the smallest true concentration which is reliably detected. These two concentrations are quite different. If one makes several assumptions about the statistical properties of analytical results when a linear response function is used, such as the random errors are normally distributed and *the variance is constant* over the range of interest, then L_D is about a factor of two greater than L_C when Type I and Type II error rates are equal. For this discussion, L_C is the same as $L_D(EPA)$. Methods for estimating L_D have developed from using replicate measurements at concentrations near L_D (18) to using all the results available for determining the linear calibration curve (*19–21*).

Quantitation. A related true concentration, the limit of quantitation, L_Q, should also be mentioned in this context. The limit of quantitation is the lower limit for reporting a quantitative value with a specified degree of confidence. This value is typically defined as: $L_Q = 10 \cdot \hat{s}_C$ (*22*). This is an ad hoc definition rather than a definition based on achieving a particular level of confidence. If one assumes constant variance over the concentration range $[0, L_Q]$ and a normal distribution of results, the concentration uncertainty at this value is about ±30% at the 99% confidence level (*22*). It is interesting to note that L_Q as defined above is a random variable. If one repeats the procedure one will get a different estimate of $\hat{s}_C$ (because of random error), and a different L_Q. The USEPA has selected the value: $L_Q(EPA) = 3.18 \cdot L_D(EPA)$ as the minimum level for quantification (*23*), which is relatively close to L_Q when n is small (e.g., below 10). This definition also results in a random variable.

The above definitions are not appropriate for most immunoassay work, where non-linear curves and non-constant variance are present. A more direct, operational, definition has been proposed by Adams et al. (*24*) who defined L_Q as the concentration at which the relative standard deviation is 10%. Gibbons has shown how to apply this results-oriented definition for the linear calibration case when variance stabilizing transformations are required (*25*). However, work in the area of non-linear calibration is at the stage where correct estimation and characterization of the calibration function is still of interest (*26*). Identification of limits of detection and quantitation in the presence of nonlinear calibration curves with varying error levels is at the boundary of current practices.

Assumptions and Presumptions. The lack of readily available and appropriate techniques has not prevented the reporting of various detection and quantitation limits when nonlinear response functions are present! Unfortunately, there is a tendency to apply $L_D(EPA)$ or other definitions developed for linear models or constant variance situations directly to methods with non-linear calibration functions. A typical result is a detection limit claim based on $R_D = R_B - 3 \cdot s_B$, where we have written the equations in response units and the sign change compensates for the fact that the curve has a negative slope. We demonstrate the result in Figure 3, where the claimed L_D (on the response axis) is the line labeled "[ABS]$_o$ - 3xSD$_o$". The "detection limit" falls well into the apparent linear range of the assay. The result is incorrect because the assumptions for the original formula have not translated from the linear to the nonlinear case. (As an exercise, the reader can estimate where the "limit of quantitation" is when a factor of 10 is substituted for the factor of 3.)

The main problem with the above example is the underlying assumption that the variance is constant over the range from the blank to L_D; however, the variance is not constant in the nonlinear case. Figure 4 shows the precision profile for the assay used for Figure 3. The precision profile graphs relative error versus analyte concentration (*27*). Because the response curve has a sigmoidal shape, the relative error is smallest at the low concentration range of the linear portion of the curve and becomes large at both low and high ends of the analytical range.

Precision Goals. Characterizing the precision profile is necessary in order to use the information in it as a weighting function for calibration curve fitting or variance estimates. The precision profile has been used for a long time in the clinical immunoassay field (*28*). A recent article by Sadler and Smith utilized 100 degrees-of-freedom as their lower bound when estimating error in precision profiles (*29*). For duplicate sample runs, this requires 200 analysis results. Their conclusion was that "a reliable picture of assay precision requires large quantities of data." One hundred duplicate runs distributed equally across the linear range gave relatively high uncertainties for about a third of the range. A higher number of replicates or runs provided much more reliable estimates across the quantitative range (*29*).

Since the above $L_D(EPA)$ example actually corresponds to the more traditional L_C, the estimated value is still appropriate for identifying where Type I error levels exist. This value is important in identifying when the percent of false positives from blank sample analyses exceed a particular limit. However, without knowing the response variance above L_C, one cannot estimate what the Type II error is at any higher concentration. Based on a typical precision profile (Figure 4, also see Sadler and Smith (*29*)), the Type II error for samples at concentrations above $2 \cdot L_C$ is expected to be much less than the Type I error. This suggests that an important task in the development of immunoassays with low detection limits is minimizing the variability of the blank. This is especially true with immunoassay formats which have a large response at low concentrations and a small response at high concentrations.

Upper Quantiative Limits. Another, often neglected, problem is the identification of an upper quantitative limit. At high concentrations the immunoassay response may have a low absolute error, but a correspondingly low slope for the standard curve may lead to poor resolution (figures 3 and 4). We can define the concentration resolution as:

$$R_C = k \cdot \sigma(r) / b \tag{8}$$

where $\sigma(r)$ is the standard deviation for the response, b is the slope of the curve, and k is an integer. The constant k is related to the level of confidence one is willing to accept. For the case of a linear curve with constant $\sigma(r)$ in the neighborhood of interest,

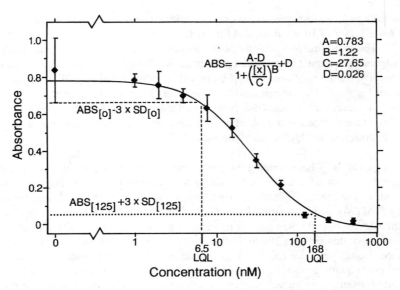

Figure 3. Calibration curve fit to a four-parameter model. The lower quantitative limit (LQL) and upper quantitative limit (UQL) are 3 standard deviations from the 0 and 125 nM standards, respectively. The lack of fit at high concentrations is greater than the replicate analysis error, which invalidates the assumptions for using this definition of the UQL.

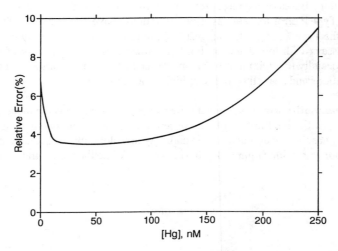

Figure 4. Precision profile example; the coefficient of variation as a function of concentration across the analytical range for a 4-parameter immunoassay response curve.

$k = 3$ corresponds to an approximate 95% CI. While Equation 8 is simple to apply to linear curves, it is difficult to use with the four-parameter model. Not only does the slope change across the curve, but the standard deviation changes non-uniformly across the quantitative range.

Several problems affect the development of reasonable precision profiles. One factor is the nonlinear functional behavior. A second problem is the fact that the calibration function only approximates the true response-concentration behavior. When only one set of data is present for analysis, it is difficult to distinguish this last problem from additional factors which increase the lack of fit. For instance, dilution error for a particular standard concentration may produce results with excellent replicate (within run or within plate) precision which are biased from the expected true result. This may introduce uncertainty in the fitted standard curve which overshadows any effect due to replicate measurement error. This can easily happen with environmental immuno-assays run in a 96-well plate format, where standards are run in replicate wells and calibration curves are produced on a per plate basis. When this is the case, precision profiles based on replicate standards used to characterize error across the calibration curves will under-estimate the error at high concentrations.

Figure 3 shows an example where lack-of-fit error to the standard curve dominates the total error at higher concentration levels. For discussion purposes we have plotted an upper quantitation limit generated at three times the standard deviation from a concentration near the upper linear range. Due to bias to the least-square fit at this concentration, we find the estimated upper quantitative limit at a higher concentration than where we began (Figure 3).

In theory, methods have been developed for solving each of the above problems. Estimating variability across non-linear functions has been demonstrated (*30*). Solving for statistical confidence intervals with non-linear calibration functions has been addressed (*31*). It has even been noted in the analytical chemistry literature that most of these problems have been addressed long before by applied mathematicians and statisticians (*32*). The question thus arises, "Why isn't existing theory utilized to solve these problems?" To answer this question one has to evaluate the requirements for applying known methods.

Limitations and Barriers. There are several requirements which inhibit or prevent the use of the so-called known methods for addressing the above problems. Often the requirements for the proposed methods are not specified. For example, Schwartz outlined three different methods for fitting a non-linear calibration curve (*30*), but didn't provide guidance on which method was most appropriate under any specific circumstance. In addition, this paper leaves many operational decisions in doubt. Guidance for the number of standards required by Schwartz was "as many analyses as possible." This suggests that there is a high cost associated with developing appropriate statistics for these non-linear curves.

If standards are run on a per 96-well plate basis, one ends up with a correspond-ingly low throughput. Five to eight concentration levels have been suggested for the generation of standard curves with linear response relationships and more than this for non-linear situations (*33*). If there were eight concentration levels, a zero concentration standard, and a blank without antibody to check for non-specific binding, with each analysis done in triplicate, then 30 analyses would be needed for the calibration curve. The estimation of quantitation limits for linear response functions by Gibbons et al., (25) suggested 50 to 100 determinations in order to avoid complex measurement error components in the calculation. This is certainly possible for a 96-well plate format or where large numbers of samples are run in highly controlled conditions, but rather formidable for field analysis of a few samples or immunoassay formats accommodating few standards. One might expect that even more determinations would be required for the nonlinear case. These requirements are unattainable for most field kits.

An additional problem with per plate or per batch evaluations is the possibility of shifting standard curve positions. This is the norm with field immunoassay methods and even for methods run occasionally in a laboratory setting. The net result is that information from multiple runs of standard curves can't be combined because the calibration model shifts from run to run.

An operational barrier to using the so called traditional methods is that they involve relatively complex mathematical statistics compared to data treatments most analysts are used to. These treatments also tend to be less available commercially and more difficult to implement by the analyst. This leads to lack of familiarity with the methods, loss of opportunity for analysts to experiment with the methods, and low comfort levels when immunoassay experts encounter results from appropriate data treatment methods.

Improvements and Direction. What can be done to improve the use of more appropriate data treatments? First, one must recognize that traditional data treatments based on linear (or linearized) calibration curves are often inappropriate. Detailed tutorials are an excellent way to introduce users to methods for nonlinear calibration methods (10). Second, analysts need to understand the limitations of each data analysis procedure. The scientific assumptions associated with the experimental results need to match the mathematical assumptions upon which the statistics were based. The requirements for applying the published methods for nonlinear immunoassay calibration curves are associated with assays which are repeatedly run in identically controlled conditions, or where one runs a large numbers of standards per batch of samples. These requirements are only met with high volume work in laboratory settings and/or where robotic control is utilized. In these circumstances, performance characteristics may be estimated by the application of appropriate statistical analysis methods. In the absence of adequate control and reproducibility, the analyst is left with few options and large uncertainties associated with the quality of the results. Studies providing guidance on how to assess these low information situations would be helpful. Meanwhile, use of quality assurance techniques, performance standards, and periodic confirmation by better characterized methods can provide an indirect monitor on field method performance.

Experimental Design Factors

All field evaluations involve both controlled and uncontrolled experimental factors. Controlled factors are explicitly specified during the study. For instance, 3 particular soil types might be selected to investigate matrix effects. However, there are many other conditions and factors which may affect the results which aren't specified or controlled. An example of an uncontrolled factor is the analyst. Analysts aren't usually chosen by design. Whether one or more analysts performs the method over the course of an evaluation may not be controlled either.

This type of information is needed in order to decide whether particular sources of error consist of random effects or fixed effects (*34*). A fixed effect is one where factor levels are specifically chosen, while a random effect is when factor levels are chosen at random. Whether fixed or random effects are at work may not change the statistical calculations, but they are very important in terms of understanding and interpreting the results. Suppose each analyst in an evaluation imparts a different, but uniform, level of uncertainty to results for samples they analyze over the timeframe of the study. If only one analyst/site was involved in the study, then the analyst effect would become a fixed bias confounded with other site characteristics. If several analysts/site were used for the study, then a random error component has been introduced. If there was a bias between field and laboratory sample results, the cause could have been analyst bias in the one analyst/site design, while it is less likely that analyst bias was the cause when multiple analysts/site were involved. Interpretation of numerous other types of factors associated with a field study, such as sample sources, sampling times, analysis times, reagent lots, confirmatory analysis laboratory, audit sample characteristics, etc. are also colored by fixed and random components of the design. The correct analysis of the data and interpretation of results depends on a comprehensive understanding of each of these potential variance factors (*35*).

In terms of design issues for field evaluations, most studies try to include expected sources of variability with the assumption that almost all factors are random effects. One assumes a randomly selected analyst is performing the field evaluation. However, unless a design includes a large number of sites, traditional "random" labels don't apply. Evaluation of the data requires one to assess whether the assumed random factors should be interpreted as random or fixed components. Fixed components are of interest because they may be a source of bias. Random components imply less of a chance for significant bias, though the possibility still exists. Due to the multiplicity of factors and limitations for most evaluations, these assumptions often cannot be tested. However, potentially significant factors should at least be identified. For instance, temporal day-to-day effects are assumed to be random when analyses are carried out over several weeks. However, this assumption may not be appropriate if analyses were performed in two periods separated by 6 months. In this latter case, changes between each period might dominate temporal variability compared to day-to-day changes.

The statistical evaluation and interpretation of summary statistics depends on knowledge of both controlled and uncontrolled factors in the experimental design. Design related information is needed to understand each component of a study. For the precision profile discussed above, Sadler and Smith (29) have noted the need for 1) whether the profile estimates within- or between-assay variability, 2) how much data was used, 3) how the data was obtained, 4) how the estimate was calculated, and 5) whether the profile estimates variability for a sample analyzed singly or with some other degree of replication. This type of information is needed for each estimated parameter in order to fully understand and utilize the results.

Conclusion

The above discussions touch on only some of the issues which play a role in analyzing data from immunoassay field evaluations. The use and evaluation of analytical methods can be partitioned into three activities: optimization of the analytical technique, development of an experimental design used when applying it, and the interpretation of the resulting data set. Results from a method evaluation can only be understood by connecting information from all three phases. The development of method evaluation data and statistical results are necessary, but not sufficient for assessing performance. It is the information in the data that is used to make decisions, information that results from understanding the scientific principles behind the method, the experimental design factors which isolate individual characteristics, and the assumptions required by each statistical analysis technique. It is hoped that the development of new evaluation and data analysis procedures will follow the high activity levels producing new environmental immunoassays in order to advance their acceptance.

Acknowledgments

We thank Richard J. White of Lockheed Environmental Systems & Technologies Company, Las Vegas, NV, for valuable discussions about field immunoassays and Steve C. Hern of U.S. EPA's Characterization Research Division, Las Vegas, NV for enthusiastic support and encouragement in the development of this work. NOTICE: The U.S. Environmental Protection Agency (EPA), through its Office of Research and Development (ORD), funded and collaborated in the research described here. It has been subjected to the Agency's peer review system and has been approved as an EPA publication. Mention of trade names or commercial products does not constitute endorsement or recommendation for use.

Literature Cited

1. Silverstein, M.E.; White, R.J.; Gerlach, R.W.; Van Emon, J.M. *Superfund Innovative Technology Evaluation (SITE) Report for the Westinghouse Bio-Analytic Systems Pentachlorophenol (PCP) Immunoassays*; EPA/600/R-92/032; U.S. Environmental Protection Agency: Las Vegas, NV, 1992.
2. Natrella, M.G. *Experimental Statistics*; NBS Handbook 91; U.S. Government Printing Office: Washington, DC, 1963, p 2-3.
3. Pitzer, K.S.; Brewer, L. *Thermodynamics, Second Ed.*; McGraw-Hill: Englewood Cliffs, NJ, 1972, pp 204-205.

4. Fleiss, J.L. *Statistical Methods for Rates and Proportions, Second Ed.;* Wiley: New York, NY, 1981.
5. Zar, J.H. *Biostatistical Analysis, Third Ed.*; Prentice-Hall: Upper Saddle River, NJ, 1996.
6. Dixon, W.J.; Massey, F.J., Jr. *Introduction to Statistical Analysis, Third Ed.*; McGraw-Hill: New York, NY, 1969.
7. Blythe, C.R. "Approximate binomial confidence limits," *J. Amer. Statist. Assoc.*, **1986**, *81*, 843-855.
8. Gerlach, R.W.; White, R.J.; O' Leary, N.F.D.; Van Emon, J.M. *Superfund Innovative Technology Evaluation (SITE) Program Evaluation Report for Antox BTX Water Screen (BTX Immunoassay)*; EPA/540/R-93/518; U.S. Environmental Protection Agency: Las Vegas, NV, 1993.
9. Gerlach, R.W.; White, R.J.; Deming, S.N.; Palasota, J.A.; Van Emon, J. M. "An evaluation of five commercial immunoassay data analysis software systems," *Anal. Biochem.*, **1993**, *212*, 185-193.
10. O'Connell, M.A.; Belanger, B.A.; Haaland, P.D. "Calibration and assay development using the four-parameter logistic model," *Chemometrics Intell. Lab. Syst.*, **1993**, *20*, 97-114.
11. Brady, J.F. "Interpretation of immunoassay data," In *Immunoanalysis of Agrochemicals: Emerging Technologies;* Nelson, J.O.; Karu, A.E.; Wong, R.B., Eds.; ACS Symposium Series 586; American Chemical Society: Washington, DC, 1995, pp 266-287.
12. Meyers, R.H. *Classical and Modern Regression with Applications*; Duxbury Press: Boston, MA, 1986.
13. Gallant, A.R. *Nonlinear Statistical Models*; Wiley: New York, NY, 1987.
14. Currie, L.A. "Nomenclature in evaluation of analytical methods including detection and quantification capabilities," *Pure & Appl. Chem.*, **1995**, *67*, 1699-1723.
15. U.S. EPA. "Definition and procedure for determination of the method detection limit, Revision 1.11," Title 40, Part 136, Appendix B; *Code Fed. Reg.*; U.S. Government Printing Office: Washington, DC, July 1, 1995, pp 882-884.
16. Glaser, J.A.; Foerst, D.L.; McKee, G.D.; Quane, S.A.; Budde, W.L. "Trace analyses for wastewaters," *Environ. Sci. Technol.*, **1981**, *15*, 1426-1435.
17. Currie, L.A. Detection: "Overview of historical, societal, and technical issues," In *Detection in Analytical Chemistry: Importance, Theory, and Practice;* Currie, L.A., Ed.; ACS Symposium Series 361; American Chemical Society: Washington, DC, 1988, pp 1-62.
18. Currie, L.A. "Limits for qualitative decision and quantitative determination," *Anal. Chem.*, **1968**, *40*, 586-593.
19. Hubaux, A.; Vos, G. "Decision and detection limits for linear calibration curves," *Anal. Chem.*, **1970**, *42*, 849-855.
20. Clayton, C.A.; Hines, J.W.; Elkins, P.D. "Detection limits with specified assurance probabilities," *Anal. Chem.*, **1987**, *59*, 2506-2514.
21. Gibbons, R.D.; Jarke, F.H.; Stoub, K.P. "Detection limits for linear calibration

curves with increasing variance and multiple future detection decision," In *Waste Testing and Quality Assurance;* Friedman, D., Ed.; Vol. 3, ASTM STP 1075; American Society for Testing and Materials: Philadelphia, PA, 1991, pp 377-390.

22. Keith, L.H.; Crummett, W.; Deegan, J.Jr.; Libby, R.A.; Taylor, J.K.; Wentler, G. "Principles of Environmental Analysis," *Anal. Chem.*, **1983**, *55*, 2210-2218.

23. U.S. EPA. *Guidance on Evaluation, Resolution, and Documentation of Analytical Problems Associated with Compliance Monitoring;* EPA/821/B-93/001; Office of Water, U.S. Environmental Protection Agency: Washington, DC, June, 1993.

24. Adams, P.B.; Passmore, W.O.; Campbell, D.E., paper No. 14, *Symposium on Trace Characterization – Chemical and Physical*; National Bureau of Standards: Washington, DC, October, 1966.

25. Gibbons, R.D.; Grams, N.E.; Jarke, F.H.; Stoub, K.P. "Practical quantitation limits." *Chemometrics Intell. Lab. Sys.*, **1992**, *12*, 225-235.

26. Davidian, M.; Haaland, P.D. "Regression and calibration with nonconstant error variance." *Chemometrics Intell. Lab. Sys.*, **1990**, *9*, 231-248.

27. Feldkamp, C.S.; Smith, S.W. "Practical guide to immunoassay method validation," In *Immunoassay: A Practical Guide*; Chan, W.C., Ed.; Academic Press: Orlando, FL, 1987, pp 49-95.

28. Rodbard, D. "Statistical quality control and routine data processing for radioimmunoassays and immunoradiometric assays," *Clin. Chem.*, **1974**, *20*, 1255-1270.

29. Sadler, W.A.; Smith, M.H.; "Use and abuse of imprecision profiles: Some pitfalls illustrated by computing and plotting confidence intervals," *Clin. Chem.*, **1990**, *36*, 1246-1250.

30. Schwartz, L.M. "Nonlinear calibration," *Anal. Chem.*, **1977**, *49(13)*, 2062-2068.

31. Long, G.L.; Wineforder, J.D. "Limit of detection," *Anal. Chem.*, **1983**, *55(7)*, 712A-724A.

32. Porter, W.R. "Proper statistical evaluation of calibration data," *Anal. Chem.*, **1983**, *55(13)*, 1290A.

33. Shah, V.P.; Midha, K.K.; Dighe, S.; McGilveray, I.J.; Skelly, J.P.; Yacobi, A.; Layloff, T.; Viswanathan, C.T.; Cook, C.E.; McDowall, R.D.; Pittman, K.A.; Spector, S. "Analytical methods validation: Bioavailability, bioequivalence, and pharmacokinetic studies," *J. Pharm. Sci.*, **1992**, *81(3)*, 309-312.

34. Montgomery, D.C. *Design and Analysis of Experiments, Second Ed.*, Wiley, New York, NY, 1984.

35. Evans, J.C.; Coote, B.G. "Matching sampling designs and significance tests in environmental studies," *Environmetrics*, **1993**, *4(4)*, 413-437.

HUMAN EXPOSURE ASSESSMENT

Chapter 24

Biomonitoring for Occupational Exposures Using Immunoassays

Raymond E. Biagini, R. DeLon Hull, Cynthia A. Striley,
Barbara A. MacKenzie, Shirley R. Robertson, Wendy Wippel,
and J. Patrick Mastin

National Institute for Occupational Safety and Health,
Centers for Disease Control and Prevention, Public Health Service,
U.S. Department of Health and Human Services,
4676 Columbia Parkway, Cincinnati, OH 45226

Biomonitoring for occupational exposures involves measurement of parent compounds or metabolites in excreta (usually urine), sera or exhaled breath. Classically, biomonitoring involves collection of the matrix, separation and/or isolation of the compounds of interest from the matrix, followed by identification and quantification of the analytes. In most cases this procedure is labor intensive and involves the need for specialized high capital expenditure equipment. Alternative methods for biomonitoring exist that use immunochemical techniques rather than classical chemical techniques for quantification. Immunochemical methods have advantages over classical chemical techniques in speed and cost of analyses, and capital expenditure for equipment, and in most cases are more sensitive than chemical techniques. In the present monograph we review the use of enzyme linked immunosorbent assay (ELISA) immunochemical techniques for the detection and quantitation of pesticides and/or metabolites with comparison to classical analytical techniques. In addition, the use of circulating antibodies developed in response to xenobiotic exposure are also discussed as potential biomarkers. These "legacy biomarkers" of exposure have potentially far reaching medico-legal and other ramifications inherent in their use as they can serve as biomarkers of exposure in the absence of any chemically detectable analyte in excreta or blood.

Exposure to a xenobiotic does not necessarily mean the existence of a body burden of an agent. In order for a xenobiotic to gain entrance it must be absorbed. Occupational absorption can occur via dermal, inhalation, ingestion or a combination of routes. Whether or not absorption occurs depends on the chemical properties of the xenobiotic, in general, related to its lipid solubility. Once absorbed, a chemical is distributed and partitioned into various tissues due to tissue variations in pH, permeability, etc. More water soluble chemicals are absorbed throughout the total body water, while more lipophilic substances may be distributed totally in body fat. The loss of chemical from the body can loosely be defined as elimination, which depends on metabolism and excretion. Chemicals may be eliminated by numerous routes including fecal, urinary, exhalation, perspiration and lactation. A chemical can be excreted from the body without metabolism, in which case the parent chemical will be detectable in the urine or other excreta. In other cases, the chemical may be metabolized, which is the process of chemical alteration of the xenobiotic in the body. Metabolism may occur in numerous body tissues including liver, kidney, brain, etc. In most cases, metabolism is the result of oxidation, reduction, hydrolysis, or combination of these processes followed by conjugation, however, direct conjugation with an endogenous substrate is also a pathway for excretion. Most important conjugation reactions include glucuronidation, amino acid conjugation, acetylation, sulfate conjugation and methylation. Glucuronidation is the most common metabolic pathway. Metabolism/excretion and the rate of metabolism/excretion can be affected by age, diet, general health status, race, as well as other factors. In general, the metabolized chemical will be more water soluble than the parent. Also, there may be more than one type of metabolite produced from exposure to one parent (e.g., parent-glucuronide, parent-sulfate, etc.). The amount and ratios of parent-metabolites produced are affected by an individual's general health status, diet, nutrition, degree of hydration, time after chemical exposure, etc. In general, the kidney is the major organ of excretion and is the primary route of excretion of water soluble substances.

Biological Monitoring

Biological monitoring has the potential to assess worker exposure to industrial chemicals by all routes including skin absorption and ingestion. However, biological monitoring is not without its limitations. One limitation is the lack of detailed information on the metabolic fate of industrial chemicals in humans. Most of the toxicological/pharmacological (absorption, distribution, metabolism, excretion) information available is from experimental animals and not easily applied to humans. Another concern is the apparent wide variability seen between individuals in response to a toxicant exposure. The human response to the same exposure of a particular chemical may vary widely. This variability has two sources, 1) variability associated with differences in the penetration of the chemical from the environment to the target organ where the enzyme or biochemical system is affected, and 2) variability associated with differences among individuals in the response and delay of the response of the target organ itself. Working conditions in industry are likely to vary considerably from day to day as well as within the shift due to the fluctuation of the exposure concentration. Inhalation exposure and consequently uptake is not constant. The fluctuation of environmental exposure results in fluctuation of concentrations in the target organ(s). The effect of fluctuation of exposure intensity on biological concentrations depends on the kinetic behavior of the chemical in the body. Target organ concentrations of chemicals with short biological half-lives closely follow the environmental concentration and therefore have a larger variability. On the other hand, levels of chemicals with long biological half-lives fluctuate very little in the target organs in the majority of biological monitoring data (*1*). Proper knowledge of the fate of a chemical, its pharmacokinetic properties, specific methodologies and other factors described later can control such variability; however, specific methodology and data interpretation remain an obstacle to widespread use of biological monitoring (*2*).

Biological Monitoring for Occupational Exposures. Biological monitoring for occupational exposures usually involves the detection of analytes in matrices such as blood, urine or exhaled air. The classical procedure used to identify and quantitate the analyte of interest includes isolation of the analyte, separation of the analyte from other potentially interferring substances and quantitation by instrumental or other methods. These classical methods have many shortcomings including being highly labor intensive, requiring capital expenditures for expensive equipment, (e.g., gas chromatographs [GC], liquid chromatographs [LC], mass spectrometers [MS] or combinations of these instruments [e.g., GC-MS, LC-tandem-MS]). In addition, recoveries during the separation and isolation phases of the procedure may not be constant, and in some cases be associated with the level of analyte in the original sample, potentially yielding confounding systematic errors. Despite these shortcomings, when adequately controlled, classical chemical biological monitoring has the capacity to quantitate the body burden of substances to the sub-ppb level.

Immunoassays for Urinary Biomarkers

Numerous investigators have described urinary immunoassay as a screen for occupational exposure to a variety of compounds including pesticides (*3-4*). Screening assays are generally quite sensitive, specific and accurate. However, unique patterns of sensitivity and cross-reactivity appear to be assay dependent and a detailed knowledge of assay performance characteristics is necessary for accurate interpretation of urine testing data (*15*). Screening immunoassays have been shown to have high concordance with instrumental methods for the analysis of clinical specimens (*6*). Immunoassays have also been proposed for use in the screening of large numbers of water samples for the analysis of pesticide residues, as they are cost effective and highly efficient (*7*).

Although many examples of the use of immunoassay for biological monitoring could be cited, (e.g., other investigators (*4*) have also shown the usefulness of ELISA techniques as biomarkers of pesticide (atrazine) exposure), a detailed description of one such use in our laboratory will be reviewed comprehensively to give the reader an appreciation of the some of the inherent difficulties and advantages of these techniques. We recently reported on the use of a commercial enzyme linked immunosorbent assay (ELISA) to qualitatively evaluate the body burdens of alachlor or alachlor metabolites in urines collected from pesticide applicators (*8-10*). Twenty pesticide applicators and seven hauler/mixers participated in the study. Also, eight employees of the pesticide application companies, who were thought to have limited exposure to pesticides, submitted urine samples for estimates of alachlor dose. All study participants were male. Participants in the study were asked to provide three urine voids over a 24-hr period: one on the morning of the exposure survey before they began work; one at the end of the application period; and one as the first-void sample the morning following the exposure survey. Each void was collected separately in a wide-mouthed 500-mL polyethylene bottle, and the time and volume of the void were noted. Two 25- to 50-mL aliquots of each void were transferred to 60-mL high density polyethylene bottles and immediately frozen on dry ice. To estimate possible contamination of urine samples during voiding, a second uncapped 500 mL high density polyethylene bottle containing 50 mL of distilled water was taped to the side of the urine collection bottle (N=4). The urine samples were analyzed by both high-performance liquid chromotography (HPLC) and ELISA techniques.

High-performance Liquid Chromatography (HPLC). Briefly, putative alachlor metabolites present in the urine are alkaline-hydrolyzed at 150°C, and the resultant diethylaniline (DEA) produced is quantitated by HPLC. The urine samples were hydrolyzed in methanol/ sodium hydroxide, and hydrolysis performed for 1 hr in a bath consisting of sand and aluminum oxide which was fluidized by compressed air and maintained at 150°C. In order to control for hydrolyses and systematic losses, normal volunteer control urine samples were spiked with a DEA-yielding pseudo-metabolite of alachlor [(2-[2,6-diethylphenyl)-(methoxymethyl)-amino]-2-oxo-ethane acid, pseudo-DEA] and analyzed by HPLC as above.

Enzyme Linked Immunosorbent Assay (ELISA). A commercially available immunoassay kit (EnviroGard™, ImmunoSystems, a subsidiary of Millipore Corp, Scarborough, ME), designed for the analysis of alachlor in water, was also modified for the urinary analyses. The immunoassay format is a competitive solid phase ELISA method which is based on the inhibition of the reaction of enzyme-labelled (horseradish peroxidase) alachlor with immobilized polyclonal anti-alachlor antibodies by free alachlor present in the standard or test sample. Briefly, 80 μL standardized sample or a diluted urine sample were added (in triplicate) to each of the wells of the pre-coated 96-well microtiter plates. Eighty (80) μL of alachlor-enzyme-conjugate were then added to each well and the plates were covered and mixed on an orbital shaker (200 rpm) for 2 hr at room temperature. Plates were then thoroughly washed, substrate (hydrogen peroxide)added followed by chromogen (tetramethylbenzidine) followed by incubation at room temperature for 1 hr, again, with shaking. Forty (40) μL of stop-solution (H_2SO_4) were then added and the plate solutions were agitated again (200 rpm). The absorbance of the solutions in the wells were read on an automatic microplate reader at 450 nm against an air blank.

The ELISA and HPLC analytical methods gave statistically significantly ($P<0.0001$, one way ANOVA) different results when applied to the 82 specimens that were above the analytical LOD for both methods. The mean result for the samples, analyzed by ELISA (N=82) with results above the analytical LOD was 22.6 ± 1.79 μmole/L as alachlor equivalents (± standard error, [SE]), while the HPLC method gave a mean result of 3.23 ± 0.38 μmole/L DEA (Figure 1). When correlation between the two methods was investigated using simple orthogonal regression techniques, a highly significant ($P<0.0001$; r=0.89) linear association was observed. The relationship between the two methods was ELISA results (as alachlor equivalents [μmole/L]) = 4.12 HPLC (as DEA, [μmole/L]) + 9.25 (Figure 2). These results demonstrate a positive bias

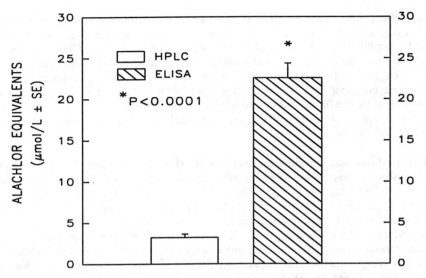

Figure 1. Results for alachlor equivalents determinations by HPLC (μmole/L DEA and ELISA μmole/L), N=82. Adapted from ref. 8.

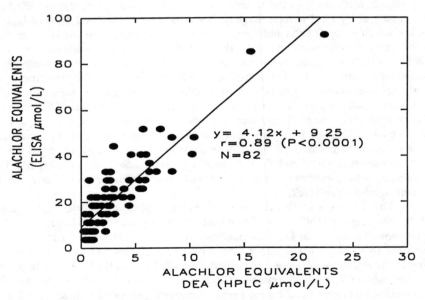

Figure 2. Orthogonal regression analysis of alachlor equivalents measured by HPLC and ELISA. The equation for the regression line is given on the Figure. Adapted from ref. 8.

by the ELISA method when compared with the hydrolysis/HPLC method (8-10). The basis of this systematic bias is unknown, but probably is related to similarities in structure between putative human alachlor metabolites excreted in the urine and the primary immunogen used to produce the polyclonal antibodies used as the basis of commercial kits.

Alachlor, MW 270, is too small to be immunogenic in its own right. To overcome this, most antibodies for alachlor and other chloroacetanilide herbicides are raised against a derivatized chloroacetanilide that is coupled to a carrier macromolecule (usually a protein) with a thioether linkage (7,15). Polyclonal antisera to these alachlor-protein-thioethers would be expected to contain antibodies to numerous antigenic determinants on the immunogen molecule, including the thioether region, probably with differing affinities and avidities for each antigenic determinant. We hypothesized that higher affinity of the putative thiolated human urinary metabolites of alachlor present in the operators' urine was the reason for the discrepancy between our observed HPLC and ELISA results. In order to test this hypothesis, alachlor mercapturate (a known human metabolite of alachlor metabolism in humans, personal communication, Jack Driscoll, CDC/CEH) was synthesized. Briefly, alachlor was reacted with N-acetyl cysteine in the presence of sodium methoxide as a base. The resulting reaction mixture is adjusted to pH ~ 7 with aqueous sodium phosphate followed by continuous liquid/liquid extraction with methylene chloride. The existence of the mercapturate was verified by fast atom bombardment mass spectrum (FAB-MS) MH+ peak at 397 atomic mass units (amu) and the electron impact mass spectrum (EI-MS) M+/. peak at 396 amu.

When alachlor mercapturate was evaluated for binding in the Millipore kit, it was found that essentially parallel standard curves were observed, with the alachlor mercapturate curve shifted to lower concentrations yielding greater (lower optical density) responses. Linear interpolation of the two curves suggests that that if alachlor mercapturate concentrations were interpolated from a standard curve produced with alachlor parent, an approximate 5X overestimation in concentration would occur (see Figure 3).

Immunoassays for Circulating Antibodies

The measurement of circulating antibodies for biomonitoring of exposure is not new, and is commonly used in clinical medicine for assistance in the diagnosis of diseases where exposure to a pathogen has caused an antibody response. The use of anti-xenobiotic antibodies as biomarkers of exposure, has also been reported (11-13). And in some cases, such as immediate hypersensitivity diseases, the biomarker (IgE) serves as both a marker of exposure and is pathognomic in diagnosis. Again, to describe these phenomena, a detailed example from our laboratory will be reviewed. An interesting concept regarding the use of circulating antibodies for detecting exposure is that the half-life of the antibody may be longer than the half-life for elimination of the parent compound or metabolite. In this situation, a specific biomarker of exposure is present in the absence of chemically detectable parent or metabolite in serum or excreta. This is more of a "legacy" or history of exposure rather than a direct biomonitoring technique. Also, "legacy biomonitoring" is subject to false negatives, but not to false

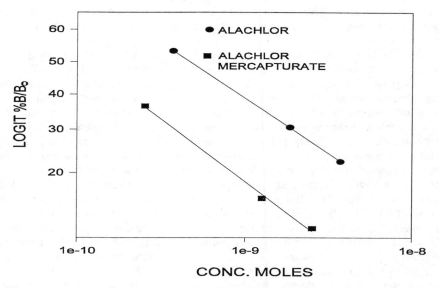

Figure 3. Comparison of standard curves of alachlor vs. alachlor mercapturate. For clarity in direct comparisons, concentrations are given in moles. Adapted from ref. 8.

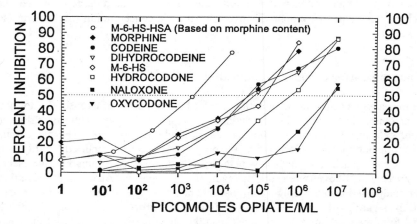

Figure 4. ELISA inhibition studies with opiate nucleus containing compounds. Fifty percent inhibition is indexed by a dotted line. Adapted from ref. 16.

positives, i.e., one can only make antibodies to substances to which they are exposed (with the caveat of antibody cross-reactivity to similar substances due to the polyclonal nature of the human antibody response) . In the case of environmental exposures, numerous individuals are exposed to numerous environmental immunogens daily, (e.g., molds and pollen). Not all individuals will make the same type, amount or specificity of antibody from these exposures as there are individual genetic and other factors which control antibody production. However, in the case of an occupational or environmental agent to which there are few confounding environmental exposures (e.g., soluble platinum halide salts, opiates), the existence of antibodies indicates, with some exceptions, a positive exposure history to the compound or class of compounds.

Some reactive small molecular weight molecules, while not immunogenic in their own right because of size and other limitations, may bind to constitutive polymers (such as host proteins) and become immunogenic, causing the production of specific antibodies. Alternatively or in addition, exposure to some small molecular weight proteins may cause the production of new antigenic determinants (NADs) formed by interaction of these relatively reactive small molecular weight compounds with selected protein carrier molecules. Antibodies can be made to these NADs of constitutive proteins or to the parent hapten-conjugate.

In an investigation of factory workers who extract morphine and other related alkaloids from opium gum or related opium poppy (Papaver somniferum) concentrates, antibodies to opiates were observed. Morphine-6-hemisuccinate (M-6-HS) was prepared by heating morphine (Morphine Alkaloid Powder U.S.P.) with succinic anhydride. The M-6-HS was then conjugated to human serum albumin, dialyzed under reduced pressure and purified by gel filtration. Specific IgG antibodies to M-6-HS-HSA were measured by a modified indirect microtitre plate ELISA method. In order to determine the specificity of human antimorphine antibodies, inhibition studies with morphine nucleus containing pharmaceuticals were performed (see Figure 4). Varying concentrations of M-6-HS-HSA, morphine sulfate, codeine phosphate, dihydrocodeine bitartrate, oxycodone HCl, hydrocodone and naloxone HCl solutions were incubated with a positive serum known to have high levels of specific IgG antibodies to M-6-HS-HSA, for two hours at 37°C. Following this 2 hr preincubation, the sera were analyzed by ELISA as above. The amount of specific IgG binding (in triplicate, represented by optical density) contained in the inhibited serum was compared to that in the uninhibited serum, and the percent inhibition calculated. As can be seen, there was cross- reactivity between the different opiate compounds tested, indicating, as would be expected, a polyclonal antibody response from opiate exposure. These findings have been essentially corroborated by other investigators studying heroin addicts (*14*).

The usefulness of "legacy biomonitoring" becomes most apparent when one wants to evaluate the results of engineering controls for reducing exposures. It would be impractical or impossible to daily monitor the urine or other excreta of someone exposed to a xenobiotic in order to evaluate if an engineering control methodology was effective. However, within certain limitations, antibody levels are related to exposure levels. In general, if antibody levels are reduced, exposures have been reduced. For example, in the study of workers exposed to opiates previously described, engineering controls were initiated to control exposures in February, 1988, and sera were collected

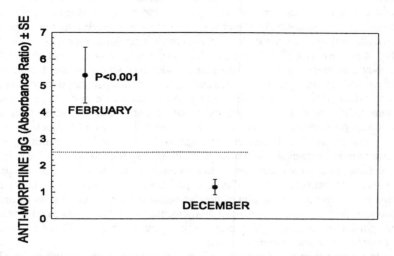

Figure 5. Morphine specific antibodies in workers' sera in February and December of the same year after implementation of an improved respiratory protection program. The dotted line indicates the mean absorbance ratio value of normal non-exposed controls. See text for details.   Adapted from ref. 16.

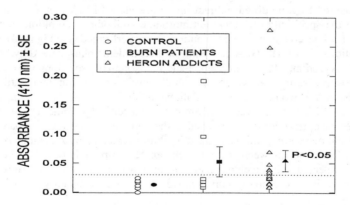

Figure 6. Morphine antibody levels in heroin abusers, burn patients, and non-opiate exposed controls. Individual absorbance values as well as group mean absorbance values (filled symbols ± standard error; variance of the control patient means is within the symbol) are given on the Figure. The dotted line indicates 2.5 X the mean absorbance value of normal non-exposed controls. Adapted from ref. 16.

for antibody levels. Ten months later in December, 1988, sera were again collected and evaluated for antibody levels. As can be seen from Figure 5, results from workers who provided sera at both the February and December, 1988, testing periods showed statistically significant reductions in antibody levels at the December 1988 testing period, coinciding with the improved engineering controls and putative reductions in opiate exposures. Data are presented as mean $\pm$ SE (standard error) absorbance ratios (5.4 ± 0.95 in February, 1988 and 1.19 ± 0.95 in December, 1988). ELISA results were normalized by calculating an absorbance ratio (absorbance of experimental sera/absorbance of volunteer negative control sera [N=6]). The negative control sera were obtained from individuals in the Cincinnati, OH, area with no known chronic or abusive opiate exposure except for possible sporadic therapeutic exposure to low levels of codeine-containing analgesics and antitussives.

It seemed apparent that anti-morphine antibodies could be used to determine morphine exposure in the absence of identifiable parent or metabolite in excreta. We obtained 28 samples from the National Institute on Drug Abuse (NIDA) for use in a double-blind study. Eighteen of the samples were obtained from 8 healthy male, HIV negative, individuals admitted to NIDA for treatment. These subjects identified heroin as their drug of choice, had been drug-free for a minimum of three days, and all had non-detectable levels of heroin or morphine (opiates) in their urine. Seven other samples were obtained from burn patients prior to surgery, who were considered drug-free except for perhaps sporadic use of morphine for analgesia. Three control samples were obtained from NIDA staff members. Two of 8 of the burn patient samples and 7/18 of the heroin abuser samples were positive (≥ 2.5 times mean control absorbance). Mean absorbance values for the heroin abusers were significantly greater than control values ($P<0.05$). Sera from burn patients, while having positive evidence of antibodies in 2 individuals, was statistically indistinguishable from control sera (Figure 6).

Summary

In the present monograph, we reviewed some of the methods we have used with immunoassays to detect exposure to xenobiotics in occupationallly and other exposed individuals. We have divided our discussion into two distinct parts, immunochemical measurements of parent or metabolite in excreta (urine) or measurements of specific antibodies produced from exposure to selected xenobiotics (legacy biomonitoring). Examples were given from our past work and describe the benefits and disadvantages of both methods when applied to real field study paradigms.

Literature Cited

1. Droz, P. O. *Appl Ind Hyg*, **1989**, 4, 1-20.
2. Hull, R.D., Lowry, L/K. **1994**. *Special Considerations for Biological Samples*.In Eller, P.M. (ed). NIOSH Manual of Analytical Methods, 4th ed., DHHS (NIOSH) Publication No. 94-113, National Institute for Occupational Safety & Health, Cincinnati, OH.
3. Feng, P.C.C., Sharp, C.R., Horton, S.R., **1994**, *J Agric Food Chem* 42, 316-319.

4. Lucas, A.D., Jones A.D., Goodrow M.H., Saiz S.G., Blewett C., Seiber J.N., Hammock   B. D., **1993**, *Chem Res Toxicol* 6, 107-16.

5. Cone E.J., Dickerson S., Paul B.D., Mitchell J.M., **1992**, *J Anal Toxicol* 16: 72-78

6. Cone E.J., Yousefnejad D., Dickerson S.L. **1990**, *J Forensic Sci* 4: 786-91

7. Feng P.C.C., Wratten S.J., Horton S.R., Sharp C.R., Logusch E.W. , **1990**, *J Agric Food Chem* 38, 159-163.

8. Biagini, R.E., Henningsen, G.M., MacKenzie, B., Sanderson, W.T. **1995**, *Bull Env Contam Toxicol*, 54,245-250.

9. Sanderson, W.T., Biagini, R.E., Henningsen, G.M., Ringenburg, V., MacKenzie, B.M., **1995**, *Amer Ind Hyg J*, 56,883-889.

10. Sanderson, W.T., Ringenburg V., Biagini,R.E., **1995**, *Amer. Ind. Hyg. J*, 56,890-897.

11. Biagini, R.E., Bernstein, I.L., Gallagher, J.S., Moorman, W.J., Brooks, S.M., Gann, P., **1985**, *J Allergy Clin Immunol* 76, 794-802.

12. Biagini, R.E., Gallagher, J.S., Moorman, W.M., Knecht, E.A., Smallwood, W., Bernstein, I.L., Bernstein, DI, **1988**, *J Allergy Clin Immunol*, 82, 23-29.

13. Biagini, R.E., Driscoll, R., Henningsen, G.M., et al., **1995**, *J Appl Tox*, in press.

14. Gamaleya N., Tagliaro F., Parshin A., Vrublevskii A., Bugari G., Dorizzi R., Ghielmi S., Marigo M., **1993**, *Life Sci* 1993, 53,99-105.

15. Rittenberg, J.H., Grothaus, G.D., Fitzpatrick, D.A., Lankow, R.K., **1991**, *Rapid on-site Immunoassay System, Agricultural and Environmental Applications*. In Vanderlann M, Stanker LH, Watkins BE, Roberts, DW (ed) Immunoassays for Trace Chemical Analysis, American Chemical Society Symposium Series 451, American Chemical Society, Washingtion, DC, pp, 28-39.

16. Biagini, R.E., Klincewitz, S.L., Henningsen, G.M., Mackenzie, B., Gallagher, J.S., Bernstein, I.L., and Bbernstein, D.M., **1990** *Lif Sci* 47:897-908.

Chapter 25

Application of an Immunomagnetic Assay System for Detection of Virulent Bacteria in Biological Samples

Hao Yu[1] and Peter J. Stopa[2]

[1]Calspan SRL Corporation and [2]SCBRD-RTE, Building E 3549, U.S. Army Edgewood Research, Development and Engineering Center, Aberdeen Proving Ground, MD 21010

Virulent pathogenic bacteria could pose a serious health threat in contaminated food and water resources. Traditional bacterial culture methods or enzyme linked immunosorbent assay (ELISA) for identification of the bacteria are time consuming and labor intensive. Some new technologies are very sensitive but analysis time can be lengthy. For example, the polymerase chain reaction (PCR) can be used to amplify small quantities of genetic material to determine the presence of bacteria. This is a sensitive method that requires pure samples and considerable laboratory time. Alternative methods should be quicker and retain sensitivity. The magnetic separation technique appears promising for rapid bacterial isolation from the media prior to the detection. In this work, immunomagnetic assay system (IMAS) has been coupled to an electrochemiluminescent (ECL) technology for rapid and sensitive bacterial detection of biological samples within an hour. The sensitivity of the IMAS-ECL for *Bacillus anthrax* spores (sterne), *Escherichia coli* O157:H7 and *Salmonella typhimurium* detection is about 1000 cells/mL in biological samples. In addition, IMAS can also be coupled to a flow cytometer or any analytical instruments for target agent monitoring. Results of this study strongly suggest that IMAS methodology is useful for rapid and sensitive detection.

Rapid and sensitive screening methods for *Escherichia coli* O157:H7, *Salmonella typhimurium* and other virulent bacteria such as *Bacillus anthracis* in contaminated food, water and other biological samples are important to prevent the spread of bacteria. Traditional methods for bacterial identification and detection are time consuming (e.g., membrane filtration onto eosin-methylene blue agar or culture takes 24-48 hours). The immunoblotting technique is very sensitive and the detection levels of 1-10 colony forming unit/g (CFU/g) for *E. coli* O157 can be obtained by using capture on hydrophobic grid membranes (1, 2, 3), or by overnight culture.

The polymerase chain reaction and restriction fragment mapping identification techniques can be definitive and extremely sensitive, but require hours of processing and expertise in molecular biology. Enzyme linked immunosorbent assay (ELISA) can be rapid (less than a few hours), but it is labor intensive and multiple pipetting is required. An immuno-latex bead based agglutination assay has been used to detect *E. coli* O157:H7 (4), but the assay is involved a pre-enrichment culture procedure which limits the application for rapid screening.

An immunomagnetic assay system (IMAS) has been developed for effective magnetic particle capture followed by rapid detection techniques. This approach can speed up the assay time to less than one hour and also increase the sensitivity by significantly reducing biological sample interference. The IMAS includes a magnetic separator for capturing the antigen and an electrochemiluminescent (ECL) detector for detection. Detection sensitivities of as low as picogram or attomogram levels can be achieved for bacteria and toxoid, as well as ds-DNA (5, 6, 7) using ECL. Alternatively, in IMAS configuration, a fluorescence microscope can be used for bacterial identification or a continuous fluorimeter or a flow cytometer can be coupled to IMAS for routine positive antigen screening.

The current work illustrates the utility of easy, rapid, and sensitive detection for *B. anthrax* spores, *E. coli* O157:H7 and *Salmonella sp.* in biological samples by use of IMAS. The extension of current methodologies could apply to environmental and clinical needs by isolating specific molecules and soluble antigens in biological samples.

Experimental procedures

1. Bacteria, antibodies and magnetic particles

Heat-killed *E. coli* O157:H7, *Salmonella typhimurium* and anti-*E. coli* O157, -*Salmonella sp.* antibodies were obtained from Kirkegaard Perry Labs.(KPL; Gaithersburg, MD). Irradiated *E. coli* O157:H7 and *Salmonella sp.* and ATCC-11775 (*E. coli*) were obtained from USDA (Philadelphia, PA). Nonpathogenic *E. coli* O111:B4 strain and *B. anthrax* spores (Sterne strain) were obtained from Sigma Chemical Co.(St. Louis, MO) and USAMRIID (Ft. Detrick, MD), respectively. Anti-*Salmonella sp.* antibody is broadly reactive with all sero group D of *Salmonella sp.*. Goat anti-*B. anthracis* GT-576, -577 and -578 antibodies were obtained from Antibodies Inc. (Davis, CA). These antibodies were very specific to surface antigens of the spore coating rather than vegetative cells. Bacterial cell counts were performed with a hemacytometer and the stock bacterial suspensions in phosphate buffered saline (PBS, 10 mM phosphate, pH 7.4) contained 10^9 cells/mL. Streptavidin-coated magnetic beads, Dyna M-280 and MACS microbeads were obtained from Dynal, Inc. (Lake Success, NY.) and Miltenyi Biotec Inc. (Sunnyvale, CA.), respectively.

2. Biotin, fluorescein and Ru(bpy)$_3^{2+}$-antibody conjugations

N-hydroxy-succinimide (NHS)-biotin and fluorescein-NHS ester reagents from Molecular Device Corp. were used to label polyclonal anti-*B. anthracis*, *E. coli* and *Salmonella* antibodies. Ru(bpy)$_3^{2+}$- NHS ester was obtained from IGEN Corp.(Gaithersburg, MD) and used for antibody-conjugation (8). Both fluorescein- and biotin-antibody conjugation assays were performed for one hour. Unreacted labels were removed by using a G-25 Sephadex (PD-10) size exclusion chromatographic column (Pharmacia, Sweden). All protein concentrations were determined by the Bradford Protein Assay (BioRad Corp., Hercules, CA). A sandwich immunoassay, with biotinylated antibody as the primary capture antibody and fluorescein- or Ru(bpy)$_3^{2+}$-labeled antibodies as Tag-antibodies, was used in these studies.

3. Sample processing

Environmental water samples including bay water, pond water, stream water and tap water were collected along the Chesapeake Bay area in Maryland. Fresh ground beef and poultry samples were purchased from retail grocery stores. Liquid samples and minced solid samples (5 gram with their liquids) were placed in 15 mL polypropylene tubes and agitated for 30 minutes. One mL of supernatant from these samples was inoculated with varying amounts of target bacteria. Undiluted juice and water samples were similarly inoculated with bacteria.

4. IMAS and immunoassay

A diagram of the IMAS is shown in Figure 1. One mL of biological samples mixed with 100 μL (100 ng) biotinylated antibody plus 100 μL (100 μg) streptavidin-coated beads were processed by IMAS at a rate of 2 minutes/sample. Following the IMAS procedure, 100 μL of sample were collected for further measurement. An ORIGEN® analyzer from IGEN was used for ECL assay. The principle of ECL has been previously described (5, 6). Previously collected samples were incubated for 30 minutes with 100 mL of Tag-antibody (200 ng) prior to the ECL assay. Following the sampling process, an optical or fluorescence microscope can be employed for cell identification or other analytical instruments can be used for more definitive analysis such as a flow cytometer.

The IMAS was built in house. It works with four subsystem procedures which sequentially mix, magnetically capture, rinse and collect particles from the sample though a flow cell. Downstream from the flow cell, a peristaltic pump (Cole-Parmer, Chicago, IL) is used at a flow rate of 2 mL/minute.

Bacteria in artificially inoculated food supernatant fluids, environmental water and biological samples were subjected to the separator prior to the ECL assay. IMAS

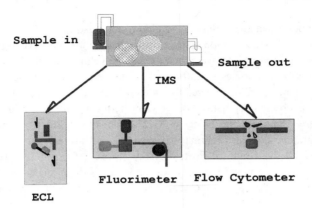

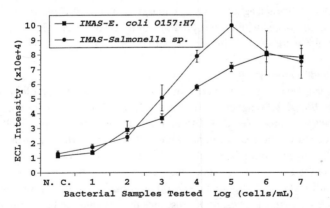

Figure 1.   A diagram of the IMAS. Samples are
initially processed by the IMAS, then analyzed by an
ECL, a fluorimeter and a flow cytometer. Samples
also can be collected on a micron-sized membrane
filter for identification by a fluorescence
microscope.

Figure 2.   ECL detection for *E. coli* O157:H7 and
Salmonella sp. assays in buffer.   Results indicate
the detection limits approximately 100 to 1000
cells/mL.   ECL values represent the mean and
standard deviation of three independent
measurements. Values in amount of antigens indicate
the log of cells/mL used in each assay.

was accomplished by adding 100 μL (200 ng) of biotinylated antibody to 100 to 200 μg of streptavidin-coated magnetic beads. This was followed by adding the antibody-coated magnetic beads to various concentrations of bacteria in 1 mL of sample supernatant for 15 minutes with continuous shaking. Samples were collected after magnetic separation. Magnetic particles were resuspended in PBS and washed again followed by resuspension in 50 μL of PBS. The ECL analyzer was operated at a photomultiplier tube (PMT) setting of 650V (or 1000x gains) and a carousel vortexing speed of 100 rpm.

In the fluorescence sandwich immunoassay, biotinylated antibody at 5 ng/μL was used for antigen (100 μL) capture and 100 μL of fluorescein-conjugated antibody at 5 ng/mL were used to enable detection. One hundred μL samples with or without bacteria were flowed through a Jasco-920 fluorometer (Jasco, Inc., Japan). The Bio-Rad Bryte HS® (Microscience Ltd., England) was used for flow cytometry studies. Forward light scattering and fluorescent data were obtained from the flow cytometer. The data were collected under the 1.1 μL/minute flow rate and 0.7 Bar pressure conditions.

Results

This experiment was designed to determine internal standards for *bacterial* detection. In Figure 2 the detection limits of ECL results for *E. coli* and *Salmonella* were approximately between 100 and 1000 cells/mL in buffer. Both results showed a dynamic range over four orders of magnitude. *E. coli* O111:B4 and ATCC-11775 at 10^6 cells/mL were selected as negative control antigens.

E. coli and *Salmonella sp.* were inoculated in various food and water samples. The final concentration of bacteria in these samples was 2000 cells/mL. Figure 3 shows the results of ECL assays with and without *E. coli*. The ECL intensities with different samples should be compared to the intensity in PBS buffer condition. The signal to noise ratios of biological samples that spiked with 2000 bacteria cells/mL in ECL determinations were between 5 and 10.

In Figure 4, the ECL results of *Salmonella* detection in different water and food samples are illustrated. Even though some interference can be seen in fish, beef and juice samples, the signal to noise ratios were still significant. Detection for *B. anthrax* spores in biological sample is shown in Figure 5.

In flow cytometry assays, a negative control assay was performed by injection of immunoparticles into the flow cytometer without bacteria present. The results of the negative control assays showed that there were no fluorescence (FL) or forward light scattering (LS) peaks in both cases (Figures 6a and 6b). Miltenyi MAC magnetic particles used in these assays were too small to be detected in the current assay scale. On the other hand, there were distinguishable peaks revealed in the LS result (Figure 6c) and the FL (Figure 6d) when *E. coli* antigen at 10^6 cells/mL was introduced into

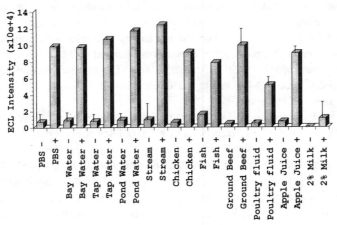

Figure 3. IMAS-ECL assay of *E. coli* O157:H7(2000 cells/mL) in various biological samples. ECL values represent the mean and standard deviation of three independent measurements.

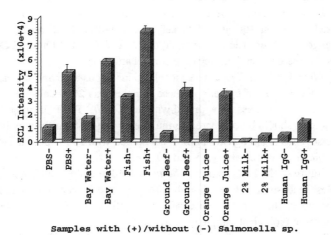

Figure 4. IMAS-ECL detection of *Salmonella sp.*(2000 cells/mL)in various food supernatant fluids and environmental samples. ECL values represent the mean and standard deviation of three independent measurements.

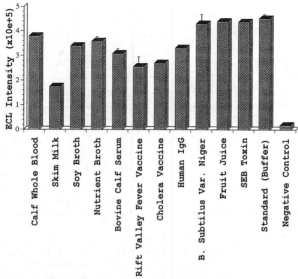

Inoculated B. anthrax Spores (500 cells/mL)

Figure 5.   IMAS-ECL detection of *B. anthracis* (500 cells/mL) in various biological samples.   ECL values represent the mean and standard deviation of three independent measurements.   All ECL values should be compared to the ECL value at buffer condition. Negative control is PBS without *B. anthrax* antigen.

the flow system. The shape and the location of the peaks in both histograms indicated that the particle size distribution of the bacteria was approximately a few micrometers in diameter. The peak height in both LS and FL (Figures 6b and 6d) windows corresponds to the quantity of the bacterial cells detected by flow cytometry.

In addition, pathogenic bacterial antigens from various samples were further examined by other techniques. Results from fluorescence microscopy showed that about 60-80% of the bacterial cells were captured by immunoparticles (data not shown). A distinguishable fluorescent signal was detected at $>10^5$ cells/mL by fluorimetric measurement (data not shown) in the fluorescence assay.

Discussion

IMAS provides a rapid, sensitive and facile technique for virulent bacterial detection that is sensitive to at least >1000 cells/mL in biological samples. Total IMAS assay procedure is less than one hour. The main advantages using magnetic separation prior

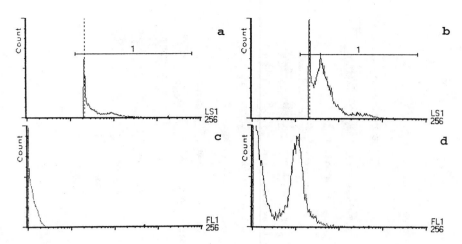

Figure 6. Forward light scattering (LS) and particle fluorescence (FL) histograms of IMAS-flow cytometry showed distinguishable peaks (b and d), respectively, in the presence of *E. coli* O157:H7 antigens (10^6 cells/mL) compared to the controls (without *E. coli* O157:H7 antigens) as a and c.

to the ECL or other assays are in reducing the direct interference from the biological samples; concentrating the target antigens from large volume to small volume; and avoiding the operator's direct contact with the samples which contain the pathogenetic bacteria. IMAS in conjunction with ECL is a powerful approach to perform very sensitive and rapid assays. The IMAS technique is capable of broader applications to any type of prokaryotic and eukaryotic cell assay, as well as soluble antigens, and nucleic acids(5, 6, 7). The advantages of non-radioisotope labels, non-intrinsic fluorescence background and controlled electric potential make ECL technology more sensitive and effective than EIA, chemiluminescence and fluorescence in clinical and biological applications.

In the *E. coli* and *Salmonella sp.* immunoassays, both ECL responses were not linear over a broad range. This is probably related to the immunologic "Bell Effect" (9) which is apparent beyond $10^5, 10^6$ cells/mL or a simple physical over-loading of the ECL flow cell. In the latter hypothesis, increased amounts of captured bacteria do not lead to proportionately higher levels of ECL signal because bacterial absorption of light emitted by the ECL labels diminishes the ECL signal intensity reaching the PMT.

Results of the IMAS-ECL assay for *B. anthracis*, *E. coli* and *Salmonella sp.* showed that detection limits were about >100 cells/mL in PBS and >1000 cells/mL in most biological samples. These results may not be as sensitive as the immunoblotting technique; however, this is sensitive enough to detect the presence of bacteria in biological samples. More importantly, the early detection (within an hour) of this rapid assay could prevent these virulent bacteria from spreading and could save lives. Both heat-killed and irradiated bacterial *E. coli* cells were tested in my lab. It is believed that the heating process can cause bacterial cell membrane break-down. Therefore, cell surface antigens may not be in a condition for recognition by the antibody which was produced against normal surface antigen. Our results indicated that there were no difference in ECL assays whether heat-killed or irradiated bacterial cells were used. Results from other researchers (10, 11) using the same polyclonal antibodies from KPL (not necessary the same lot) demonstrated a similar detection sensitivity for both live and heat-killed *E. coli* O157:H7 and *Salmonella sp.*

Bacterial detection in biological samples constitutes authentic sample detection. In the case of solid meat samples, only the interference from their fluids was considered. The efficiency of bead collection from solid meat remains unknown. In order to perform quantitative ECL assays, positive detections in buffer were done as standard procedure prior to assays of real biological samples. Decreased sensitivity of ECL detection in biological samples was expected because of the sample interference. It was also anticipated that the competition between the primary and secondary antibodies could reduce the ability to detect bacteria. However, this was apparently not a significant problem in this protocol. The limitation of the ECL assay like any immunoassays is largely dependent upon the antibody affinity to the antigens; however, the advantage using IMAS-ECL instead of ELISA and culture-based methods is that IMAS-ECL is more sensitive than ELISA or fluorescence (6) and the total assay time is only one hour.

Many hand-held test kits and commercial products, including Dynal antibody-based test kits (Dynal, Inc.) have been developed for *E. coli* O157 and *Salmonella sp.* detection. These testing kits could be very rapid; however, they are not suitable for quantitative detection. In the present work, the IMAS method is useful for both sensitive detection and relative quantitation. The IMAS-ECL assay has shown the most sensitive detection for these bacteria, however, the results of IMAS-fluorimetry and IMAS-flow cytometry assays currently are not promising. Improvement of these assays could potentially increase sensitivity and make the IMAS more useful in biological detection as well as environmental monitoring.

Acknowledgement: This work was funded under U.S. Air Force HSC scholarship.

Literature cited

1.   Doyle, M.P. and Schoeni, J.L. Isolation of *Escherichia coli* O157 from retail fresh meats and poultry. Appl. Environ. Microbiol. 53, 2394-2396, (1987).

2.   Todd, E.C.D., Szabo, R.A., Peterkin, P., Sharpe, A.N., Parrington, L.,
        Bundle, D., Gidney, M.A.J. and Perry, M.B.   Rapid hydrophobic grid
        membrane filter-enzyme-labeled antibody procedure for identification and
        enumeration of *Escherichia coli* 0157 in foods. Appl. Environ. Microbiol.
        54, 2536-2540, (1988).
3.   Szabo, R., Todd, E., Mackenzie, J., Parrington, L. and Armstrong, A.
        Increased sensitivity of the rapid grid membrane filter enzyme-labeled
        antibody procedure for   *Escherichia coli* O157 detection in foods and bovine
        feces. Appl. Environ. Microbiol. 56, 3546-3549, (1990).
4.   Notermans, S., Wernars, K., Soentoro, P.S., Dufrenne, J. and Jansen, W.
        DNA- hybridization and latex agglutination for detection of heat-labile and
        shiga-like toxin-producing *Escherichia coli* in meat. Int. J. Food Microbiol.
        13, 31-40, (1991).
5.   Gatto-Menking, D.L., Yu, H., Bruno, J.G., Goode, M.T., Miller, M. and
        Zulich, A.W. Sensitive detection of biotoxoids and bacterial spores using an
        immunomagnetic electrochemiluminescence sensor.Biosensors Bioelectronics,
        10:501-50, (1995).
6.   Yu, H., Bruno, J.G., Cheng, T., Calomiris, J.J., Goode, M.T. and Gatto-
        Menking, D.L. A comparative study of PCR product detection and
        quantitation by electrochemiluminescence and fluorescence. J. Biolum.
        Chemilum., 10, 239-245,   (1995).
7.   Yu, H. Enhancing Immunoelectrochemluminescence for Sensitive Bacterial
        Detection, J. Immunol. Methods, in press, (1996).
8.   Blackburn, G.F., Shah, H.P., Kenten, J.H., Leland, J., Kamin, R.A., Link, J.,
        Peterman, J., Powell, M.J., Shah, A., Talley, D.B., Tyagi, S.K., Wilkins, E.,
        Wu, T.. and Massey, R.J.   Electrochemiluminescence detection for
        development of immunoassays and DNA probe assays for clinical diagnostics.
        Clinical Chem. 37, 1534-15, (1991).
9.   Ugelstad, J., Kilaas, L., Aune, O., Bjorgum, J., Herje, R., Schmid, R.,
        Stenstad, P.   and Berge, A. Monodisperse polymer particles: Preparation
        and new biochemical   and   biomedical applications. In: M. Uhlen, E.
        Hornes and O. Olsvik (Eds.) Advances in Biomagnetic Separation. Eaton
        Publishing, Natick, Mass., p. 6-7, (1994).
10.   Tison, D.L. Culture confirmation of *Escherichia coli* serotype O157:H7 by
        direct immuno-fluorescence, J.Clin. Microbiolgy 28, 3546-3549, (1990).
11.   Park, C.H., Hixon,D.L., Morrison, W.L. and Cook, C.B. Rapid diagnosis of
        enterohemorrhagic *Escherichia coli* O157:H7 dirctly from fecal specimens
        using immuno-fluorescence stain. Am. J. Clin. Pathol. 101, 91-94, (1994).

Chapter 26

Application of Enzyme-Linked Immunosorbent Assay for Measurement of Polychlorinated Biphenyls from Hydrophobic Solutions
Extracts of Fish and Dialysates of Semipermeable Membrane Devices

James L. Zajicek[1], Donald E. Tillitt[1], James N. Huckins[1],
Jimmie D. Petty[1], Michael E. Potts[1], and David A. Nardone[2]

[1]Midwest Science Center, National Biological Service,
U.S. Department of the Interior, 4200 New Haven Road,
Columbia, MO 65201
[2]Ohmicron Corporation, 375 Pheasant Run, Newtown, PA 18940

Determination of PCBs in biological tissue extracts by enzyme-linked immunosorbent assays (ELISAs) can be problematic, since the hydrophobic solvents used for their extraction and isolation from interfering biochemicals have limited compatibility with the polar solvents (e.g. methanol/water) and the immunochemical reagents used in ELISA. Our studies of these solvent effects indicate that significant errors can occur when microliter volumes of PCB containing extracts, in hydrophobic solvents, are diluted directly into methanol/water diluents. Errors include low recovery and excess variability among sub-samples taken from the same sample dilution. These errors are associated with inhomogeneity of the dilution, which is readily visualized by the use of a hydrophobic dye, Solvent Blue 35. Solvent Blue 35 is also used to visualize the evaporative removal of hydrophobic solvent and the dissolution of the resulting PCB/dye residue by pure methanol and 50% (v/v) methanol/water, typical ELISA diluents. Evaporative removal of isooctane by an ambient temperature nitrogen purge with subsequent dissolution in 100% methanol gives near quantitative recovery of model PCB congeners. We also compare concentrations of total PCBs from ELISA (ePCB) to their corresponding concentrations determined from capillary gas chromatography (GC) in selected fish sample extracts and dialysates of semipermeable membrane device (SPMD) passive samplers using an optimized solvent exchange procedure. Based on Aroclor 1254 calibrations, ePCBs (ng/mL) determined in fish extracts are positively correlated with total PCB concentrations (ng/mL) determined by GC: ePCB = 1.16 * total-cPCB - 5.92. Measured ePCBs (ng/3 SPMDs) were also positively correlated ($r^2 = 0.999$) with PCB totals (ng/3 SPMDs) measured by GC for

0097–6156/96/0646–0307$15.00/0

dialysates of SPMDs: ePCB = 1.52 * total PCB - 212. Therefore, this ELISA system for PCBs can be a rapid alternative to traditional GC analyses for determination of PCBs in extracts of biota or in SPMD dialysates.

Polychlorinated biphenyls (PCBs) are ubiquitous hydrophobic contaminants that continue to be of environmental concern. PCB exposures from contaminated areas continue to be linked to adverse effects in fish and wildlife (1) and humans (2). For this reason, a quick and simple method for measuring concentrations of total PCBs in biological sample extracts and other integrative samplers, such as SPMDs, is needed.

Enzyme-linked immunosorbent assay (ELISA) measurement of total polychlorinated biphenyls (ePCB) in water samples and methanolic extracts of soil is currently enabled by commercial kits (Ohmicron Corp., Newtown, PA, Millipore Corp., Bedford, MA, and EnSys, Inc., Research Triangle Park, NC). The kits from Ohmicron Corp. are configured for rapid PCB quantitation, while those kits from Millipore Corp. and EnSys, Inc. are configured for rapid semi-quantitative screening of PCBs. The latter kits can be used for PCB quantitation by inclusion of multilevel calibration and appropriate quality control materials (3). In these ELISA methods water samples are simply, quickly and inexpensively either analyzed directly, or PCBs are concentrated in methanolic eluates of solid-phase extraction (SPE) columns and analyzed. For ELISA analysis of soils, samples are collected, extracted with methanol, filtered, diluted with either additional methanol or an equal volume of aqueous buffered diluent, and analyzed. ELISA determination of PCBs in biological tissue extracts can be problematic, however. PCBs must be separated from non-polar biogenic compounds, such as lipids, which have limited compatibility with the polar solvents and the immunochemical reagents used in ELISA techniques. Although some have attempted to extract PCBs from tissues with polar solvents (4), they have met with only limited success due to problems associated with matrix co-extractables. Because of their hydrophobicity, PCBs have been traditionally extracted from biological tissues, separated from co-extracted lipids, and concentrated for instrumental analyses in nonpolar solvents, such as isooctane (**2,2,4**-trimethylpentane) and hexane (5). These solvents and PCBs have limited solubility in aqueous-methanolic solutions (6-9). Hydrophobic solvents in the methanolic ELISA solutions can cause errors in the ELISA results. Recently, it has been observed that low percentages of isooctane (0.5% [v/v] in methanolic sample solutions) can diminish the apparent ePCB concentration, by as much as 94% (3). Therefore, it was important to examine several approaches for transferring PCBs dissolved in hydrophobic solvents into PCB ELISA diluents, such as methanol and 50% (v/v) methanol/water.

The objectives of this work were: 1) to study the recovery and homogeneity of PCBs in solutions prepared for ELISA by direct dilution of isooctane solutions; 2) to evaluate nitrogen evaporation as a quick and efficient means for solvent removal; 3) to examine the efficacy of methanol and aqueous/methanol mixtures for quantitative recovery of residual PCBs, following evaporation of the hydrophobic solvent; and 4) to compare concentrations of ePCBs with their corresponding concentrations determined by gas chromatography (GC) in selected fish extracts and in dialysates of semipermeable membrane devices (SPMDs) used to monitor laboratory air.

Experimental

Materials.

PCBs. Solutions of Aroclors 1242, 1248, 1254, and 1260 (neat materials provided by Monsanto Corp., St. Louis, MO), ^{14}C-TePCB (18 mCi/mmol ^{14}C-radiolabeled **2,2',5,5'**-tetrachlorobiphenyl, PCB-52 [*10*], Pathfinders Labs, St. Louis, MO), and ^{14}C-HxPCB (**2,2',4,4',5,5'**-hexachlorobiphenyl, PCB-153, Pathfinders Labs, St. Louis, MO) were prepared in isooctane and methanol.

Hydrophobic Blue Dye. A 0.1-mg/mL solution of a nonpolar blue dye was prepared by dissolving a measured amount of Solvent Blue 35 (**1,4**-bis[butylamino]-**9,10**-anthraquinone, Sudan Blue II, CAS No. 17354-14-2, Sigma Chemical Company) in hexane.

Solvent Exchange. Microliter aliquots of isooctane solutions of two carbon-14 radiolabeled PCB congeners were diluted directly into 1.00 mL and 5.00 mL of 50% (v/v) methanol/water (50% MeOH/H$_2$O) or Ohmicron diluent (Ohm-diluent, a 50% [v/v] methanol/buffered aqueous solution with stabilizers). In some direct dilution experiments, after mixing, the undissolved isooctane droplets were evaporated by gently purging with nitrogen. In other experiments, microliter aliquots of these same radiolabeled solutions were transferred into 1.1-mL conical-tipped glass vials. The isooctane was evaporated to dryness at ambient temperature by gentle nitrogen purge, and the resulting PCB residues were redissolved in 1.00 mL of Ohm-diluent, 50% MeOH/H$_2$O, or pure methanol. The purge rate was adjusted such that 40 μL of isooctane evaporated in 1.5 to 2.5 min. Selected extracts in isooctane, or isooctane solutions of ^{14}C-HxPCB and ^{14}C-TePCB, were treated as above with nitrogen purge to remove the solvent and then redissolved with Ohm-diluent, 50% MeOH/H$_2$O, or methanol. Solvent Blue 35, a hydrophobic blue dye, was used in some of the experiments to facilitate visual monitoring of the mixing and solvent exchange processes. In each of the above types of experiments, after thorough mixing, replicate 200-μL aliquots (sub-samples) were transferred from each dilution (samples) into separate vials containing scintillar. Their PCB content was determined by liquid scintillation counting (LSC).

Radioactivity Measurements. Liquid scintillation counting (LSC) analyses used a Model LS 3801 scintillation counter (Beckman Instruments, Fullerton, CA). Microliter volumes of Ohm-diluent, 50% MeOH/H$_2$O, methanol, or isooctane solutions of ^{14}C-TePCB or ^{14}C-HxPCB were mixed with 10 mL of Ecolume liquid scintillation cocktail (ICN Biomedicals, Costa Mesa, CA) and counted.

Sample Preparation.

Fish. Fish samples were a subset (23 of 117 monitoring stations) of the U.S. Department of Interior's National Contaminant Biomonitoring Program (NCBP) fish collected in 1988 (Table I). Extracts of whole fish composites were prepared as previously described (*11*) for mayflies (*Hexagenia bilineata*). Briefly, whole fish

Table I. Collection Locations, Species, PCB Concentrations as Determined by GC and ELISA, and Associated QA/QC Samples

Station #[b]	River or Lake	Species Code[c]	(geq/mL)[d]	Total-cPCB[e] (μg/mL)	ePCB[f] (μg/mL)
			Fish Primary ELISA Samples[a]		
2	Connecticut R.	CHC	0.130	0.229	0.310
2	Connecticut R.	CHC	0.114	0.229	0.301
3	Hudson R.	WSU	0.056	0.152	0.124
4	Delaware R.	CHC	0.115	0.134	0.190
8	Cape Fear R.	CHC	0.264	0.126	0.123
18	L. Ontario	WSU	0.114	0.148	0.232
19	L. Erie	WSU	0.220	0.097	0.031
20	Saginaw Bay	C	0.114	0.084	0.133
21	L. Michigan	LT	0.065	0.134	0.212
22	L. Superior	LT	0.536	0.161	0.148
46	Columbia R.	BRB	1.21	0.045	0.028
53	Merrimack R.	WSU	0.116	0.070	0.052
54	Raritan R.	WSU	0.114	0.167	0.218
66	St. Lawrence R.	WSU	1.19	0.253	0.243
68	Wabash R.	C	0.116	0.072	0.072
70	Ohio R.	C	0.114	0.153	0.155
70	Ohio R.	CHC	0.117	0.181	0.162
102	L. Superior	LT	0.065	0.234	0.232
104	L. Michigan	LT	0.059	0.065	0.062
105	L. Michigan	LT	0.057	0.124	0.120
105	L. Michigan	LT	0.057	0.124	0.180
105	L. Michigan	LT	0.057	0.124	0.199
106	L. Huron	LT	0.122	0.209	0.185
107	L. St. Clair	C	0.067	0.263	0.330
108	L. Erie	C	0.108	0.068	0.062
109	L. Ontario	LT	0.036	0.167	0.236
111	Mississippi R.	C	0.117	0.230	0.339
Quality Control	Procedural		7.07	0.012	0.007
	Matrix Blank		6.00	0.091	0.074
	Matrix Spike		1.25	0.184	0.075
	Matrix Spike		1.27	0.208	0.166
	Matrix Spike		1.21	0.182	0.214
	Matrix Spike		1.17	0.156	0.234

[a] Methanolic dilutions prepared from sample extracts in isooctane by complete nitrogen-evaporation of the isooctane and subsequent dissolution of PCBs with methanol (nitrogen-evaporation solvent-exchange).
[b] Collection sites as designated by the NCBP (5).
[c] CHC, channel catfish (*Ictalurus punctatus*); WSU, white sucker (*Catostomus commersoni*);　C, common carp (*Cyprinus carpio*); LT, lake trout (*Salvelinus namaycush*); BRB, brown bullhead (*Ictalurus nebulosus*).
[d] grams fish tissue (wet weight)/mL methanol.
[e] μg Total-cPCB/mL methanol as determined by GC after correction for the dilution resulting from the nitrogen-evaporation solvent-exchange procedure.
[f] μg ePCB/mL methanol as determined by ELISA after correction for dilution(s) with Ohm-diluent.

samples were ground and mixed with sodium sulfate, column extracted with methylene chloride, applied to multi-layer reactive chromatography columns (sodium sulfate, 40% sulfuric acid silica gel [SA/SG], potassium silicate [KS], silica gel [SG], and sodium sulfate), and extracted with methylene chloride. The partially purified extracts were concentrated, and applied to a final reactive chromatography column (sodium sulfate, SA/SG, SG, and sodium sulfate), and were eluted with 0.5% benzene/99.5% hexane (v/v). Together these chromatography steps removed ≥ 99.5% of the co-extracted lipids from the tissue extracts. During the subsequent solvent reduction the clean extracts were exchanged into isooctane, which is a preferred solvent for GC analysis and for sample storage.

For separate ELISA determinations, small aliquots (40-200 μL) of the concentrated fish extracts (in isooctane) were brought to dryness in 1.1-mL conical glass vials, and the PCB residues were redissolved in 200 μL to 1000 μL of methanol to give solutions referred to as the "Primary ELISA Samples" (Table I). Aliquots of each "Primary ELISA Sample" were subsequently diluted with Ohm-diluent to give concentrations in the appropriate range defined by the ELISA calibration standards (0.25 ng to 5.0 ng Aroclor 1254/mL).

SPMDs. Semipermeable-membrane devices were prepared and exposed to room air as previously described (*12*). Briefly, SPMDs were exposed to laboratory air for varying lengths of time up to 28 days and then dialyzed with hexane. Concentrated dialysates were then subjected to gel permeation chromatography, Florisil adsorption chromatography, and exchanged into hexane. Separate aliquots of the clean hexane extracts were brought to dryness in 1.1-mL conical glass vials, and the PCB residues were redissolved in 1000 μL of pure methanol. These solutions were referred to as the "Primary ELISA Samples". Aliquots of each "Primary ELISA Sample" were subsequently diluted at least 100-fold with Ohm-diluent to give concentrations in the appropriate range defined by the ELISA calibration standards (0.25 ng/mL to 5.0 ng Aroclor 1254/mL). The less than 1% increase in the methanol content of the resulting sample dilutions was considered to be trivial with respect to ELISA interference.

GC Analysis.

Fish. Aliquots of the concentrated fish extracts were analyzed by capillary GC with electron capture detection to determine the concentrations of 105 individual PCB congeners (cPCBs,[*11*]). The sum of the individual congeners is referred to as total-cPCBs.

SPMDs. Aliquots of the concentrated dialysates were analyzed by a separate GC method to measure concentrations of total PCBs (*12*). This GC method used to screen for total PCBs in SPMDs required less time per sample injection (45 min vs. 160 min), required less time for data reduction (minutes vs hours/sample), but measured only about one half the number of cPCBs measured by the GC method for fish extracts. Total PCBs measured by this GC screening method are generally similar to those obtained by the more rigorous GC method used for fish, but the relationship between the two methods has not been determined statistically.

Enzyme-Linked Immunosorbent Assay.

ELISA. The PCB-competitive ELISA (Ohmicron PCB RaPID Assay), a tube format assay employing paramagnetic particles that are covalently coated with anti-PCB antibodies, was supplied by Ohmicron Corporation and used according to the manufacturer's recommendations. Briefly, individual samples (200-μL aliquots) were analyzed singly or in duplicate together with four duplicate calibration standards: 0, 0.25, 1.0, 5.0 ng/mL. Up to thirty individual sample dilutions (or 15 duplicates) were analyzed with one Ohmicron positive control (Ohmicron-control) per 20 tubes in batches of 20 to 40 tubes. The immunochemical reagents, 250 μL of PCB enzyme conjugate (a horseradish peroxidase-labeled PCB analog) and 500 μL of the particle suspension, were added to aliquots of each sample, standard, and control in test tubes. The tubes were vortexed, and the competitive binding reaction was allowed to occur at room temperature for 15 min. The paramagnetic particles containing bound PCBs and/or PCB enzyme conjugate were isolated using the vendor supplied magnetic separation rack and two washes. The magnetic rack was separated from the tube rack, 500 μL of substrate/chromogen solution (hydrogen peroxide/3,3',5,5'-tetramethylbenzidine) was added to each tube, and the color was developed at room temperature. After 20 minutes the color reaction was stopped by adding 500 μL of 2 M H$_2$SO$_4$. It is important to note that although, some of the ELISA measurements were performed over one year after the reference GC measurements, it has been our experience that repeat GC analysis of similar fish extracts stored in isooctane for comparable periods of time show no measurable losses of PCBs (J.L. Zajicek, unpublished data).

ELISA Reader. ELISA absorbances at 450 nm were measured with a Model RPA-1 spectrophotometer (Ohmicron Corp., Newtown, PA). Measured absorbance data (B) were related to that of the zero standard (B$_0$), and concentrations were calculated from the relation:

$$\text{Logit B = Slope * Ln [PCB] + Intercept} \qquad (1)$$

where

$$\text{Logit B = Ln ([B/B}_0\text{]/[1-B/B}_0\text{])} \qquad (2)$$

Concentration calculations were either performed automatically using the RPA-1 or by using Microsoft Excel (Microsoft Corporation, Redmond, WA).

Parallelism. Selected samples (one enriched extract each from fish and SPMDs) were serially diluted and analyzed by ELISA to compare the resulting dose-response curves of PCBs in these samples to similar curves resulting from dilutions of the Aroclor 1254 calibration standards. Lack of parallelism between standard and sample curves gives an indication of matrix effects associated with the sample, solvents, or the reagents (*13,14*).

Results and Discussion

Direct Dilution Studies. The most straightforward procedure for transferring the PCBs, contained in a hydrophobic solvent into an ELISA-compatible media, would be to directly dilute an aliquant of the sample into the ELISA diluent. To examine this approach, aliquots of isooctane ($\leq 10~\mu L$) were added directly to the ELISA media (1.0 mL of Ohm-diluent) and the recovery of ^{14}C-PCB (TePCB or HxPCB) was measured in the media. Recovery of the added radioactive ^{14}C-HxPCB was 72% for the 1001-fold dilution (1 μL of isooctane), but rapidly decreased as the volume of isooctane increased (Table II). Similarly, the variation (%RSD) among sub-samples was lowest for the sample prepared from 1.0 μL of isooctane and rapidly increased to an asymptote between 40% and 50% for all samples prepared with larger volumes of isooctane (Table II).   Visual examination of the corresponding samples suggested that only the samples prepared from 1.0 μL of isooctane appeared to be homogenous, while those prepared from larger volumes of isooctane appeared to be biphasic with small (barely visible) droplets of isooctane at the surface of and floating throughout the methanol-water phase.

Table II. Dependence of Recovery from Ohmicron ELISA Diluent on Isooctane Volume and ^{14}C-PCB Solute; Direct Dilution of Isooctane Solutions of ^{14}C-HxPCB and ^{14}C-TePCB with 1.0 mL of ELISA Diluent in 1.1-mL Conical Vials

Isooctane[a] ($\mu L/mL$)	^{14}C-HxPCB[b] Recovery (%)	RSD[c] (%)	^{14}C-TePCB[b] Recovery (%)	RSD[c] (%)
1.0	72	5	65	2
2.0	32	9	69	1
3.0	17	49	66	10
5.0	29	35	60	5
10.0	7	50	42	5

[a] μL of isooctane solution of ^{14}C-PCB/mL of Ohmicron ELISA diluent.
[b] Mean sub-sample recovery (n=3).
[c] For n=3 sub-samples.

The phase separation was more easily observed in samples by addition of microliter amounts of a solution of Solvent Blue 35 (blue dye). Previously, this dye was used by Vanderlaan et.al. (*15*) to follow the processing of dioxin residues in sample extracts and their dissolution into an ELISA-compatible solvent system. In our aqueous-methanol-isooctane samples, the blue dye was preferentially concentrated in the isooctane phase. Using the dye, blue microdroplets were observed adhering to the walls of the vial and at the meniscus-glass interface when only 1.0 μL of isooctane was added. This, together with the low recoveries of HxPCB shown in Table II, suggest that the solubility limit for isooctane is less than 1.0 μL isooctane/mL of Ohm-diluent. When the above experiment was repeated with microliter volumes of an isooctane solution of ^{14}C-TePCB, higher recoveries with greater precision were obtained (Table II). The improved recoveries and decreased variances of the ^{14}C-TePCB are likely due to its greater solubility in Ohm-diluent, relative to ^{14}C-HxPCB.

We had observed that the blue dye, concentrated in the isooctane phase, readily dissolved in the aqueous methanol phase as the isooctane droplet was evaporated by a gentle purge of nitrogen. Therefore, evaporation of the residual isooctane phase was tested for its ability to increase the amount of ^{14}C-PCB recovered relative to the above direct dilution method. The removal of the overlying isooctane phase, in fact, resulted in a significant improvement in the percentage of radioactivity recovered from the resulting aqueous-methanolic phase. Also, the variability of replicate sub-samples was reasonably low for all samples (Table III).

Table III. Effect of Blue Dye and Isooctane Volume on Recovery of ^{14}C-HxPCB from Ohmicron ELISA Diluent Following Nitrogen Evaporation of Undissolved Isooctane; Direct Dilution of Isooctane Solutions with 1.0 mL of ELISA Diluent in 1.1-mL Conical Vials

Isooctane[a] (μL/mL)	Without Dye[b] Recovery (%)	RSD[c] (%)	With Dye[b] Recovery (%)	RSD[c] (%)
1.0	55	1	58	0.3
2.0	38	1	70	0.4
3.0	62	1	64	1.1
5.0	44	4	73	0.4
10.0	33	2	50	1.1

[a] μL of isooctane solution of ^{14}C-HxPCB/mL of Ohmicron ELISA diluent.
[b] Mean sub-sample recovery (n=3).
[c] For sub-samples (n=3).

The 1.1-mL conical glass vials were convenient for sample storage and quantitative removal of isooctane from small sample aliquots ($\leq$ 10 μL, see **Evaporation of Isooctane and Dissolution of PCB Residues** below). However, the vials were not optimum for vigorous and efficient sample mixing, due to their low head space (about 0.1 mL for 1.0-mL dilution volumes) and their near-capillary conical tips. To allow for greater sample dilution (up to 5001-fold), and to simultaneously increase the vigor of the sample mixing, we diluted 1.0-μL to 50-μL aliquots of isooctane solutions (^{14}C-HxPCB and blue dye) with 5.00 mL of 50% MeOH/H$_2$O in 10-mL glass culture tubes. After mixing and evaporative removal of the overlying residual isooctane, these dilutions resulted in nearly quantitative recoveries of ^{14}C-HxPCB ($\geq$ 80%) with dilutions between 5001- and 334-fold (Table IV). Sample homogeneity was also good; relative standard deviation among sub-samples over the range of dilutions from 0.2 to 5.0 μL isooctane/mL diluent was

%RSD ≤ 4. Apparently, better mixing was achieved in the 10-mL culture tubes than in the 1.1-mL conical glass vials. Therefore, direct dilution of PCB congeners into either Ohm-diluent or 50% MeOH/H_2O can give nearly quantitative results for dilutions greater-than-or-equal-to 0.1% (v/v) isooctane in diluent, when the sample container allows for adequate mixing (e.g. 5.0 μL of isooctane diluted into 5.00 mL of diluent in a 10-mL tube) and the undissolved isooctane is evaporated.

Table IV. Effect of Isooctane Volume on Recovery of ^{14}C-HxPCB from 50% MeOH/H_2O ELISA Diluent Solutions Following Nitrogen Evaporation of Undissolved Isooctane; Direct Dilution of Isooctane Solutions with 5.0 mL of ELISA Diluent in 10-mL Culture Tubes (Blue Dye Added)

Isooctane (μL/mL)[a]	Mean Recovery[b] (%)	RSD[c] (%)	Mean RSD[d] (%)
0.2	93	14	1
0.4	81	17	1
0.6	88	19	1
0.8	94	13	1
1.0	101	2	1
2.0	85	12	2
3.0	83	7	2
5.0	78	9	4
10.0	73	34	18

[a] μL of isooctane solution of ^{14}C-HxPCB/mL of 50% MeOH/H_2O ELISA diluent.
[b] Mean recovery for samples (n=4).
[c] For samples (n=4).
[d] Measure of sample homogeneity; mean of sample %RSDs (n=4), where each sample %RSD was based on n=3 sub-samples.

Evaporation of Isooctane and Dissolution of the PCB Residues. Evaporation of the solvent to dryness was tested as an alternative transfer technique. In this method, isooctane was directly evaporated from aliquots (≤ 10 μL) of ^{14}C-PCBs solutions, and 50% MeOH/H_2O diluent was used to redissolve the PCB residues. Under these conditions, the recoveries of ^{14}C-PCBs were incomplete (Table V).

The low PCB recoveries from the solvent evaporation technique resulted from either incomplete dissolution of the PCB residues or evaporative losses. The solvent evaporation step was known to provide near quantitative recoveries of PCBs (*3*), so a better understanding of the dissolution step was needed. Up to this point attempts had been made to transfer ^{14}C-PCBs from isooctane into the polar ELISA diluents composed of 50% (v/v) methanol/H_2O. These diluents are most compatible with the requirements of the targeted Ohmicron PCB RaPID Assay. It is generally known that PCBs are readily soluble in organic solvents such as alkanes (e.g. hexane and isooctane), aromatics (e.g benzene), and many other solvents including alcohols (e.g. methanol). In sharp contrast, PCBs are poorly soluble in water, and their solubility in methanol-water mixtures up to 50% (v/v) is considerably less than in pure methanol (see the discussion below and [6]). Some PCB ELISAs, such as the Millipore EnviroGard Assay, are compatible with microliter volumes of samples in 100% methanol. Results of a previous study showed that isooctane could be completely evaporated from PCB solutions and the PCB residues recovered in 100% methanol in a precise quantitative ($\geq$ 90%) manner (3). By inference, these earlier results using 100% methanol showed that losses due to PCB volatility were minimal, and the incomplete recoveries reported here must be associated with the dissolution step using 50% MeOH/H$_2$O.

Table V. Evaporative Solvent Exchange in 1.1-mL Conical Vials; Effect of Blue Dye and Isooctane Volume (Amount of ^{14}C-HxPCB) on the Recovery of ^{14}C-HxPCB from 50% MeOH/H$_2$O ELISA Solutions

Isooctane[a] (μL)	Without Dye[b] Recovery (%)	RSD[c] (%)	With Dye[b] Recovery (%)	RSD[c] (%)
1.0	84	10	91	7
2.0	86	4	93	5
3.0	84	9	89	6
5.0	75	18	89	5
10.0	72	13	80	13

[a] μL of isooctane solution of ^{14}C-HxPCB added to 1.1-mL vials.
[b] Mean sample recovery (n=6).
[c] For samples (n=6).

Adding blue dye to sample aliquots made it possible to clearly observe when the samples had gone to dryness and to visually follow the process of dye (and presumably the ^{14}C-HxPCB) dissolution with the 50% MeOH/H$_2$O diluent. Evaporation of the solvent to dryness deposited the blue dye residues around the walls of the conical vial tip at the point where the isooctane evaporated under nitrogen purge. When 50% MeOH/H$_2$O was added to the vials, it was easy to see that only part of the dye residue (and presumably the ^{14}C-HxPCB) was redissolved in the ELISA diluent, even after repeated vigorous mixing. Radioactivity measurements confirmed that incomplete recoveries were obtained when 50% MeOH/H$_2$O ELISA diluent was used to redissolve ^{14}C-PCB residues (Table V).

When methanol was used to redissolve the blue dye and the PCBs, near quantitative recoveries of both ^{14}C-HxPCB and ^{14}C-TePCB were obtained (Table VI). In general, evaporative removal of the isooctane solvent, followed by dissolution of the dye and associated PCB residues into pure methanol, consistently gave near quantitative mean recoveries of radioactivity associated with both radiolabeled PCB congeners. Use of the blue dye improved the precision of the solvent exchange procedure by making it easier to visualize when the solvent had gone to dryness, and thus decreased variability due to evaporative losses. Such losses should be important for PCB mixtures composed of lower average chlorination of the biphenyl molecules (e.g. Aroclor 1242), due to their greater volatility.

Table VI. Effects of Dilution Treatment, Solvent Exchange Treatment, and Isooctane Volume on the Recovery of ^{14}C-HxPCB and ^{14}C-TePCB from Three ELISA Diluents

Isooctane-Sample Treatments (in 1.1-mL Conical Vials)	^{14}C-HxPCB Recovery, % (Isooctane Volume)		^{14}C-TePCB Recovery, % (Isooctane Volume)	
	(2 μL)	(10 μL)	(2 μL)	(10 μL)
Direct Dilution				
Ohm-Diluent (No Dye or N$_2$-Evap.)	33	8	66	42
Ohm-Diluent (N$_2$-Evap., But No Dye)	38	33	NA[a]	NA[a]
Ohm-Diluent (Dye and N$_2$-Evap.)	70	50	NA[a]	NA[a]
50% MeOH/H$_2$O[b] (Dye and N$_2$-Evap.)	85	73	NA[a]	NA[a]
Solvent Exchange				
50% MeOH/H$_2$O (N$_2$-Evap., But No Dye)	86	72	NA[a]	NA[a]
50% MeOH/H$_2$O (Dye and N$_2$-Evap.)	93	80	NA[a]	NA[a]
Ohm-Diluent (N$_2$-Evap., But No Dye)	80	53	76	93
Ohm-Diluent (Dye and N$_2$-Evap.)	78	45	78	78
Methanol (N$_2$-Evap., But No Dye)	91	90	83	90
Methanol (Dye and N$_2$-Evap.)	91	90	86	79

[a] Not analyzed.
[b] Diluted to 5.0 mL in 10-mL culture tubes.

Solubilities of Hydrophobic Chemicals in ELISA Diluents. Hydrophobic solvents impart some of the same problems associated with solubility as do hydrophobic analytes. Although these solvents allow complete solvation of hydrophobic analytes, in an ELISA they can lead to a variety of detrimental effects including solvation of hydrophobic areas of the antibody or the enzyme label and miscibility problems by development of separate liquid phases. In contrast, ELISA diluents are hydrophilic and solubility (solvation) of hydrophobic chemicals can be a significant issue. For example, the solubilities of PCB 3, PCB 30, and PCB 155 were examined in a series of methanol/H_2O mixtures (6). The solubility of PCB 30 was found to be 0.2 mg/L in H_2O, 73 mg/L in 50% MeOH/H_2O, and 56,600 mg/L in 100% methanol. This demonstrates that the solubility of a hydrophobic analyte like PCB 30 in methanol/H_2O mixtures is not linear, that PCB 30's solubility in 50% MeOH/H_2O is closer to that of H_2O than pure methanol, and thus methanol/H_2O ELISA diluents may not have adequate solvation capacity for high concentrations of the intended hydrophobic analytes. Furthermore, the water solubilities of hydrophobic solvents, such as n-heptane, isooctane, and n-octane (range from 2.9 mg/L to 0.66 mg/L [8]) are similar to those of PCB 3 and PCB 30. Therefore, when working with hydrophobic chemicals one must be aware of the solubility of the solvents and analytes in the ELISA media, and should consult solubility data (e.g. [6-9]) or use estimation methods (e.g. [7]) to guarantee that you are working within this range.

ELISA Determinations.

Method Validation. ELISA analyses were performed according to the manufacturer's recommendations, and the performance met or exceeded the acceptability criteria established by the manufacturer (Table VIIa and VIIb).

Table VIIa.   Results for ELISA[a] Measurements of Ohmicron Control (n=13)

	This Study	Reference Values
Mean (ng/mL)	3.13	3.0
SD (ng/mL)	0.425	0.3
Acceptable Range (ng/mL)		2.4 - 3.6

[a]Ohmicron RaPID Assay calibration curve-fit correlations (n=7) were: mean (r) = 0.9964, SD = 0.0049 compared with the target minimum (r) = 0.990.

PCB concentrations measured by ELISA (ePCB) in the procedural blank, the matrix blank, and the clean fish spiked with Aroclor 1254 had nearly a one-to-one correspondence with those PCB concentrations obtained by GC measurements (Figure 1, Table I). The Aroclor 1260 and 1248 matrix spikes lie above and below a one-to-one correspondence, respectively, in approximate proportion to their ELISA

Table VIIb.   Results for ELISA[a] Measurements of Ohmicron Proficiency Samples (n=4)

	Sample A	Sample B
Mean (ng/mL)	0.42	1.65
SD (ng/mL)	0.135	0.222
Ohmicron Target Value (ng/mL)	0.44	1.70
± 2SD Target Range (ng/mL)	0.22 - 0.66	1.18 - 2.22

[a]Ohmicron RaPID Assay calibration curve-fit correlations (n=7) were: mean (r) = 0.9964, SD = 0.0049 compared with the target minimum (r) = 0.990.

responses relative to the Aroclor 1254 calibrations (Figure 1, and PCB RaPID Assay documentation, Ohmicron Corp.). In other words, the mixture of PCB congeners of Aroclor 1248 have a lower binding efficiency than the congeners of Aroclor 1254, and thus the ELISA response of the Aroclor 1248 spike fell below a one-to-one relationship. Conversely, Aroclor 1260, which has a greater proportion of the strongest binding congeners, has a greater binding efficiency than Aroclor 1254, and the predicted concentration by ELISA was greater for the Aroclor 1260 spike than would be expected. For example, according to the manufacturer, the 50% B/B_0 concentrations for Aroclor 1254 and Aroclor 1248 are 0.93 ng/mL and 2.1 ng/mL, respectively. The normalized ELISA response (or bias) = (50% B/B_0 concentration of the standard)/(50% B/B_0 concentration of the competing Aroclor), and is equal to 0.44 for Aroclor 1248. Thus, bias can be introduced by mixtures of PCB congeners not representative of the calibration mixtures.

The materials used for extraction and sample processing, and the biological matrix (fish) do not appear to bias the ELISA procedure. To further investigate the possibility of a matrix bias we prepared serial dilutions of a NCBP sample and compared the ELISA dose-response curves to that of the Aroclor 1254 calibration standards (Figure 2). Dose-response curves for both samples parallel the curve based on dilutions of Aroclor 1254. This indicates that neither procedure contributed significant amounts of interfering co-extractables into subsequent dilutions (13,14). Although, we did not test this fish extract for parallelism using our highest yield solvent-exchange method (e.g. N_2-evaporation, methanol dissolution, and water dilution), we routinely found that the ELISA responses of dilutions of fish samples, so prepared, were parallel with those of standards.

PCB Determinations in NCBP Fish Samples. A more rigorous test of the use of the PCB ELISA method on extracts of biological samples was made with environmental samples taken from the NCBP (5). A wide range of total-cPCB concentrations was previously measured in these fish (≤200 ng/g to 6700 ng/g, wet

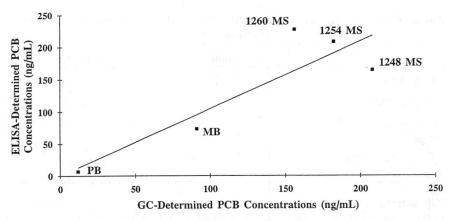

Figure 1.   Comparison of ELISA and GC determinations of total PCB concentrations (ng/mL) in selected QA/QC sample extracts. Extracts include: (PB) - procedural blank, (MB) - matrix blank (clean fish), and fish spiked with known amounts of (1260 MS) - Aroclor 1260, (1254 MS) - Aroclor 1254, and (1248 MS) - Aroclor 1248. Regression line: [ePCB] = 1.05 * [total-cPCB] + 0.25. The curve correlation (R^2 = 0.78) was significantly less than 1.0, since the ELISA responses of Aroclors 1248 and 1260 differ from that of Aroclor 1254 calibration standards.

weight) (5). Additionally, the fish were taken from a number of locations and represent several species, which increases the possibility of having interferences and affords a more robust test of the ELISA procedure. PCB concentrations derived from ELISA and GC measurements on the same NCBP fish sample extracts were compared (Table I and Figure 3). The average ePCB concentrations appear to be about 16% greater than the average GC total-cPCB concentrations. Considering that these fish extracts contained PCB mixtures that were significantly different from the parent technical Aroclors or their combinations (16), the relationships and correlations were quite good. In the earlier studies with these same samples, we used a commercial antibody-coated tube ELISA (EnviroGard, Millipore Corp., Bedford, MA), and calibrations were based on dilutions of Aroclor 1248 (3). The coated tube ELISA was based on polyclonal antibodies that had responses with technical Aroclors very similar to those of the Ohmicron ELISA used for these studies. The ELISA results from the Millipore ELISA on these same fish samples indicated a positive bias (slope = 4) as compared to GC determinations of total PCBs. Because of this relationship, it was our recommendation that future PCB ELISA determinations of unknown fish samples be calibrated against dilutions of Aroclor 1254, which has a 2.7-fold greater response relative to Aroclor 1248. This recommendation was also supported by our finding that the majority of the PCB mixtures in these fish extracts, although statistically different from technical Aroclors or their combinations based on principal component analysis, are most closely related to the technical Aroclors 1254 and 1260 (3).

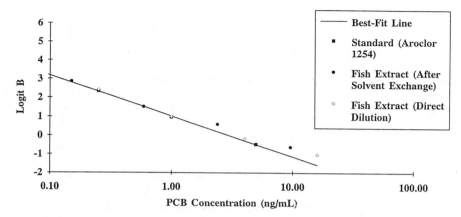

Figure 2. Test for parallelism by comparison of the dose vs. response curves for Aroclor 1254 standards (■) and serially diluted samples of a fish tissue extract. Sample dilution series prepared by directly diluting 2 μL of an isooctane extract into 1.0 mL of Ohm-diluent (◊). The response of the fish extract dilution, whose ELISA measured concentration was 0.59 ng ePCB/mL diluent, was assumed to be correct. The logit transformed responses for the remaining sample dilutions were paired with their corresponding nominal concentrations that, by definition, differed from 0.59 ng/mL by factors of four. The second sample dilution series prepared after evaporation of the isooctane and re-dissolution of the extract PCBs with Ohm-diluent (●). The response of the fish extract dilution, whose ELISA measured concentration was 1.0 ng ePCB/mL diluent, was assumed to be correct. The logit transformed responses for the remaining sample dilutions were paired with their corresponding nominal concentrations that, by definition, differed from 1.0 ng/mL by factors of four. Matrix effects appear to be minimal as shown by the close correspondence of the three curves.

PCB Determinations in SPMDs. SPMDs have been developed to passively sample hydrophobic compounds from environmental media such as water and air (*17,12*). In a recent study, SPMDs were used to sample volatile PCBs in a laboratory room for up to 28 days (*12*). We measured ePCB concentrations in these same dialysates for comparison to the total-PCB concentrations measured in the previous study by GC. The ePCB concentrations were positively correlated ($R^2=0.999$, n=3) to the total-PCB concentrations with a slope relative to GC measured concentrations of about 1.52 (Figure 4). To investigate the nature of the higher ELISA measurement results relative to those by GC, we examined the dose-response curve of one of the dialysates of SPMDs collected after 28 days (Figure 5). The curve for the PCBs of the dialysate dilutions is essentially parallel to that of the Aroclor 1254 calibration standards, except for the least dilute SPMD sample (Figure 5). The most likely

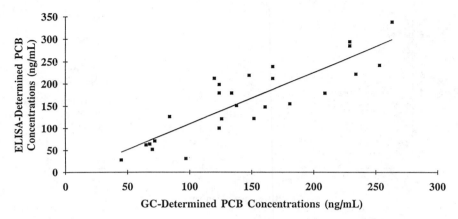

Figure 3.   Comparison of ELISA and GC determinations of total PCB concentrations (ng/mL) in fish extract primary ELISA solutions. Regression line: [ePCB] = 1.16 * [total-cPCB] + 0.7. The curve correlation (R^2 = 0.74) is similar to that of Figure 1, since the cPCB mixtures, and the ELISA responses of the fish extracts differ from that of Aroclor 1254 used for calibration.

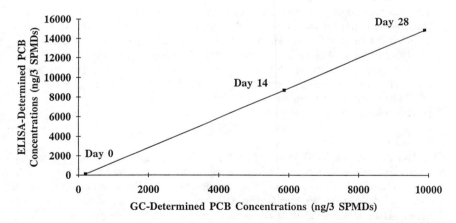

Figure 4.   Comparison of ELISA and GC determinations of total PCB concentrations in hexane dialysates of SPMDs exposed to laboratory air. The best-fit line for SPMD samples exposed to laboratory air for zero, 14, and 28 days is: [ePCB] = 1.52 * [PCB] - 212; concentration units are ng/3 SPMDs. The slope of the line and the curve correlation (R^2 = 0.999) are consistent with the fact that, the mixtures of PCBs in these dialysates are from a single source, whose ELISA response is 1.52 times that of the Aroclor 1254 calibration standards.

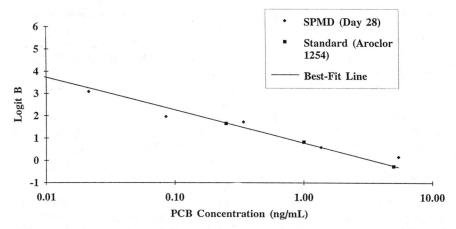

Figure 5. Test for parallelism by comparison of the dose-response curves of a four-fold serially diluted day-28 SPMD sample (♦) and that of Aroclor 1254 standards (■) . The response of the SPMD sample dilution, whose ELISA measured concentration was 1.42 ng ePCB/mL diluent, was assumed to be correct. The logit transformed responses for the remaining sample dilutions were paired with their corresponding nominal concentrations that, by definition, differed from 1.42 ng/mL by factors of four. Matrix effects appear to be minimal due to close correspondence of the two curves.

cause for this negative bias was that the sample solution was 54.5% (v/v) methanol/aqueous buffer, which according to the manufacturer, is at the PCB ELISA's solvent tolerance of methanol (Scott Jourdan, Ohmicron Corporation, personal communication). The positive deviations (low Logit B values) for the sample dilutions that were extrapolated below the ELISA detection limit of 0.1 ng/mL were within the expected experimental error of this part of the dose-response curve.

Conclusions and Recommendations

ELISA can be used to accurately measure PCBs in hydrophobic extracts following complete removal of the nonpolar solvent and quantitative dissolution of the PCB residues with methanol. After solvent evaporation to dryness during traditional solvent exchange, PCB residues should first be redissolved with methanol, and then be diluted with an equal volume of water to provide the 50% aqueous methanolic solution most compatible with the Ohmicron PCB RaPID Assay or any other ELISA that employs methanol/H_2O as the sample medium.

The accuracy of PCB ELISA measurements can be maximized by grouping samples with a common source of PCB contamination and by using a mixture of cPCBs in the calibration standards that is representative of the mixture of cPCBs in the samples.   Since this is not practical for large screening studies on a state, regional or national scale, such as screening environmental fish samples collected from PCB contaminated sites across the United States, calibration with Aroclor 1254 standards appears to be appropriate.

Fish extracts and SPMD dialysates traditionally prepared for GC are free of ELISA interferences for the analysis of PCBs.

Dissolution of PCB residues with Ohmicron-diluent or 50% MeOH/H_2O after complete removal of lipophilic solvents may be less than quantitative, due to the limited solubility of PCBs in these polar solvent systems.

Acknowledgments

The authors would like to thank Ted R. Schwartz and Jon A. Lebo of the Midwest Science Center (MSC) and Jeffre C. Johnson of the USEPA, Characterization Research Division, Las Vegas for their thoughtful comments on earlier drafts. We would also like to thank Christopher J. Schmitt of the MSC, Coordinator of the NCBP, for supplying the fish samples and for his continued support of this research.

References to trade names or manufacturers do not imply Federal government endorsement of commercial products.

Literature Cited

1.   Tillitt, D. E., Ankley, G. T., Giesy, J. P., Ludwig, J. P., Kurita-Matsuba, H., Weseloh, D. V., Ross, P. S., Bishop, C. A., Sileo, L., Stromborg, K. L., Larson, J., & Kubiak, T. J. *Environ. Toxicol. Chem.* **1992**, *11*, 1281-1288.

2.   Rogan, W. J.   In *Halogenated Biphenyls, Terphenyls, Naphthalenes, Dibenzodioxins and Related Products*, R. D. Kimbrough & A. A. Jensen Eds., Elsevier Science Publishers (Biomedical Division), New York, NY, 1989, 401-415.

3.   Zajicek, J. L., Harrison, R. O., & Tillitt, D. E. *Options 2000, Eighth IUPAC International Congress of Pesticide Chemistry, Biotechnology Based Approaches Session*, Washington, DC, 1994, Poster No. 66.

4.   Richter, C. A., Drake, J. B., Giesy, J. P., & Harrison, R. O. *Environ. Sci. & Pollut. Res.* **1994**, *1*, 69-74.

5.   Schmitt, C. J., Zajicek, J. L., & Peterman, P. H. *Arch. Environ. Contam. Toxicol.* **1990**, *19*, 748-781.

6.   Li, A. & Andren, A. W. *Environ. Sci. Technol.* **1994**, *28*, 47-52.

7.   Lyman, W. J., Reehl, W. F., & Rosenblatt, D. H. *Handbook of Chemical Properties Estimation Methods: Environmental Behavior of Organic Compounds*; American Chemical Society: Washington, DC, **1990**, 1.0-2.52.

8.   Riddick, J. A. & Binger, W. B. *Organic Solvents, Physical Properties, and Methods of Purification*; Techniques of Chemistry, Wiley-Interscience: New York, NY, **1970**; Vol. II, 19-551.

9.   Shiu, W. Y. & Mackay, D. *J. Phys. Chem. Data* **1986**, *15*, 911-929.

10.   Balschmiter, K. & Zell, M. Z. *Z. Anal. Chem.* **1980**, *320*, 20-31.

11.   Steingraeber, M. T., Schwartz, T. R., Weiner, J. G., & Lebo, J. A. *Environ. Sci. Technol.* **1994**, *28*, 707-714.

12.   Petty, J. D., Huckins, J. N., & Zajicek, J.L. *Chemosphere* **1993**, *27*, 1609-1624.

13.   *Immunoassay A Practical Guide*; Chan, D. W. & Perlstein, M. T., Eds.; Academic Press, Inc.: San Diego, CA, **1987**, 157-159.

14.   Hayes, M. C., Dautlick, J. X., & Herzog, D. P. *Quality Control of Immunoassays for Pesticide Residues*; Ohmicron Corporation: Newtown, PA, **1992**, 18 pages.

15.   Vanderlaan, M., Stanker, L. H., Watkins, B. E., Petrovic, P., & Gorbach, S. *Environ. Toxicol. and Chem.* **1988**, *7*, 859-870.

16.   Schwartz, T. R., Stalling, D. L., & Rice, C. L. *Environ. Sci. Technol.* **1987**, *21*, 72-76.

17.   Huckins, J. N., Manuweera, G. W., Petty, J. D. Mackay, D., & Lebo, J. A. *Environ. Sci. Technol.* **1993**, *27*, 2489-2496.

Chapter 27

Immunochemical Methods for Fumonisins in Corn

Mary W. Trucksess

Division of Natural Products, Center for Food Safety and Applied Nutrition, U.S. Food and Drug Administration, 200 C Street Southwest, Washington, DC 20204

An enzyme-linked immunosorbent assay (ELISA) and an immunoaffinity column (IAC) cleanup procedure have been successfully applied to the determination of fumonisins in corn. The performance of the ELISA was evaluated by comparison to a reference high-performance liquid chromatographic (HPLC) method. The IAC procedure was coupled with HPLC determination. The recoveries of fumonisin B_1 from corn spiked at the 1,2, and 4μg/g levels were 73-106, 79-83, and 64-92% for the ELISA, IAC, and HPLC methods, respectively. The accuracy and precision of the methods compared favorably. In the comparative studies using naturally contaminated corn samples, the ELISA results were 2-100% higher than those determined by HPLC. The immunoaffinity procedure results were about 71% of the levels observed using HPLC.

The fumonisins are a group of structurally related mycotoxins, which are secondary metabolites produced on corn by *Fusarium moniliforme* (1,2), *Fusarium proliferatum* (3), and several other fungi (4,5). Of the known naturally occurring fumonisins, fumonisins B_1 (FB_1) and B_2 (FB_2) are the most abundant (6). In the United States the FB_1/FB_2 ratio in corn is about 3:1 (7). FB_1 was found to cause equine leukoencephalomalacia (2), porcine pulmonary edema (8), and rodent hepatotoxicity (9). Cattle and poultry can also be affected, but are not as susceptible to the mycotoxin as horses and swine. The Mycotoxin Committee of the American Association of Veterinary Laboratory Diagnosticians recommends that the total intake of FB_1 be limited to less than 5 μg/g in the non-roughage diet of horses, 10 μg/g in the total diet of swine and 50 μg/g in the feed for cattle and poultry (10). FB_1 has also been implicated in human esophageal cancer on the basis of epidemiological data (11,12). The International Agency for Research on Cancer, Working Group on the Evaluation of Carcinogenic Risks in Humans, has classified the toxins derived from *F. Moniliforme*, which include FB_1 and FB_2, as possible carcinogens to humans (13). Subsequently many methods have been developed for the determination of these toxins.

Methods for fumonisin determination include thin-layer, liquid, and gas chromatography, as well as mass spectrometry (14-16). All these methods entail sample preparation, extraction, solid phase purification, chromatographic separation, and derivitization prior to quantitation. Mycotoxin testing by chromatographic methods is thus relatively slow and costly. With the advance of biotechnology, antibodies have been produced against the fumonisins and have been used in both enzyme-linked immunoassays (ELISAs) and immunoaffinity columns. Immunochemical methods can produce results more quickly than the traditional methods (17-22). Immunochemical methods are often the methods of choice for both mycotoxin monitoring and surveillance studies, which require rapid analysis of a large number of samples. Monitoring programs are useful for measuring the effectiveness of milling or food processing in controlling fumonisins in human and animal food. Surveillance studies can identify the incidence and occurrence of the fumonisins as well as the geographical areas where fumonisin contamination is a problem. Data can also be used to estimate human exposure for risk assessments. The validity of the data depends on several factors, including the method of analysis. It is extremely important that the immunochemical methods used in these studies are evaluated for their ability to produce accurate results when compared with a reference method. The objective of the current paper is to compare the performance of immunochemical methods with an Association of Official Analytical Chemists (AOAC) International first action high performance liquid chromatographic (HPLC) method.

Commercial Immunochemical Kits

Several immunochemically based commercial kits have been marketed in the United States for detection and cleanup of fumonisins in corn (Table I). Two formats have been used: a competitive ELISA for determination of fumonisins and an immunoaffinity column cleanup procedure . The ELISA is a microtiter-well format. The anti-fumonisin antibodies are bound to the polystyrene microtiter wells. The free fumonisins in the extract and the fumonisin B_1-horseradish peroxidase (FB_1-HRP) conjugate compete for the antibody binding sites. After incubation, washing, and addition of substrate, the color that develops in the wells is inversely related to the amount of toxin in the test sample. In the immunoaffinity column procedure the antibodies are attached to an agarose bead support. The fumonisins can be bound to the specific antibodies conjugated to the column support. The column is then washed, resulting in removal of unbound impurities. The fumonisins can then be desorbed and eluted with a strong organic solvent such as methanol, resulting in purification. The purified fumonisins are then derivatized. The fluorescent fumonisin derivatives are quantitated by HPLC.

Table I. Commercial Immunochemical Methods for Fumonisins in Corn

Kit	Format	Detection Limit (ng/g)	Analysis Time/min.	Cost $/Test
Veratox[a]	ELISA Microwell	500	35	7.00
Fumonitest[b]	Affinity Column	25	50	10.00
Ridascreen Fumonisin Fast[c]	ELISA Microwell	9	65	7.00

[a]Neogen Corp, Lansing, MI 48912
[b]Viacam, Somerville, MA 02145
[c] Bio-Tek Instruments, Inc., Winooski, VT 05404

Methods.

Sample Extraction. For all three methods, corn samples (50 g) were extracted with 250 mL methanol water (70/30).

Analytical Procedures.

HPLC. The reference HPLC method was carried out according to an Association of Official Analytical Chemists first action method (25). The HPLC method uses a strong anion exchange column (SAX) for purification.

ELISA Procedure. The ELISA (Veratox, Neogen, Corp.) was performed as follows: Corn extract (100μL) were diluted with 3.9 mL methanol-water (10/90). The diluted extracts and standard solutions were added to microwell plates and mixed with FB_1-horseradish peroxidase conjugate. After mixing, the contents of the wells were transferred to anti-fumonisin antibody coated plates. The plates were incubated at room temperature for 20 minutes with mixing for 30 seconds at 5 minute intervals.The wells were washed 5 times with water, and 100 μL tetramethylbenzidine substrate solution was added to each well. After 10 minutes, the reaction was stopped by addition of 100 μL dilute sulfuric acid. Absorbance of the wells was measured at 650 nanometers. Standard curves were generated using a log/logit fit.

Immunoaffinity Column Cleanup and Derivatization.   Ten mL of corn extract was diluted with 40 mL of diluting solution (12.5 g sodium chloride, 2.5 g sodium bicarbonate, 2 drops Tween 20 in 500 mL water). After filtration, 5 mL of the diluted extract was placed on the column. The column was washed with 5 mL diluting solution and 5 mL water. The fumonisin was eluted (2 x 0.8 mL) with methanol-water (80/20). The eluate was evaporated and then redissolved in 200 μL methanol.

A Waters model 710 Plus autoinjector was used to deliver 100 μL of derivatization reagent (40 mg o-phthaldialdehyde, 1 mL methanol, 5 mL 0.1 M sodium tetraborate, and 50 μL mercaptoethanol) to 25 μL of extract prior to injection onto the HPLC).

Results

Initial studies (23) were aimed at determining cross-reactivity with structurally related fumonisins as well as other mycotoxins, spike recovery, and immunoaffinity column capacity. In addition, immunochemical results were compared with results generated by the reference HPLC method.

Cross Reactivity. For the ELISA method, the relative cross-reactivities of FB_1, FB_2, and FB_3 were found to be 100, 24, and 30%, respectively. The relative cross-reactivity of each fumonisin was calculated by comparing the fumonisin concentration necessary to inhibit the ELISA response by 50%. The cross-reactivities of the fumonisins on the immunoaffinity columns were determined indirectly by comparing the recoveries of added FB_1 and FB_2 from columns. The average recoveries of FB_1 and FB_2 simultaneously added to the columns at various combinations totaling 1 μg/g were 89 and 79%, respectively. Similar recoveries were obtained in a previous study (26). The results suggested that the monoclonal antibodies in the immunoaffinity column have similar binding affinities with both FB_1 and FB_2. The antibodies of both procedures do not react with other Fusarium toxins, such as deoxynivalenol and zearalenone.

Analytical Range. The applicable range for the ELISA method was 0.1-2.5 ng/mL. This is equivalent to 0.5-10 μg/g of corn. The immunoaffinity column has a maximum binding capacity of 1 μg/g FB_1 according to the manufacturer. The manufacturer recommends applying the equivalent of 0.2 g corn extract to the column. The upper limit of determination was 5 μg/g. However, the maximum sample loading capacity of the column was determined to be about 2 g corn extract (27). The lower limit of determination was 25 ng/g. The limit of determination of the HPLC method depends on the SAX purification cartridge. On the basis of loading 5 g equivalent of test portion extract onto the SAX cartridge the applicable range of the method was 25-15,000 ng/g.

Recoveries of Fumonisin B_1 from Spiked Corn. Each corn extract was analyzed by the three methods to eliminate sampling and extraction variability. Recoveries of FB_1 from corn spiked over the range of 1- 4 μg/g were 73-106, 79-83, and 64-92% for the ELISA, IAC, and HPLC methods, respectively;   the respective relative standard deviations were 4.9-7.2, 0.9-3.5 and 4.7-8.9%.

Analysis of Naturally Contaminated Corn. Ten naturally contaminated corn samples were extracted and each extract was analyzed by ELISA, the immunoaffinity based method and by HPLC. The results obtained by the ELISA were about 38% higher than those obtained by the HPLC method. This difference probably was caused by the cross-reactivity of the antibodies with compounds structurally related to FB_1 or the loss of FB_1 in the solid-phase purification prior to HPLC analysis. The slope of the ELISA concentration vs. the HPLC concentration was 1.3765 (Y = 1.3765X + 0.0456); the correlation coefficient was 0.9964. Results obtained by the IAC method were about 71% of those obtained by the HPLC method. The slope of the IAC concentration vs. the HPLC concentration was 0.7135 (Y = 0.7135X + 0.1169); the correlation coefficient was 0.9674. Since the HPLC method (including the IAC extract) quantitatively determined FB_1 alone, and the ELISA quantitatively determined "total" fumonisins, the overall agreement of the results obtained by the three methods was considered acceptable.

The polyclonal antibody ELISA method was further evaluated (24) with an additional 18 naturally contaminated corn samples that contained total fumonisin (B_1 +B_2 ,+B_3) levels ranging from 0.1 to > 5 µg/g. The samples were extracted as described and analyzed by both ELISA and HPLC. The correlation coefficient between the results generated by HPLC and ELISA was 0.967. In 3 of 18 samples the fumonisin levels determined by ELISA were 85-100% higher than those determined in the same extracts by HPLC; in 13 of 18 samples ELISA results were 2-53% higher than those by HPLC; and in 2 of 18 samples ELISA results were 10% and 20% lower than those by HPLC. No recovery data were given in this study.

Conclusion

Results indicated that the polyclonal antibody ELISA method is suitable for use as a screening method for fumonisins in corn and that the immunoaffinity column method can be used for the determination of fumonisin B_1 in corn. The immunoaffinity column method has the added advantage over the HPLC method because it can be used to determine FB_1 in canned corn and frozen corn. The HPLC method gave poor recoveries for added fumonisin B_1 (<40%) in these commodities (27). The level of fumonisin B_1 in these commodities were too low for the ELISA method to detect.

The ELISA test is the simplest and quickest of the three methods. However, the ELISA results for contaminated corn tend to be biased high relative to HPLC. This phenomenon may be due to the compound specific nature of the HPLC method versus the ELISA method, which responds to some extent to the presence of other fungal metabolites in the corn. Additional studies are needed to resolve this issue.

The development and improvement of solid phase extraction (SPE) methods, including columns (25, 28), discs, and membrane filters, is progressing rapidly.

The price of these devices ranges from $1 to $3/test. The methods using these devices are solvent-efficient and rapid. The immunochemical methods are usually more expensive, ranging from $7 to $10/test. In the future, unless the prices of the immunochemical devices are made more competitive with the SPE methods, they will be at a disadvantage with the solvent-efficient rapid methods.

Some of the immunochemical methods require dedicated instrumentation such as ELISA readers and scanners. The traditional methods require the use of HPLC, GC, or TLC instrumentation; however, this expensive instrumentation can be used for many other analytes.

In summary, the immunochemical methods for fumonisins at the present time have advantages over the traditional and SPE methods for their speedy results and on-site field tests. It is hoped that future developments will include immunochemical methods for multianalytes including other mycotoxins in single or multiple matrices; the development of reusable biosensors; and the use of recombinant antibodies.

Literature Cited

1. Bezuidenhout, S.C.; Gelderblom, W.C.A.; Gorst-Allman, C.P.; Horak, R.M.; Marasas, W.F.O.; Spiteller, G.; Vleggaar, R. J. Chem. Soc., Chem. Commun. 1988, 743-745.
2. Gelderblom, W.C.A.; Jaskiewicz, K.; Marasas, W.F. O.; Thiel, P.G.; Horak, R.M.; Vleggaar, R.; Kriek, N.P. J. Appl. Environ. Microbiol. 1988, 54, 1806-1811.
3. Ross, P.F.; Nelson, P.E.; Richard, J.L.; Osweiler, G.D.; Rice, L.G.; Plattner, R.D.; Wilson, T. M. Appl. Environ. Microbiol. 1990, 56, 3225-3226.
4. Nelson, P.E.; Plattner, R.D.; Shackelford, D.D.; Desjardins, A.E. Appl. Microbiol. 1992, 58, 984-989.
5. Thiel, P.G.; Marasas, W.F.O.; Sydenham, E.W.; Shephard, G.S.; Gelderblom, W.C.A.; Nieuwenhuis, J. Appl. Environ. Microbiol. 1991, 57, 1089-1093.
6. Branham, B.E.; Plattner, R.D. J. Nat. Prod. 1993, 56, 1630-1633.
7. Pohland, A.E. In The Toxicology Forum; Caset Assoc. Ltd.: Fairfax, VA, 1994; pp 186-196.
8. Harrison, L.R.; Colvin, B.M.; Greene, J.T.; Newman, L.E.; Cole, J.R. J. Vet. Diagn. Invest. 1990, 2, 217-221.
9. Gelderblom, W.C.A.; Kriek, N.P.J.; Marasas, W.F. O.; Thiel, P.G. Carcinogenesis 1991, 12, 1247-1251.
10. Food & Chemical News, May 29, 1995, 57.

11. Sydenham, E.W.; Thiel, P.G.; Marasas, W.F.O.; Shephard, G.S.; van Schalkwyk, D.J.; Koch, K.R. J. Agric. Food Chem. 1990, 38, 313-318.

12. Thiel, P.G.; Marasas, W.F.O.; Sydenham, E.W.; Shephard, G.S.; Gelderblom, W.C.A. Mycopathologia 1992, 117, 3-9.

13. International Agency for Research on Cancer, Monograph 56, 1993; pp 445-466.

14. Rottinghaus, G.E.; Coatney, C.E.; Minor, H.C. J. Vet. Diagn. Invest. 1992, 4, 326-329.

15. Sydenham, E.W.; Shephard, G.S.; Thiel, P.G. J. AOAC Int. 1992, 75, 313-318.

16. Plattner, R.D.; Branham, B.E. J. AOAC Int. 1994, 77, 525-532.

17. Azcona-Olivera, J.I.; Abouzied, M.M.; Plattner, R.D.; Norred, W.P.; Pestka, J. J. Appl. Environ. Microbiol. 1992, 58, 169-173.

18. Usleber, E.; Straka, M.; Terplan, G. J. Agric. Food Chem. 1994, 42, 1392-1396.

19. Shelby, R.A.; Rottinghaus, G.E.; Minor, H.C. J. Agric. Food Chem. 1994, 42, 2064-2067.

20. Fukuda, S.; Nagahara, A.; Kikuchi, M.; Kumagai, S. Biosci. Biotech. Biochem. 1994, 58, 765-767.

21. Abouzied, M.M.; Pestka, J.J. J. AOAC Int. 1994, 77, 495-501.

22. Chu, F.S.; Huang, X.; Maragos, C.M. J. Agric. Food Chem. 1995, 43, 261-267.

23. Trucksess, M.W.; Abouzied, M.M. in Immunochemical Detection of Residues in Foods, Beier, R.C., Ed.; ACS Symposium Series 621, American Chemical Society, Wahington, D.C., 1996, 358-367.

24. Sydenham, E.W.; Stockenstrom, S.; Thiel, P.G.; Doko, M.B.; Bird, C.; Miller, B.M. J. Food Prot. (in press).

25. Sydenham, E.W.; Shephard, G.S.; Thiel, P.G.; Stockenstrom, S.; Snijman, P.W. JAOAC Int. (in press).

26. Ware, G.M.; Umrigar, P.P.; Carman, A.S.; Kuan, S. S. Anal. Lett., 1994, 27, 693-715.

27. Trucksess, M.W.; Stack, M.E.; Shantae, A.; Barrion, N. JAOAC Int., 1995, 78, 705-710.

28. Rice, L.G, Ross, P.F., J. Food Protect., 1994,57,536-540.

Author Index

Adkisson, Scott M., 56
Andreas, Christine M., 254
Balduf, Thomas J., 155
Biagini, Raymond E., 286
Blake, Diane A., 10
Carlson, Larry, 23
Chakrabarti, Pampa, 10
Coakley, William A., 254
Coulter, Stephen L., 103
Dawson, Gary N., 10
Dill, Kilian, 89
Dombrowski, Tonya R., 148
Ellis, Richard L., 227
Fan, Titan S., 161
Fare, Thomas L., 240
Gee, Shirley J., 110
Gerlach, Clare L., 2
Gerlach, Robert W., 265
Gilman, S. Douglass, 110
Goodrow, Marvin H., 110
Goolsby, Donald A., 170
Hammock, Bruce D., 110
Harris, Adam S., 110
Hatcher, Frank M., 10
Heider, Patrick J., 155
Herzog, David P., 240
Holmquist, Bart, 23
Huckins, James N., 307
Hull, R. DeLon, 286
Humphrey, Alan, 191
Jacobowitz, Susan M., 254
Jaeger, Lynn L., 110
Kido, Horacio, 110
Klainer, Stanley M., 103
Kuenzli, Brent, 155
Kusterbeck, Anne W., 46
Ligler, Frances S., 46
Lopez-Avila, Viorica, 74
MacKenzie, Barbara A., 286
Mastin, J. Patrick, 286
Matt, Jonathan J., 161

Mohrman, Greg B., 148
Nardone, David A., 307
Parsons, Andrew, 183
Petty, Jimmie D., 307
Pitts, J. Terry, 161
Pomes, Michael L., 170
Potts, Michael E., 307
Riddell, Mal, 23
Ridgill, Tadd E., 56
Robertson, Shirley R., 286
Sadik, Omowunmi A., 37,127
Sanborn, James R., 110
Sandberg, Robert G., 240
Schweitzer, Craig, 23
Shriver-Lake, Lisa C., 46
Skoczenski, Brian A., 161
Stearman, G. Kim, 56
Stopa, Peter J., 297
Stoutamire, Donald W., 110
Strahan, Johanne, 65
Striley, Cynthia A., 286
Sutton, Donna W., 216
Szurdoki, Ferenc, 110
Thurman, E. M., 148,170
Tillitt, Donald E., 307
Trucksess, Mary W., 326
Van Emon, Jeanette M.,
    2,74,127,216,265
Vampola, Christine, 183
Wander, Jeff, 155
Wang, Hong, 183
Wells, Martha J. M., 56
Wengatz, Ingrid, 110
Williams, Philip A., 155
Wippel, Wendy, 286
Wortberg, Monika, 110
Wylie, Dwayne, 23
Young, Barbara Staller, 183
Yu, Hao, 297
Zajicek, James L., 307
Zettler, J. Larry, 161

Affiliation Index

BioNebraska Inc., 23
Calspan SRL Corporation, 297
DuPont Agricultural Products, 65
EnSys Inc., 183
Environmental Health Laboratories, 183
FCI Environmental, Inc., 103
Gustafson, Inc., 161
Lockheed Martin Environmental
    Systems, 2,216,265
Meharry Medical College, 10
Midwest Research Institute, 74
Millipore Corporation, 161,183
Molecular Devices Corporation, 89
Naval Research Laboratory, 46
Ohio Environmental Protection
    Agency, 155
Ohmicron Corporation, 240,307

Rocky Mountain Arsenal, 148
Tennessee Technological University, 56
Tulane University School of
    Medicine, 10
U.S. Army Edgewood Research,
    Development and Engineering
    Center, 297
U.S. Department of Agriculture, 161,227
U.S. Department of Health and Human
    Services, 286
U.S. Department of the Interior, 307
U.S. Environmental Protection Agency,
    2,74,127,191,216,254,265
U.S. Food and Drug Administration, 326
U.S. Geological Survey, 25,148,170
University of California—Davis, 110
University of Wollongong, 37

Subject Index

A

Alachlor automated immunoassay system
    for quantitative analysis in water samples
    assay performance, 184–186,187t
    experimental objective, 183
    kit description, 183–184
Analyte specificity, core indicator of
    confidence, 260–261
Antibodies
    circulating, 291–295
    definition, 127
    description, 192–193
    immobilization, 76
    purification, 75
    to metal chelate complexes, 10–21
    use in electrochemical sensing
        technologies, 37
Antibody–analyte interaction, use of
    immunochemistry, 4
Antibody–antigen interactions, generation
    of electrochemical signal, 37–44

Antibody titer, determination, 67
Assay, heavy metals using antibodies to
    metal chelate complexes, 10–21
Assay design, calibration considerations,
    245–246,247f
Atmospheric pressure chemical
    ionization, use as detector for
    immunoaffinity extraction with
    on-line LC–MS, 81–85
Atrazine
    automated immunoassay system for
        quantitative analysis in water
        samples
    assay performance, 186–189
    experimental objective, 183
    kit description, 183–184
    detection using threshold immunoassay
        system, 96,98–100
    structure, 96,97f
Automated immunoassay system for
    quantitative analysis in water
    samples, atrazine and alachlor, 183–189

B

Bacillus anthrax, detection using immunomagnetic assay system in biological samples, 297–305

Bacteria, detection using immunomagnetic assay system in biological samples, 297–305

Bacterial cells, monitoring using fiber-optic biosensor, 51

Bayes's rule, use in microtiter-plate enzyme-linked immunosorbent assay, 170–181

Benlate 50 DF, enzyme-linked immunoassay for determination on same microwell plate, 65–73

BiMelyze immunoassay, 23–24

Biological samples, virulent bacteria detection using immunomagnetic assay system, 297–305

Biomonitoring for occupational exposures using immunoassays

human response variability, 287–288

immunoassays

circulating antibodies, 291–295

urinary biomarkers, 288–292*f*

limitations, 287

procedure, 288

Biosensor development, use of immunochemistry, 4

C

Cadmium

assay using antibodies to metal chelate complexes, 10–21

environmental contamination problem, 10

Calibration, use for field immunoassay evaluation, 274–280

Calibration standards, generic indicator of confidence, 259

Capillary electrochromatography, use in immunoaffinity extraction with on-line LC–MS, 84–86

Carbendazim, immunoaffinity extraction with on-line LC–MS, 78,81–82

Chelate assay, detection of heavy metals, 117,119,120*f*

Chip level waveguide sensor, 106*f*,107–108

Chloroacetanilide herbicide microtiter-plate enzyme-linked immunosorbent assay for storm runoff samples

advantage, 170

Bayes's rule application, 178–181

comparison to GC–MS, 174,177–178

cross-reactivity, 174–176*f*

experimental description, 170–173

Chlorpyrifos-methyl residue screening immunoassay on grain, 161–169

Chromium, environmental contamination problem, 10–11

Circulating antibodies, use for biomonitoring of occupational exposures, 291–295

CLU-IN, description, 218

Comparability, generic indicator of confidence, 259

Competitive immunoassay optical sensor, description, 105–107

Concentration, calculation, 198–199

Concentration resolution, definition, 277

Conducting electroactive polymers

electrochemical conversion, 38

immunological detection, 128,139–141

use in electrochemical immunoassays, 37–44

Confirmation analysis, generic indicator of confidence, 259

Conjugate reagent, description, 193

Conjugation, description, 287

Contact Laboratory Program, sediment screening survey of Maumee area of concern, 157–159

Continuous flow immunosensor, 47–50

Core indicators of confidence, 260–262

Corn, fumonisin detection using immunochemical methods, 326–331

Curve fitting, role in immunoassay calibration, 241–245

Cyclic quartz voltammetry, generation of electrochemical signal with antibody–antigen interactions, 37–44

Cyclodiene insecticide(s), 149,150*f*

Cyclodiene insecticide screening in groundwater using enzyme-linked immunoassay, 149–153

D

Data processing, immunoassay calibration considerations, 251,253
"Data Quality Objectives for Superfund," 255
Detection, use for field immunoassay evaluation, 274–280
Dialysates of semipermeable membrane devices, polychlorinated biphenyl measurement using enzyme-linked immunosorbent assay from hydrophobic solutions, 308–323
Dilutions, core indicator of confidence, 261–262
O,O-Dimethyl *O*-3,5,6-trichloro-2-pyridyl phosphorothionate, residue screening, 161–169
Diode array, use as detector for immunoaffinity extraction with on-line LC–MS, 78,81
DuPont sulfonylurea herbicides, enzyme-linked immunoassay for determination on same microwell plate, 65–73
Dynal antibody-based test kits, bacteria detection, 306

E

Electroactive polymers, use in electrochemical immunoassays, 37–44
Electroanalytical techniques for immunological detection
advantages, 127
conducting electroactive polymers, 139–141
development, 128
electrochemical enzyme immunoassay
capillary, 132
competitive, 129,131*f*
homogeneous, 132–133
nonenzymatic, 133
sandwich, 132
electrochemical immunosensors
amperometric immunosensors, 138
capacitive immunosensors, 138–139
conductometric immunosensors, 139
features, 137–138
potentiometric immunosensors, 138,140*f*

Electroanalytical techniques for immunological detection—*Continued*
electrochemiluminescence immunoassay, 135–137
flow injection immunoassay, 135
immunoaffinity chromatography with electrochemical detection, 133–135
immunosensors, 137
performance improvement, 142–145
use of conducting polymer membranes, 128
Electrochemical enzyme immunoassay, description, 129,131–133
Electrochemical immunoassays using conducting electroactive polymers
cyclic voltammograms for electrode, 39,40*f*
experimental procedure, 38–39
flow injection analysis system, 39,40*f*
pulsed potential wave form, 39,41*f*
reversibility, 43,44*f*
selectivity, 39,42–43
signal response, 39,42*f*
Electrochemical immunosensors, immunological detection, 137–138,140
Electrochemical quartz crystal microbalance, generation of electrochemical signal with antibody–antigen interactions, 37–44
Electrochemiluminescence immunoassay, description, 135–137
Electron ionization–chemical ionization, description, 84
Electrospray, description, 84
Elimination, 287
ELs 1000 kit, automated immunoassay system for quantitative analysis of atrazine and alachlor in water samples, 183–189
EnviroGard chlorpyrifos-methyl screening kit
quantitative analysis of atrazine and alachlor in water samples, 183–189
validation, 161–169
Environmental immunoassays, optical sensing technology, 103–108
Environmental immunochemistry
antibody–analyte interaction, 3
biosensor development, 4

Environmental immunochemistry—
Continued
food and drug purity monitoring, 4–5
human exposure monitoring, 3
immunoaffinity chromatographic
extraction coupling with on-line
LC and MS, 3
need for current information, 216–217
quality assurance considerations, 5
quantitative evaluations, 5–6
screening applications, 5
xenobiotic exposure assessment, 4
Environmental immunosensing at Naval
Research Laboratory
applications, 54
continuous flow immunosensor, 47–50
fiber-optic biosensor, 50–54
Environmental matrices using enzyme-
linked immunoassay, mercury
detection, 23–36
Environmental monitoring, field
requirements, 103
Enzyme-linked immunoassay
cyclodiene insecticide screening in
groundwater, 148–153
determination on same microwell plate,
sulfonylurea herbicide, 65–73
mercury detection in environmental
matrices, 23–36
Enzyme-linked immunoassay analysis
applications, 56
cross-reactivity problem, 56–57
Enzyme-linked immunoassay coupled
with supercritical fluid extraction,
analysis of soil herbicides, 56–63
Enzyme-linked immunosorbent assay
fumonisin detection in corn, 326–331
measurement of polychlorinated
biophenyls in hydrophobic solutions,
308–323
triazine and chloroacetanilide
herbicides, 170–181
Escherichia coli, detection using
immunomagnetic assay system in
biological samples, 297–305
Ethylenediaminetetraacetic acid (EDTA),
use in heavy metals assay using
antibodies, 10–21

Exposures, biomonitoring using
immunoassays, 286–295
Extraction, *See* Immunoaffinity extraction
with on-line LC–MS

F

False negatives and false positives, use
for field immunoassay evaluation,
271–274,278
Fiber-optic biosensor
bacterial cell monitoring, 51
chamber, 50–51
evanescent wave, 50,52*f*
field tests, 51,53
improvements, 53
reusability of probe, 53
small molecule detection, 51
storage stability, 51,53,54*f*
TNT detection, 51,52*f*
Fiber-optic chemical sensors, 104–105
Field immunoassay(s), 265–282
Field immunoassay evaluation
calibration and detection, 274–280
conflicting results, 270–271,272*f*
experimental design factors, 281–282
minimizing false negative and false
positive rates, 271–274,278*f*
multiple statistical estimates, 266–268
proportion estimates, 268–270
Field screening techniques, immunoassay
test kit, 254–264
File Transfer Protocol, source for
immunochemistry Web site, 219
Fish, polychlorinated biphenyl
measurement using enzyme-linked
immunosorbent assay from hydrophobic
solutions, 308–323
Fish, Tissue, Bottom Sediment, Surface
Water, Organic and Metal Chemical
Evaluation and Biological Community
Evaluation, function, 156
Food purity, monitoring using
immunochemistry, 4–5
Food Safety and Inspection Service
HACCP system for meat and poultry
products, 228–238
pollution prevention strategy, 231–233

Fumonisin(s)
  detection using immunochemical methods
    in corn, 327–331
  toxicity, 326–327

G

GC comparisons
  to enzyme-linked immunosorbent assay
    for polychlorinated biphenyl
    measurement from hydrophobic
    solutions, 308–323
  to immunoassay for chlorpyrifos-methyl
    residue analysis on grain, 161–169
GC–MS, comparison to microtiter-plate
    enzyme-linked immunosorbent assay
    for triazine and chloroanilide
    herbicides, 170–181
Generic indicators of confidence, 258–259
Gopher, source for immunochemistry
    Web site, 219
Grain, chlorpyrifos-methyl residue
    screening immunoassay, 161–169
Groundwater, cyclodiene insecticide
    screening using enzyme-linked
    immunoassay, 148–153

H

Hapten, role in antibody binding
    affinity, 19
Hazard Analysis and Critical Control Point
    (HACCP) system for meat and poultry
    products, 228–238
Heavy metals
  assay using antibodies to metal chelate
    complexes, 11–20
  detection using chelate assay, 117,119,120f
  environmental contamination problem,
    10–11,23
Herbicide(s), microtiter-plate enzyme-
    linked immunosorbent assay for storm
    runoff samples, 170–181
Herbicide enzyme-linked immunoassay
    coupled with supercritical fluid
    extraction, See Soil herbicide enzyme
    immunoassay coupled with supercritical
    fluid extraction

Herbicide enzyme-linked immunoassay
    for determination on same microwell
    plate, See Sulfonylurea herbicide
    enzyme-linked immunoassay for
    determination on same microwell plate
High-performance liquid chromatography,
    comparison to immunochemical methods
    for fumonisin detection in corn, 326–331
Human exposure
  assessment, 286–331
  monitoring using immunochemistry, 3
Hydrophobic solutions, polychlorinated
    biphenyl measurement using enzyme-
    linked immunosorbent assay, 308–323

I

Immuno–ligand assay format, 94–96,97f
Immunoaffinity chromatographic extraction
    coupling with on-line LC and MS, 3
Immunoaffinity chromatography
  advantages, 74–75
  role of antibodies, 74
  with electrochemical detection, 133–135
Immunoaffinity column cleanup,
    fumonisin detection in corn, 326–331
Immunoaffinity extraction with on-line
    LC–MS
  analyte elution from column, 77
  antibody immobilization, 76
  antibody purification, 75
  applications, 78–82
  column packing and operating
    conditions, 76
  detectors
    atmospheric pressure chemical
      ionization–MS, 81–85
    capillary electrochromatography and
      laser-induced fluorescence, 84–86
    diode array, 78,81
  instrumental setup, 77–78
Immunoassay
  biomonitoring of occupational exposures,
    286–295
  chlorpyrifos-methyl residue screening on
    grain, 161–169
  detection using xenobiotics, 110–121
  elements, 241

Immunoassay—*Continued*
heavy metals assay using antibodies to
metal chelate complexes, 10–21
types, 127
Immunoassay calibration
assay design, 245–246,247*f*
characteristics of curves, 241
curve fitting methods, 241–245
data processing, 251,253
errors, 246,248–253*f*
influencing factors, 240
process, 240
Immunoassay system for quantitative
analysis in water samples, atrazine
and alachlor, 183–189
Immunoassay test kit
development for environmental field
applications, 254–255
ideal features, 263–264
quality assurance indicators, 254–264
use by U.S. Environmental Response
Team, 255
See also Pentachlorophenol immunoassay
soil test kit
Immunochemistry bibliography of U.S.
Environmental Protection Agency,
description, 216–226
Immunochemistry Web site
electronic bulletin board systems, 217–218
home page and links, 219–221
implementation, 226
prototype Web page, 221,222–225*f*
recommendations for improvements,
221,226
Internet tools, 218–219
U.S. Environmental Protection Agency
resources, 219
Immunogen, definition, 127
Immunological detection, electroanalytical
techniques, 127–145
Immunomagnetic assay system, detection
of virulent bacteria in biological
samples, 297–305
Immunosensors, use in immunological
detection, 137
Indium, assay using antibodies to
metal chelate complexes, 10–21

L

Laser-induced fluorescence, use in
immunoaffinity extraction with on-line
LC–MS, 84–86
Lead, environmental contamination
problems, 10
Legacy biomonitoring, 291–295
Light-addressable potentiometric sensor,
use in threshold immunoassay system,
91,93*f*
Linear regression, use in evaluation of
immunoassay soil test kit, 200
Liquid chromatography–mass spectrometry,
use with immunoaffinity extraction,
74–86
Log–linear curve fitting, role in
immunoassay calibration, 242–243
Log–logit curve fitting, role in
immunoassay calibration, 243–245
Logistic regression, use in evaluation of
immunoassay soil test kit, 200

M

Matrix spikes, generic indicator of
confidence, 259
Maumee area of concern sediment
screening survey, 156–160
Maumee Bay, pollution problems, 155–156
Mass spectrometry–liquid chromatography,
use with immunoaffinity extraction,
74–86
Mercury, problem of environmental
contamination, 23
Mercury detection in environmental
matrices using enzyme-linked
immunoassay, 23–35
Metal chelate complexes antibodies,
heavy metals assay, 10–21
Microtiter-plate enzyme-linked
immunosorbent assay for storm runoff
samples, triazine and chloroacetanilide
herbicides, 170–181
Moisture content, core indicator of
confidence, 261

Multiple statistical estimates, use for
   field immunoassay evaluation, 266–268

N

National Performance Review, roles and
   responsibilities of federal regulatory
   agencies, 228
Near-IR fluorescence, use in detection of
   xenobiotics, 117,118*f*
4-Nitrophenols, detection using
   immunoassays, 114,116*f*
Nonanalyte specificity, core indicator of
   confidence, 261

O

Occupational exposures, biomonitoring
   using immunoassays, 286–295
Optical sensing technology for
   environmental immunoassays, 103–108
Organophosphorus insecticide metabolites,
   detection using immunoassays, 113–118
Ottawa River, fish consumption–contact
   advisory, 156
Outliers, description, 248

P

Pairwise comparison, use in evaluation of
   immunoassay soil test kit, 200
PENTA RISc soil test kit, *See*
   Pentachlorophenol immunoassay soil
   test kit
Pentachlorophenol, pollution problem, 192
Pentachlorophenol immunoassay soil test
   kit, 192–196*f*,200–213
PetroSense sensors, 104–105
Pollution prevention, strategy, 231–233
Pollution Prevention Act of 1990, 231
Polychlorinated biphenyl(s), toxicity, 308
Polychlorinated biphenyl immunoassay,
   Maumee area of concern sediment
   screening survey, 157–159
Polychlorinated biphenyl immunosensor,
   performance improvement, 143–145

Polychlorinated biphenyl measurement
   using enzyme-linked immunosorbent
   assay from hydrophobic solutions
   commercial kits, 308
   direct dilution studies, 313–315
   enzyme-linked immunosorbent assay
      determinations, 318–323
   evaporation of isooctane and dissolution
      of residues, 315–318
   experimental procedure, 308–312
   solubilities of hydrophobic chemicals in
      diluents, 318
Polychlorinated dibenzo-*p*-dioxins,
   detection using immunoassays, 119–120
Polypyrrole, use in conducting polymer
   based immunochemical sensors, 139–141
Proportion estimates, use for field
   immunoassay evaluation, 268–270
Pulsed amperometric detection, 38
Pyrethroid insecticides and metabolites,
   detection using immunoassays,
   114,116–117,118*f*

Q

Quality assurance, use of environmental
   immunochemistry, 5
Quality assurance indicators for
   immunoassay test kits
   analyte quantitation procedure, 256–257
   confidence indicators, 257–262
   current status of Environmental Response
      Team activities, 263
   definitive data quality assurance–
      quality control elements, 256,257*t*
   quality assurance–quality control data
      categories, 255–256
   screening data quality assurance–quality
      control elements, 256
   Superfund data requirements, 255
Quality assurance program, environmental
   data, 255
Quantitative analysis
   atrazine and alachlor automated
      immunoassay system in water samples,
      183–189
   use of immunochemistry, 5–6

R

Random errors, description, 248
Reaction time, core indicator of
  confidence, 262
Regulatory screening, features, 235–237
Replicates, generic indicator of
  confidence, 259
Representativeness, generic indicator of
  confidence, 259
Response factor, calculation, 198–200
Rocky Mountain Arsenal, 148–153

S

Salmonella typhimurium, detection using
  immunomagnetic assay system in
  biological samples, 297–305
Sample preparation, generic indicator of
  confidence, 259
Sandwich immunoassay, 51,91
Saxitoxin, detection using threshold
  immunoassay system, 99–101
Screening, use of immunochemistry, 5–6
Screening tests in changing environment
  analyte testing options, 237–238
  evaluation parameters, 233–235
  features of regulatory screening
    methods, 235–237
  food safety responsibilities, 228
  goal, 228
  HACCP inspection issues, 229–231
  pollution prevention strategy, 231–233
Sediment screening survey, Maumee area
  of concern, 155–160
Semipermeable membrane devices,
  polychlorinated biphenyl measurement
  using enzyme-linked immunosorbent
  assay from hydrophobic solutions,
  308–323
Sensitive analyte, detection and quantitation
  using threshold immunoassay system,
  89–101
Soil herbicide enzyme immunoassay
  coupled with supercritical fluid
  extraction, 57–63
Soil test kit, *See* Pentachlorophenol
  immunoassay soil test kit

Stability, core indicator of confidence, 262
Storm runoff samples, triazine and
  chloroacetanilide herbicide microtiter-
  plate enzyme-linked immunosorbent
  assay, 170–181
Sulfonylurea herbicide enzyme-linked
  immunoassay for determination on
  same microwell plate
  Benlate 50 DF, 69–73
  components, 67
  format, 65–66
  optimization, 67–69
Supercritical fluid extraction,
  applications, 57
Supercritical fluid extraction coupled
  with enzyme immunoassay, analysis
  of soil herbicides, 56–63
Superfund, data requirements, 255
Systematic errors, description, 248

T

Temperature, core indicator of
  confidence, 260
Tenmile Creek, fish consumption–contact
  advisory, 156
Test kits, quality assurance indicators,
  254–264
Threshold immunoassay system
  advantages, 92,101
  applications, 89
  atrazine detection, 96–100
  description, 89
  detection steps, 90f,91
  immuno–ligand assay procedure format,
    94–97f
  indirect detection method, 92
  instrumentation, 90f,91
  light-addressable potentiometric sensor,
    91,93f
  pH change monitoring mechanism,
    91–92,93f
  physical parameter determination, 92
  sandwich immunoassay formats, 91
  saxitoxin detection, 99–101
  sensitivity, 92,94
s-Triazine herbicide(s), detection
  using immunoassays, 111–113,115f

Triazine herbicide microtiter-plate
   enzyme-linked immunosorbent assay
   for storm runoff samples
   advantage, 170
   Bayes's rule application, 178–181
   comparison to GC–MS, 174,177–178
   cross-reactivity, 174–176*f*
   experimental description, 170–173
3,5,6-Trichloro-2-pyridinol,
   detection using immunoassays, 114
Trinitrotoluene (TNT), detection using
   fiber-optic biosensor, 51,52*f*

U

Urinary biomarkers, use for
   biomonitoring of occupational
   exposures, 288–292
U.S. Environmental Response Team, use
   of immunoassay test kit, 255
User friendliness, core indicator of
   confidence, 262
"User's Guide to Environmental
   Immunochemical Analysis," 216–226

V

Validation, chlorpyrifos-methyl residue
   screening immunoassay on grain,
   161–169
Virulent bacteria detection using
   immunomagnetic assay system in
   biological samples
   advantages, 297–298,303–304
   Bacillus anthrax detection, 301,303*f*
   detection limits, 300*f*,301,305
   Escherichia coli detection, 301,302*f*
   experimental procedure, 298–301
   limitation, 305

Virulent bacteria detection using
   immunomagnetic assay system in
   biological samples—*Continued*
   negative control assay, 301,303,304*f*
   Salmonella detection, 301,302*f*

W

Water samples, atrazine and alachlor
   automated immunoassay system for
   quantitative analysis, 183–189
Web site for immunochemistry, *See*
   Immunochemistry Web site
Wide Area Information Server, source for
   immunochemistry Web site, 219
Wilcoxon rank sum test, use in evaluation
   of immunoassay soil test kit, 201
World Wide Web, immunochemistry
   site, 219

X

Xenobiotic(s), occupational adsorption, 286
Xenobiotic detection using immunoassays
   advantages, 110
   applications, 121
   chelate assay for heavy metals,
      117,119,120*f*
   examples, 121–122,286–295
   experimental objectives, 110–111
   near-IR fluorescence detection, 117–118*f*
   organophosphorus insecticide
      metabolites, 113–118*f*
   polychlorinated dibenzo-*p*-dioxins,
      119–120
   s-triazine herbicides, 111–113,115*f*

Z

Zinc, environmental contamination
   problem, 10